PERSPECTIVES IN
STATISTICAL PHYSICS

STUDIES IN STATISTICAL MECHANICS

VOLUME IX

EDITORS

E. W. MONTROLL
University of Rochester, New York

J. L. LEBOWITZ
Rutgers University, New Brunswick

NORTH-HOLLAND PUBLISHING COMPANY
AMSTERDAM · NEW YORK · OXFORD

PERSPECTIVES IN STATISTICAL PHYSICS

M. S. Green Memorial Volume

Editor

H. J. RAVECHÉ

National Bureau of Standards
Washington, D.C. 20234
U.S.A.

NORTH-HOLLAND PUBLISHING COMPANY
AMSTERDAM · NEW YORK · OXFORD

© North-Holland Publishing Company – 1981.
All rights reserved. No part of this publication may be reproduced, stored in a retrieval system, or transmitted, in any form or by any means, electronic, mechanical, photocopying, recording or otherwise, without the prior permission of the copyright owner.

ISBN: 0 444 86026 6

Published by:

North-Holland Publishing Company – Amsterdam, New York, Oxford

Sole distributors for the U.S.A. and Canada:

Elsevier North-Holland, Inc.
52 Vanderbilt Avenue
New York, N.Y. 10017

Library of Congress Cataloging in Publication Data
Main entry under title:

Perspectives in statistical physics.

(Studies in statistical mechanics ; v. 9)
"M. S. Green's publications": p.
Includes index.
1. Statistical physics   2. Green, Melville S.
I. Green, Melville S. II. Raveché, H. J. III. Series.
QC175.S77   vol. 9   [QC174.8]   530.1'3s   [530.1'3]
ISBN 0-444-86026-6   80-21445

PRINTED IN THE NETHERLANDS

Dedication to M. S. Green

Dedication to M. S. Green

This volume is dedicated to the memory of Melville Saul Green. He was born on June 9, 1922 in Queens, New York and he died of cancer in his home in Philadelphia on March 27, 1979. He is survived by Vivian, his wife, Aliza Miriam and Joel Moshe his children, two brothers, Bernard and Irwin, and his mother, Ella. He lived in New York City until completing his undergraduate degree at Columbia, then in Princeton, where he earned his Ph.D., in Chicago, where he was an Assistant Professor of Physics at the University of Chicago, next in Washington, D.C., where he was at first a postdoctoral associate at the University of Maryland and then Physicist and later Chief of the Statistical[1] Physics Section at the National Bureau of Standards, and finally in Philadelphia, where he was Professor of Physics at Temple University.

In a manner of contrast that was uniquely Mel Green's, he combined strength and gentleness from which he did not waiver, even in the face of hardships. The working of his mind was also unique in its originality and creativity. Mel was naturally inclined to ponder the heart of an issue and he did this with success in many different matters. Even if he tried it is doubtful that he could have ever been superficial, that quality being utterly alien to his character. Because of this, some who did not know him well found communication difficult. Actually, Mel craved meaningful discussions on a variety of subjects. While he preferred to think some things through on his own, he often invited comments and criticisms. Since he was in fact profound, he was difficult to understand for those who did not make the effort; those who did, found substantive matters to consider.

The core of Mel's life consisted of his family, his faith, and his research in statistical physics. He was a very special person because of the depth of these commitments and the great courage that he so consistently maintained in their pursuit. Mel looked forward to activities with his family and he participated with pride in the education of his children, their extracurricular activities and the writing career of his wife. He so enjoyed being with his family, that he was happy to share the warmth of his home life with friends and visitors. The genuine hospitality of the Green family is well known by the many who have visited with them. It is perhaps less well known that Mel not only enjoyed social events at his home, but also the preparations which preceded these and he was skillful in various aspects of cooking – his salads were particularly good. To Mel a dinner was a social event, an opportunity to engage in conversation in a personal way. As a host, he made every effort to ensure an atmosphere which was harmonious with lively discussion. He had a good sense of humor and he thoroughly appreciated a funny story. When Mel laughed, the expression on his face and in his eyes made it obvious that his response was sincere. His interest in people extended beyond his family and close friends. The Greens traveled extensively, especially in summers, to places where research in statistical physics is done. In free moments he would prefer to meet people rather than gaze at artifacts. When choosing a hotel or restaurant he made sure to select spots which were frequented by local people.

Although an urbanite by nature, Mel very much enjoyed walks in rural areas. He would frequently spend Sundays walking with his wife and their canine "Maxi", a Bouvier des Flandres. The problems of urban life in America were of concern to Mel and he discussed these with compassion and vision. He was particularly concerned with the survival of American cities as heterogeneous mixtures of many ethnic groups, the so-called "melting pot". Mel enjoyed visiting ethnic neighborhoods and learning about their culture. Both in Washington, D.C. and in Philadelphia, the Green family was actively involved in city life. Although the Greens experienced some of the physical dangers of urban American life while living in Philadelphia, they refused to be intimidated and continued to develop their many interests with intrepidity.

The name "Green" was acquired by Mel's grandfather when he emigrated from Russia to the United Kingdom in the second-half of the nineteenth century. The original family name was Sassoon and is of Sephardic roots. Mel's father, Maurice, was born in Manchester, U.K. and he came to live in New York City in the first decade of the twentieth century. It was here that he met Ella Prichep, a native of New York City, and here that they were married and raised their family.

Vivian Grossman and Mel first met while attending an evening course at the Jewish Theological Seminary in New York City. From the beginning of their marriage, they intertwined their family life and religious beliefs in a manner which can be an inspiration to families of all faiths. This combination no doubt contributed significantly to the strong bonds which existed between themselves and Aliza and Joel. Mel followed his faith devoutly and, even while traveling, he adhered to rigorous daily practices. He did, however, go far beyond strict observance of the Law. His total dedication led to extraordinary inner strength and optimism. Mel called upon these in facing the serious obstacles that both he and other members of the Green family encountered at various times and his ability to do this was indeed staggering.

His religious life had many dimensions, he had scholarly interests in historical matters and he spoke authoritatively on biblical subjects. Jewish mysticism fascinated Mel, so he became knowledgeable and eventually taught a course on it at the Jewish Free University in Philadelphia. Complementing his erudite pursuits, Mel was active in various organizations. In Washington, D.C., he was a member of the Board of Education of the Hebrew Academy, and in Philadelphia he was president of his synagogue for two years and a member of the Jewish Campus Activities Board, an arm of the Federation of Jewish Agencies which is responsible for distributing funds for campus projects.

Mel's religious commitments led to a way of life that was rich in tradition, yet did not preclude a broad interest in people and events. Several of his closest friends were of different religious faiths. While firmly committed to the state of Israel, Mel interacted closely with Egyptian students at Temple University. Moreover, he, Vivian, and Professor al Farouqi organized an Arab–Jewish dialogue which was begun at Temple in 1974. Mel gave serious thought to contemporary issues; he was quite knowledgeable and worried much about such topics as race riots, the Viet Nam war, and the role of science and technology in America.

Mel's broad interests probably began in childhood. He was an excellent student in science and mathematics and a member of the Science Club at Richmond Hill High School in Queens, New York where he won awards for scientific scholarship. He also demonstrated other talents; for example, he won an award for an essay on patriotism from the American Legion and he was keenly interested in classical music. In his early teenage years Mel contracted osteomyelitis, which required him to use crutches through college and, because of this illness, he was absent from high school for one academic year. During the convalescent period he read an encyclopedia thoroughly and, with the aid of textbooks, he taught himself the rudiments of calculus. His interests in mathematics continued to grow and he developed these vigorously, along with physics, throughout his life. Mel's career in science was even forecast during childhood. When he graduated from elementary school in Queens, New York, the principal referred to him as a young Edison and when his parents took him on one occasion to get his handwriting analyzed they were told that he would go into the natural sciences!

Mel graduated from Columbia University with a B.S. in physics in 1944 and then entered graduate school at Princeton University. Even at that comparatively early stage, Mel's creativity in physics was sufficiently developed that he worked effectively on his own. V. Bargmann guided Mel and oversaw his thesis which, however, was not completed until 1951. It was fortunate for science that Mel exerted independence and took more than the customary time to finish his Ph.D. degree. After fulfilling the formal graduate course work, Mel left Princeton for a few years and he began exploring many aspects of statistical physics. He went to the University of Chicago between 1947 and 1951 where he taught physics and discussed statistical mechanics with J. E. Mayer, whose seminars he attended regularly. Mel explored more; he published a paper on relaxation in polymeric media with A. V. Tobolsky [J. Chem. Phys. **14** (1946) 80] and with J. G. Kirkwood and F. P. Buff he studied a molecular interpretation of the viscosity of a liquid [J. Chem. Phys. **17** (1949) 983]. This interim period, between classroom studies and the completion of his Ph.D., proved to be very important because it gave Mel the opportunity to do what he did best – think deeply.

After arriving at the University of Maryland as a postdoctoral associate with E. W. Montroll, Mel put some of his brilliant ideas about the microscopic aspects of irreversible processes on paper. This resulted in two highly seminal articles [M. S. Green, J. Chem. Phys. **20** (1952) 1281 and **22** (1954) 398] which are now acknowledged as monumental in statistical physics. The paper in 1954 contained the widely used relations between bulk transport coefficients and microscopic auto-correlation functions now known as the Green–Kubo formulae. This work was truly pioneering, it carved out new areas of experimental and theoretical research which have made significant contributions to science and technology. A conference was organized at the National Bureau of Standards, sponsored jointly with the Institute for Physical Science and Technology at the University of Maryland and Temple University, on April 26 and 27, 1979, to honor Mel's work on time correlation functions; several of the papers in this volume were given at the symposium. His death prevented the guest of honor from attending.

After his postdoctoral position, Mel joined the staff of the National Bureau of Standards in Washington, D.C. and within a few years he was made Chief of the newly created Statistical Physics Section in the former Heat Division. In this capacity, Mel formed an internationally recognized group comprised of people such as J. L. Jackson, M. Klein, R. D. Mountain, I. Oppenheim, R. A. Piccirelli, R. J. Rubin, J. V. Sengers, J. Weinstock, and R. Zwanzig. From abroad, scientists such as

L. Garcia-Colin, F. Chaos, M. Ernst, N. G. van Kampen and J. M. J. van Leeuwen also came to interact with Mel.

His interests continued to expand while at NBS. Kinetic theory intrigued him, especially the statistical mechanical foundation of the Boltzmann equation and Bogoliubov's ansatz about the separation of time scales. He published actively in these fields and many of his ideas influenced further developments, especially in the analysis of collision phenomena, such as the divergence of the density expansion of the microscopic expressions for the transport coefficients. Mel's work in kinetic theory is reviewed in further detail in the remarks by E. G. D. Cohen which follow later in the introduction of this memorial volume.

Graphical expansions of the molecular correlation functions and of thermodynamic properties interested Mel because of their utility and also because he excelled in combinatorics. Several promising techniques and two major developments resulted from these interests. The problem of expressing entropy, which cannot be written as the expectation of a mechanical operator, in terms of only molecular correlation functions challenged Mel. In particular, he was interested in doing this for an infinite system while avoiding the problems of long-range corrections to the correlation functions defined by the canonical ensemble in that limit. By a clever trick of using Moebius functions, R. E. Nettleton and M. S. Green found the correct expression. Their paper, [J. Chem. Phys. **29** (1958) 1365], while difficult to follow, is worth the effort because it also contains an instructive variational derivation of integral equations discovered earlier by others.

The second major development from Mel's interest in graphical expansions was his derivation of the so-called hypernetted chain equation for the pair correlation function of a fluid [M. S. Green, J. Chem. Phys. **33** (1960) 1403]. This equation had been derived independently by several others. Mel's analysis, however, shed light on the asymptotic properties of the correlation function near the critical point of a fluid. The inspiration that this paper provided in critical phenomena is illustrated in the remarks which follows later by Michael E. Fisher. The hypernetted chain equation continues to be used today, and is particularly successful in dense fluids of charged particles.

Although Mel took delight in fundamental theoretical aspects of a problem, he firmly believed that theory must ultimately be closely coupled with experimental observations. This point, which he frequently emphasized, led him to begin focusing on critical phenomena in the early '60s when he suspected that future developments would make it possible to pursue the cross-fertilization between theory and experiment.

Motivated by this, he organized a conference on critical phenomena which was held at NBS in April, 1965. Approximately 200 scientists, experimentalists and theorists from America and from abroad attended this symposium, which is generally regarded as a landmark in modern critical phenomena. Requests for the proceedings are still received, some fifteen years after the event. With his natural sense of significant issues, Mel foresaw that the field would evolve to be important and therefore he understood the need for a regular series of definitive review articles. The first volume of *Phase Transitions and Critical Phenomena*, edited by C. Domb and M. S. Green, appeared in the early '70s and the series is now recognized as a major source of carefully evaluated information.

His theoretical work in critical phenomena was influenced by several factors. Building on the notion that the molecular correlation functions of a fluid contained, in their limiting behavior at large intermolecular separations, crucial information about critical point anomalies, he developed new ideas utilizing his earlier work on entropy, and his novel work on variational principles in different ensembles [M. S. Green, Lectures on Critical Phenomena, in: *Cargese lectures in theoretical physics*, edited by B. Jancovici (Gordon and Breach publishers, New York, 1966)]. Mel reasoned that some characteristic feature of the correlation functions was responsible for the anomalies in the heat capacity, as well as in other properties. He sought to express this as a type of eigenvalue problem along the lines of earlier work by L. Tisza, where the eigenvalues are related to the reciprocal of divergent quantities and the eigenfunctions to fluctuations in extensive variables. A series of studies followed [M. S. Green, J. Math. Phys. **9** (1968) 875; M. J. Cooper and M. S. Green, Phys. Rev. **176** (1968) 302; J. D. Gunton and M. S. Green, Phys. Rev. A **4** (1971) 1282] which, while left in a formal state, did lead to a novel interpretation of why C_v is less singular than C_p [Phys. Rev. **185** (1969) 176] and an exact expression for the elusive exponent η [M. S. Green and J. D. Gunton, Phys. Letters **42A** (1972) 7].

These ideas were explored initially while Mel was at NBS. They were developed further, however, after Mel joined the physics department at Temple University in 1968. In Philadelphia, as in Washington, D.C., Mel was the core of a lively and productive group in statistical physics. At Temple he was joined in the beginning by J. D. Gunton, K. Kawasaki, S. Larsen and J. M. J. van Leeuwen, and later by D. Forster.

Mel was extremely curious about renormalization methods. In 1970 he was Director of the Course on Critical Phenomena, held in Varenna, Italy. This was an important conference because of the interaction of the field theorists who attended the meeting with condensed matter theorists and because of L. P. Kadanoff's lectures on universality. Shortly after the ideas of K. G. Wilson appeared, Mel appreciated their significance and he and Michael E. Fisher, J. D. Gunton, L. P. Kadanoff, K. Kawasaki and K. G. Wilson organized a conference on the renormalization group in critical phenomena and quantum field theory, which was held at Temple in May of 1973. Mel became intrigued with the possibility of finding similar methods that were natural to fluids. This prompted him to reconsider the McMillan–Mayer theory of potentials of mean force which presented techniques for relating the potentials at one activity to those at another activity, through a renormalization process [J. E. Mayer, J. Chem. Phys. **10** (1942) 629; W. G. McMillan and J. E. Mayer, J. Chem. Phys. **13** (1945) 276]. By developing these ideas further, he demonstrated renormalization group behavior similar to the semi-group properties used with discrete interactions [M. S. Green, Phys. Rev. A **8** (1973) 1998; M. S. Green, Phys. Rev. B **15** (1977) 5367]. Unfortunately, Mel did not have time to remove some of the difficulties encountered in the equations he developed for this novel but extremely difficult problem. Interestingly, in very recent work [K. A. Green, K. D. Luks and J. J. Kozak, Phys. Rev. Letters **42** (1979) 985; K. A. Green, K. D. Luks, Eok Lee and J. J. Kozak, Phys. Rev. A, **21** (1980) 356] non-classical exponents for the square-well fluid have been reported from an integral equation for the pair correlation function, in a manner somewhat related to the general spirit of Mel's ideas. Further studies in this vein need to be completed before the real significance of Mel's thoughts on a fixed point theory for fluids can be determined.

Mel regularly sought discussions with experimentalists in critical phenomena. He had a long-standing interest in the measurement of the critical exponent η and was influential in the light scattering experiments of D. McIntyre and A. D. Wims in the late '60s. Mel thought later that neutrons might be a better probe of η, so he suggested this experiment to V. P. Warkulwiz, one of his Ph.D. students. This led to a direct experimental determination of the exponent from neutron scattering in neon [V. P. Warkulwiz, B. Mozer and M. S. Green, Phys. Rev. Letters **32** (1974) 1410]. While this was one of the earliest direct measurements of η in a fluid, the reported value is regarded as high; however, the reason for this is not yet completely understood.

Mel's research with Vicentini-Missoni and Levelt Sengers, on a scaling-law equation of state for fluids, has had major impact on important problems in industrial process and energy technology. A series of papers was completed [M. S. Green, M. Vicentini-Missoni and J. M. H. Levelt

(1966) 450]. It discussed explicit formal expressions for the transport coefficients derived from a generalized Boltzmann equation based upon the assumption of a time independent functional of the pair distribution function but not on the assumption of a density expansion of this functional, or of the transport coefficients. This paper is interesting also because of the occurrence of a non-symmetric collision operator with different left and right eigenfunctions with zero eigenvalue.

Uhlenbeck has said somewhere that there are no clear frontiers of physics, but rather that frontiers are where one feels one can make progress. With this definition of frontier, Mel was a true pioneer: someone who thrived at the frontiers of his field. No one confronted the basic problem of kinetic theory – or for that matter of statistical mechanics – on such a wide and fundamental scale, so honestly and courageously with such complete dedication, giving the best he had. He was a deep thinker and was most generous with his many ideas to his colleagues. He was clearly primus inter pares or first among his equals.

E. G. D. Cohen
The Rockefeller University
New York, New York 10021

Remarks on M. S. Green (2)*

My first knowledge of Mel Green came through the scientific literature: his deep thoughts about the validity of the famed Ornstein–Zernike theory of critical scattering and, in particular, his discovery that the hypernetted chain approximation lead to a different decay law for the critical point correlation function [J. Chem. Phys. **33** (1960) 1403], was a direct stimulus to my own work on this central topic in statistical physics.

When I came to know Mel personally, I found a warm and sympathetic person, a born leader but a gentle and considerate one, an efficient but unobtrusive administrator. As a scientist Mel never failed to impress me with his probing ideas, his concern both for the details of theory and their match to experiment, and for the formulation of a general, holistic framework of understanding. In his scientific work and in his work for science, Mel Green was a personal inspiration to me and, undoubtedly, to many others.

As a friend and colleague, I will sadly miss my contacts with Mel. Statistical mechanics and chemical physics will suffer by Mel's premature death and departure from our life.

Michael E. Fisher
Cornell University
Baker Laboratory
Ithaca, New York 14853

---

*Excerpted from a letter to the conference in honor of M. S. Green held at the National Bureau of Standards in April 1979.

M. S. Green's Curriculum Vitae

Date of Birth: June 9, 1922

Married, two children

Education

1944	Columbia College	B.A.
1947	Princeton University	M.A.
1951	Princeton University	Ph.D.

Fellowships and scholarships

1957–58 Fulbright Grant
1957–58 John Simon Guggenheim Fellowship Award
1973–74 John Simon Guggenheim Fellowship Award

Positions

1968–1979   Professor, Temple University, Philadelphia, Pa.
1960–1968   Chief, Statistical Physics Section, National Bureau of Standards, Washington, D.C.
9/59–1960   Consultant, Hughes Aircraft Company, Los Angeles, Ca.
6/58–9/59   Physicist, Thermodynamics Section, National Bureau of Standards, Washington, D.C.
1957–1958   Fellow, Institute for Theoretical Physics, University of Utrecht, Utrecht, Netherlands
1954–1957   Physicist, Thermodynamics Section, National Bureau of Standards, Washington, D.C.
1951–1954   Research Associate, University of Maryland, College Park, Md.
1947–1951   Assistant Professor, University of Chicago, Chicago, Il.
1944–1947   Assistant Instructor, Princeton University, Princeton, N.J.

Professional and Scientific Societies

Fellow, American Physical Society
Member, Washington Academy of Sciences

Member, Washington Philosophical Society
Member of the Board of Editors:
Physical Review (1964–1966)
Journal of Mathematical Physics (1963–65)
Journal of Physics of Fluids (1966–69)

Awards

U.S. Department of Commerce Gold Medal Award for Distinguished Achievement in Federal Service.

M. S. Green's Publications

[1] A New Approach to the Theory of Relaxing Polymeric Media, J. Chem. Phys. **14** (1946) 80 (with A. V. Tobolsky).

[2] The Statistical Mechanical Theory of Transport Processes. III. The Coefficients of Shear and Bulk Viscosity of Liquids, J. Chem. Phys. **17** (1949) 988 (with J. G. Kirkwood and F. P. Buff).

[3] Brownian Motion in a Gas of Noninteracting Molecules, J. Chem. Phys. **19** (1951) 1036.

[4] Markoff Random Processes and the Statistical Mechanics of Time-Dependent Phenomena, J. Chem. Phys. **20** (1952) 1281.

[5] Markoff Random Processes and the Statistical Mechanics of Time-Dependent Phenomena. II. Irreversible Processes in Fluids, J. Chem. Phys. **22** (1954) 398.

[6] The Statistical Mechanics of Electrical Conduction in Fluids, J. Phys. Chem. **58** (1954) 714.

[7] The Statistical Mechanics of Transport and Nonequilibrium Processes, Ann. Rev. of Phys. Chem. **5** (1954) 449 (with E. W. Montroll).

[8] Boltzmann Equation from the Statistical Mechanical Point of View, J. Chem. Phys. **25** (1956) 836.

[9] Thermochemistry and Thermodynamics of Substances, Ann. Rev. of Phys. Chem. **7** (1956) 287 (with C. W. Beckett and H. W. Woolley).

[10] The Non-Equilibrium Pair Distribution Function at Low Densities, Physica **24** (1958) 393.

[11] Expression in Terms of Molecular Distribution Functions for the Entropy Density in an Infinite Systems, J. Chem. Phys. **29** (1958) 1365, (with R. E. Nettleton).

[12] Bose–Einstein Condensation and the λ-Transition in Liquid Helium, Phys. Rev. Letters **1** (1958) 409.

[13] Internal Energy of Highly Ionized Gases, Proc. IXth Ann. Congress of the Int'l. Astronautical Federation, Amsterdam, 1958. (Springer-Verlag, Vienna, 120, 1959) (with J. Hilsenrath and C. W. Beckett).

[14] Statistical Mechanics and the Boltzmann Equation, Proc. of the Colloquium on Transport Properties in Statistical Mechanics (Interscience New York, N. Y., 1959).

[15] Dense Subgraphs and Connectivity, Can. J. Math. **11** (1959) 262 (with R. E. Nettleton and K. Goldberg).

[16] Moebius Function on the Lattice of Dense Subgraphs, NBS J. Res. **64B** (1960) 41 (with R. E. Nettleton).

[17] Comment on a Paper of Mori on Time-Correlation Expressions for Transport Properties, Phys. Rev. **119** (1960) 829.

[18] Topological Derivation of the Mayer Density Series for the Pressure of an Imperfect Gas, J. Math. Phys. **1** (1960) 391.

[19] On the Theory of the Critical Point of Simple Fluid, J. Chem. Phys. **33** (1960) 1403.

[20] Some Applications of the Generating Functional of the Molecular Distribution Functions, in: Lectures in theoretical physics III (Lectures delivered at the Summer Institute for Theoretical Physics. University of Colorado, Boulder, Colo., 1960) (Interscience Publishers, New York, N.Y., 195, 1961).
[21] Critical Opalescence of Polystyrene Solutions, (J. Chem. Phys. 37 (1962) 3019 (with D. McIntyre and A. Wims).
[22] Numerical Solutions of the Convolution-Hypernetted Chain Integral Equation for the Pair Correlation Function of a Fluid. I. The Lennard–Jones (12, 6) Potential, J. Chem. Phys. 6 (1963) 1367 (with M. Klein).
[23] Basis of the Functional Assumption in the Theory of the Boltzmann Equation, Phys. Rev. 132 (1963) 1388 (with R. A. Piccirelli).
[24] Studies in Scientific and Engineering Manpower, U.S. Department of Commerce, Office of the Assistant Secretary of Commerce for Science and Technology, Staff Report: 63-1, 1-50, 1963.
[25] Surface Integral Form for Three-Body Collision in the Boltzmann Equation, Phys. Rev. 136 (1964) A905.
[26] Lectures on Critical Phenomena, Proc. of summer school of theoretical physics, Cargese, Corsica, Summer 1964 (Gordon and Breach, New York, 1966).
[27] The Chapman–Enskog Solution of the Generalized Boltzmann Equation, Physica 32 (1966) 450 (with L. S. Garcia-Colin and F. Chaos).
[28] Critical-Point Phenomena, Report on conf. on phenomena in the neighborhood of critical points, Science, 150 (1965) 229.
[29] Lectures on Critical Phenomena, Latin American school of physics, Mexico, 1965 (Gordon and Breach, New York).
[30] Critical Phenomena, Proc. of a conf. held in Washington, D.C., April 1965, NBS Misc. Publ. 273 (U.S. Government Printing Office, Washington, D.C., 1966) (edited with J. V. Sengers).
[31] Definition of Temperature in the Kinetic Theory of Dense Gases, Phys. Rev. 150 (1966) 153 (with L. S. Garcia-Colin).
[32] Scaling-Law Equation of State for Gases in the Critical Region, Phys. Rev. Letters 18 (1967) 1113 (with M. Vicentini-Missoni and J. M. H. Levelt Sengers).
[33] Generalized Ornstein–Zernike Approach to Critical Phenomena, J. Math. Phys. 9 (1968) 875.
[34] Generalized Scaling and the Critical Eigenvector in Ideal Bose Condensation, Phys. Rev. 176 (1968) 302 (with M. J. Cooper).
[35] Thermodynamic Anomalies of CO_2, Xe and He^4 in the Critical Region, Phys. Rev. Letters 22 (1969) 389 (with M. Vicentini-Missoni and J. M. H. Levelt Sengers).
[36] On the Consistency of the Yvon-Born–Green Hierarchy and its Truncations, J. Chem. Phys. 50 (1969) 5334 (with H. J. Raveché).
[37] Why is C_V Less Singular than C_p Near the Critical Point, Phys. Rev. 185 (1969) 176.
[38] Scaling Analysis of Thermodynamic Properties in the Critical Region of Fluids, J. Res., NBS 73A (1969) 563 (with M. Vicentini-Missoni and J. M. H. Levelt Sengers).
[39] Scaled Equation of State and Critical Exponents in Magnets and Fluids, Phys. Rev. B 1 (1970) 2312 (with M. Vicentini-Missoni, R. I. Joseph and J. M. H. Levelt Sengers).
[40] Equilibrium Critical Phenomena in Fluids and Mixtures: A Comprehensive Bibliography with Key-Word Descriptors, National Bureau of Standards Special Publication 327, June 1970 (with S. Michales and S. Y. Larsen).
[41] Extended Thermodynamic Scaling from a Generalized Parametric Form, Phys. Rev. Letters 26 (1971) 492 (with M. J. Cooper and J. M. H. Levelt Sengers).

[42] Critical Phenomena, Proc. of the int. school of physics Enrico Fermi, Course 51 (Academic Press, 1971).
[43] Generalized Ornstein–Zernike approach to critical phenomena. III, The critical null space, Phys. Rev. A **4** (1971) 1282 (with J. D. Gunton).
[44] Phase Transitions and Critical Phenomena, Volumes 1 and 2 (Academic Press, London, 1972) (edited with C. Domb).
[45] An Exact Expression for the Critical Exponent η, Phys. Letters, **42A** (1972) 7 (with J. D. Gunton).
[46] Critical Phenomena 1965–1972: Questions and Answers, Rev. Mex. de Fis. **21** (1972) 279.
[47] Exact Renormalization in the Statistical Mechanics of Fluids, Phys. Rev. A. **8** (1973) 1998.
[48] Dilation and Conformal Covariance of Multipoint Correlation Functions and Dimensions of Fluctuating Quantities at the Critical Point of Fluids, Phys. Rev. Letters **31** (1973) 1193 (with A. M. Wolsky and J. D. Gunton).
[49] Response Under Arbitrary Groups of Point Transformation of Multipoint Correlation Functions of Local Fluctuating Quantities, Phys. Rev. A. **9** (1974) 957 (with A. M. Wolsky).
[50] Nonlinear Extended Scaling in a Mermin-Invariant Parametric Form (preprint; with F. J. Cook).
[51] Observation of the Deviation from Ornstein–Zernike Theory in the Critical Scattering of Neutrons from Neon, Phys. Rev. Letters **32** (1974) 1410 (with V. P. Warkulwiz and B. Mozer).
[52] Renormalization Group in Critical Phenomena and Quantum Field Theory, Proceedings of a Conference (Temple University 1974) (edited with J. D. Gunton).
[53] ε-Expansion Solution of Wilson's Incomplete Integration Renormalization Group Equations, Phys. Rev. Letters **33** (1974) 1263 (with P. Shukla).
[54] Phase Transitions and Critical Phenomena, Vol. 3 (Academic Press, London, 1974) (edited with C. Domb).
[55] University of the Exponent η to Order ε for a Class of Renormalization Groups, Phys. Rev. Letters **34** (1975) 436 (with P. Shukla).
[56] Ising Model Critical Indices in Three Dimensions from the Callan–Symanzik Equation, Phys. Rev. Letters **36** (1976) 1351 (with G. A. Baker, Jr., B. G. Nickel and D. I. Meiron).
[57] Phase Transitions and Critical Phenomena, **5A** (Academic Press, London, 1976) (edited with C. Domb).
[58] The Renormalization Group and the Universality of Critical Exponents, AIP Conf. Proc., No. 27, Topics in Statistical Mechanics and Biophysics: A Memorial to Julius L. Jackson 1976 (with P. Shukla).
[59] Invariance Properties of the Renormalization Group, Lecture notes in physics–54, Critical phenomena, Sitges international school on statistical mechanics, Sitges, Barcelona, Spain, eds. J. Brey and R. B. Jones (Springer-Verlag 1976).
[60] Phase Transitions and Critical Phenomena, Volume **5B** (Academic Press, London, 1976) (edited with C. Domb).
[61] Revised and Extended Scaling for Coexisting Densities of SF_6, Phys. Rev. A, **16** (1977) 2483 (with M. Ley-Koo).
[62] Invariance of Critical Exponents for Renormalization Groups Generated by a Flow Vector, Phys. Rev. B, **15** (1977) 5367.

[63] Phase Transitions and Critical Phenomena, **6** (Academic Press, London, 1976) (edited with C. Domb).

[64] A van der Waals Fixed Point, in: Statistical mechanics and statistical methods in theory and application, ed. Uzi Landman (Plenum, New York, 1977). Dedicated to Elliott Montroll upon his 60th birthday.

[65] Phase Transitions, in: Encyclopedia of physics, eds. R. G. Lerner and G. L. Trigg (Dowden, Hutchinson & Ross, Inc., 1978).

[66] Fluctuations and Non-Linear Irreversible Processes, Phys. Rev. A. **19** (1979) 1747 (with H. Grabert).

[67] Consequences of the Renormalization Group for the Thermodynamics of Fluids Near the Critical Point (preprint with M. Ley-Koo).

[68] Fluctuations and Nonlinear Irreversible Processes II, Phys. Rev. A. (in press) (with H. Grabert and R. Graham).

CONTENTS

Dedication to M. S. Green	vii
H. J. Raveché	
Remarks on M. S. Green (1)	xv
E. G. D. Cohen	
Remarks on M. S. Green (2)	xix
Michael E. Fisher	
M. S. Green's Curriculum Vitae	xx
M. S. Green's Publications	xxii
Contents	xxvii

Part I. Non-equilibrium Processes	1
Ch. 1. The Lorentz gas	
B. J. Alder and W. E. Alley	3
Ch. 2. Some recent developments in the kinetic theory of gases	
J. R. Dorfman	23
Ch. 3. Non-equilibrium fluctuations and the hierarchy	
M. H. Ernst and E. G. D. Cohen	59
Ch. 4. Green's contributions to non-equilibrium statistical mechanics revisited	
L. S. Garcia-Colin and J. L. Del Rio	75
Ch. 5. Stochastic description of many-body systems	
N. G. van Kampen	89
Ch. 6. H-theorems for Markoffian processes	
R. Kubo	101
Ch. 7. Energy flow and thermal conductivity in one-dimensional, harmonic, isotopically disordered crystals	
R. J. Rubin	111
Ch. 8. Where do we go from here?	
R. Zwanzig	123

Part II. Phase Transitions — 135

Ch. 9. A new model Hamiltonian for a correlated electron system within the general framework of critical phenomena and phase transitions
C. Di Castro — 137

Ch. 10. Membrane flux: conditions for limit cycle oscillations
A. G. De Rocco and G. L. Clark — 155

Ch. 11. Critical phenomena – A model illustration of scientific method
C. Domb — 173

Ch. 12. Coarse-grained Helmholtz free energy functional
K. Kawasaki, T. Imaeda and J. D. Gunton — 201

Ch. 13. Exact renormalization in two dimensional ising systems
J. M. J. van Leeuwen — 225

Ch. 14. How close is "close to the critical point"?
J. M. H. Levelt Sengers and J. V. Sengers — 239

Ch. 15. The interfaces between fluid phases
B. Widom — 273

Part III. Foundations — 293

Ch. 16. On higher order WKB approximations for the calculation of energy levels
F. T. Hioe, E. W. Montroll and M. Yamawaki — 295

Ch. 17. Equilibrium density matrix for fluids
J. E. Mayer — 323

Ch. 18. The mechanisms of stochasticity in classical dynamical systems
A. S. Wightman — 343

Subject Index — 365

PART I

Non-equilibrium processes

CHAPTER 1

The Lorentz Gas

B. J. ALDER and W. E. ALLEY

*Lawrence Livermore Laboratory, University of California
Livermore, California 94550
U.S.A.*

Contents

1. Introduction — 5
2. Rigorous results — 6
3. Graph theory — 7
4. Computer results at low density — 9
5. Computer results at intermediate densities — 13
6. Theory — 15
References — 21

1. Introduction

We have chosen to discuss the Lorentz gas [1–3] in this book which is dedicated to the memory of Mel Green, for two principal reasons: Firstly, we believe that unless the behavior of the Lorentz gas is understood in detail, the theory of transport properties is in bad shape. This is because the Lorentz gas, in which a single point particle moves through a randomly arranged set of fixed scatterers, is the simplest conceivable model of transport, yet it appears not to obey, even in the low-density limit, the molecular chaos approximation, thought so generally applicable. The results for the Lorentz gas, hence, challenge the validity of even the Boltzmann equation and the fundamental belief that correlations can be ignored for any real system after a sufficiently long time has elapsed. To put it in other words, if for this model, in which the motion of the test particle can never be correlated with the behavior of the particles with which it collides, because they are held stationary, the test particle does not follow Markovian behavior at long time, even at low density, what actual system does evolve with this stochastic behavior?

This brings us to the second reason for discussing this topic. These developments were made possible primarily through a very fundamental contribution to the theory of transport properties by Mel Green, namely, the fluctuation–dissipation approach [4, 5]. That approach gives general expressions for the transport coefficients valid at all times and densities in terms of correlation or autocorrelation functions calculated from a system at equilibrium. Previously theories of transport coefficients were largely based on kinetic theory which requires, in analogy with experiments, imposition of an external gradient and which, furthermore, in order to become tractable, break correlations. In its simplest execution correlations are broken between successive collisions, so that a theory results which is rigorously valid only up to such a short time till a particle undergoes a collision a second time [3]. At low density it was thought that the second collision would be with a totally uncorrelated partner so that the kinetic theory could be extended to all times. Computer calculation of autocorrelation functions on fluids of hard disks and spheres uncovered the fallacy of that assumption [6, 7].

It is the combination of computer simulation and the autocorrelation formulation that has proved so fruitful in advancing understanding of transport theory. The autocorrelation function expressions in themselves are not much help, since they too can only be rigorously evaluated for short times. The computer, however, can numerically solve the many-body problem through many collisions per particle. This overcomes the analytical difficulties in extending the time range over which the autocorrelation functions can be evaluated. Thus, having such generally valid expressions for the transport coefficients available is an absolute necessity. The other attractive feature of this approach is that the calculations can be performed for a system at equilibrium, thus avoiding boundary complications and steep gradients that will occur if an external field is applied to the very small systems that can be simulated on computers. Such external gradient calculations have been successfully carried out by computer simulation, though they have to be extrapolated by studying successively larger systems and weaker gradients to obtain the ordinary (linear) transport coefficients [8, 9]. Beside that, the very valuable information contained in the behavior of the autocorrelation function itself is not available through the kinetic approach which yields only the value of the transport coefficient itself [3].

2. Rigorous results

We would like to discuss the autocorrelation function calculated by computer for the Lorentz gas and first of all at low density. For the Lorentz gas, the collision partners of the single moving particle are, as already mentioned, dynamically uncorrelated, and hence one would have thought the Boltzmann equation to be valid. The only possible long-time correlation effects that can arise are geometric or topological in nature, in that there could exist long trajectories that scatter the moving particle with higher than random probability back to its origin (even at low density when these scatterers are randomly distributed) [2, 10, 11]. One would have also thought, because of the relatively simple nature of this dynamic problem, that mathematically rigorous statements about this possibility can be made. Although this problem is actively pursued by eminent mathematicians, as yet this possibility cannot be discounted.

Rigorous results on the Lorentz gas so far prove the existence of transport coefficients only if the possibility of long time correlations, or long mean-free paths, is eliminated [12]. This is achieved by going to the so-called Boltzmann–Grad limit. In this limit one goes to zero density by letting the scattering particles get smaller while their number gets larger in such a way that the mean-free path stays constant [2].

Essential to the existence of a transport coefficient, for example the diffusion coefficient, would also be a proof that the distribution of beginning-to-end distances the moving particle has travelled after a long time (from many different initial conditions) is Gaussian. If that is true, the second moment of the Gaussian distribution divided by the time defines the diffusion constant. We shall return to a discussion of this very important distribution and suggest from numerical experiment that the distribution approaches the long-time Gaussian limit in a non-Gaussian manner. As far as rigorous proofs go, the Lorentz gas has been shown to obey the central-limit theorem [13], which suggests Gaussian limiting behavior, but rigorously speaking concerns itself only with the ergodic behavior of this system [14, 15].

3. Graph theory

The unexpected long-time behavior of the velocity autocorrelation function for the Lorentz gas at low density was predicted first from graph theoretical considerations [10, 11] similar to the ones that extended the kinetic theory of fluids to account for the long-time effects found there [16]. This approach is less rigorous from a mathematical point of view because a set of graphs are summed which are thought to determine the long-time correlations. In the present case these are ring graphs, representing repeated collisions of a particle which brings it back to its origin. However, consideration is not given to other, more complex sets of graphs, that could either cancel the asymptotic behavior of the ring graphs or, in fact, dominate them. Nevertheless, this graph theory is the best we have and it makes several important predictions. First, the velocity autocorrelation of a point particle, moving among a set of N disk scatterers of radius R, which are confined to an area A, is predicted to decay at long times as $-n^*/\pi t^2$, where $n^* = NR^2/A$ [17]. Note that the long-time behavior is predicted to be non-exponential or non-Markovian. Also note the negative sign which indicates an anticorrelation or a higher than random return probability. As opposed to the two-dimensional fluid, however, the diffusion coefficient for the two-dimensional Lorentz gas, the integral of the velocity autocorrelation function, is predicted to exist (that is, a finite diffusion coefficient is predicted).

A vanishing diffusion coefficient is predicted by the graph theory for a slight variant of the Lorentz gas, namely, the Ehrenfest wind-tree model [18, 19]. In this two-dimensional model squares (trees) are randomly placed such that the corners point in the x and y directions. The moving point particle (wind) is initially started in either the x or y direction and, by the placement of the squares, continues to travel only in those two directions. Since, in the random placement of squares, overlapping of some squares is possible, even in the low-density limit, a zero diffusion coefficient comes about because the moving particle sooner or later collides with two such overlapping squares exactly at their intersection, whereupon it exactly retraces its trajectory. If the squares are not allowed to overlap, a finite diffusion coefficient is predicted. This fact is worth recalling because it has also been predicted [20] that the diffusion coefficient vanishes for the overlapping Lorentz gas in the low density limit, although in that case phase space is much less restricted than in the Ehrenfest tree model. Nevertheless, these arguments make it physically plausible why these systems might have long trajectories in which particles are either trapped or return to their origin with a larger than random probability.

The graph theory predicts not only a finite diffusion coefficient in the low-density limit for the Lorentz gas, but also leads to a non-analytical density expansion of the diffusion coefficient away from that limit. The fact that the density dependence of the diffusion coefficient can not be represented by a power series in the density, but instead involves the logarithm of the density has, in the case of a fluid, been shown to be intimately related to the power law decay of the autocorrelation function [21, 22]. The same is true here.

The final prediction of the graph theory which we want to discuss here, is the divergence of the Burnett coefficients. The Burnett coefficients for diffusion represent the higher cumulants of the distribution of distances the particle travels. These cumulants are merely higher moments of the distribution defined in such a way that, for a Gaussian distribution, they vanish [23]. Thus, the first non-vanishing Burnett coefficient is the rate at which the fourth moment minus three times the second moment squared approaches its long-time limit. Physically, these Burnett coefficients measure the dependence of the transport coefficients on the distance scale or size (not amplitude) of the applied gradients or on the wave-length of the fluctuation. The graph theoretical prediction is that each successive Burnett coefficient (successive cumulant) has a long-time power-law decay of one higher power [24]. Thus, the first non-vanishing Burnett coefficient for the two-dimensional low-density Lorentz gas has an autocorrelation function that decays at long time as $1/t$ so that the time integral (the Burnett coefficient) diverges.

Since these Burnett coefficients are precisely the coefficients that appear in the Chapman–Enskog expansion for the solution of the Boltzmann equation, that expansion does not exist, according to the graph theory. The physical cause of the divergence of the Burnett coefficients, as defined by the cumulants, is the non-Markovian character of the process which must inevitably lead to a non-Gaussian long-time approach of some higher moments and hence to the conclusion of an inappropriate formulation for the coefficients.

4. Computer results at low density

The fragmentary computer studies up to the present have confirmed the predictions of the graph theory. The overlapping Ehrenfest tree model was shown to have, at a single density, a power-law decay in the velocity autocorrelation function at long times which led to a zero diffusion coefficient [25]. This abnormal diffusion was contrasted with a finite diffusion coefficient result for the non-overlapping Ehrenfest model. Two points should be made in connection with these results:

The first is that the diffusion coefficient for any overlapping system of particles will vanish at sufficiently high density, namely, beyond the percolation density. This is because at some density the randomly placed overlapping particles will always form closed regions, so that no matter where the moving particle is initially placed, it will never be able to escape from that region and hence the particle is trapped forever, leading to vanishing diffusion. The percolation density, where phase space for the first time becomes disconnected, has also been investigated for the overlapping Lorentz gas but, since it forms a separate subject, will not be discussed here [26]. The point is that the percolation density for the overlapping Ehrenfest model is at zero density.

The second point concerns the accuracy with which the computer can be made to follow trajectories since, with a finite number of digits with which positions and velocities can be calculated, round-off errors inevitably introduce irreversibility into the system. The Ehrenfest model proves, however, to be an exception, since the calculation is sufficiently simple so that it could be reduced to integer algebra. Hence, trajectories could be followed precisely through many collisions and the probability of exact retracing established [25].

For the disk Lorentz gas this is not possible; in a surprisingly few collisions the particle will not reverse its history, once its velocity is exactly reversed. It was established that in each collision about one significant number is lost, so that on a twelve digit computer, after about twelve collisions, the events are not retraceable. This casts some doubt on the validity of the asymptotic behavior of autocorrelation functions obtained from computer results at times much longer than twelve collision times. These doubts have been largely dispelled by showing that the same asymptotic behavior is obtained if the particle scatters randomly (or diffusively) instead of specularly from each stationary particle. The point is that the long time behavior of the correlation function is determined by the topology of the space, and that the velocity persistence, which is broken up by the diffusive process, is insignificant after a few elastic collisions anyway. In all cases the systems studied were made large enough so that the asymptotic behavior was not influenced by the use of periodic boundary conditions.

The previous computer study of the two-dimensional Lorentz gas at low density was able to establish consistency with the theoretical expression for the logarithmic dependence on density of the diffusion coefficient [27]. The present more extensive computer runs were able to establish consistency with a $-\alpha_D/t^2$ decay of the velocity autocorrelation function at the lowest densities that were feasible to investigate [28]. Practical considerations, primarily connected with the increasing size of the system required due to increasingly larger mean-free paths as the density is lowered, make it difficult to get precise low density information. Thus, as table 1 shows, at the highest of the three lowest densities investigated, the density is sufficiently high so that the diffusion coefficient, D, is already about half of the kinetic theory (Enskog, D_E) result. Nevertheless, this not so low density appears still to be consistent with a $-\alpha_D/t^2$ predicted autocorrelation function behavior at low-density and, furthermore, the result at this density is required, as fig. 1 shows, to

Table 1.[a] Long-time behavior at low density of the autocorrelation functions leading to the diffusion and Burnett coefficients, as well as the diffusion coefficients and the generalized Burnett coefficients themselves.

n^*	D/D_E	$-\alpha_D$	α_B	α_B^M	$D_4/D\lambda^2$
0.10	0.66_1	0.20_5	0.16_2	0.18_4	0.22_3
0.05	0.81_1	0.060_5	0.08_2	0.065_4	0.23_2
0.03	0.87_1	0.026_5	0.04_2	0.029_5	0.23_2

[a] The last small number in each entry signifies the uncertainty in the last digit given.

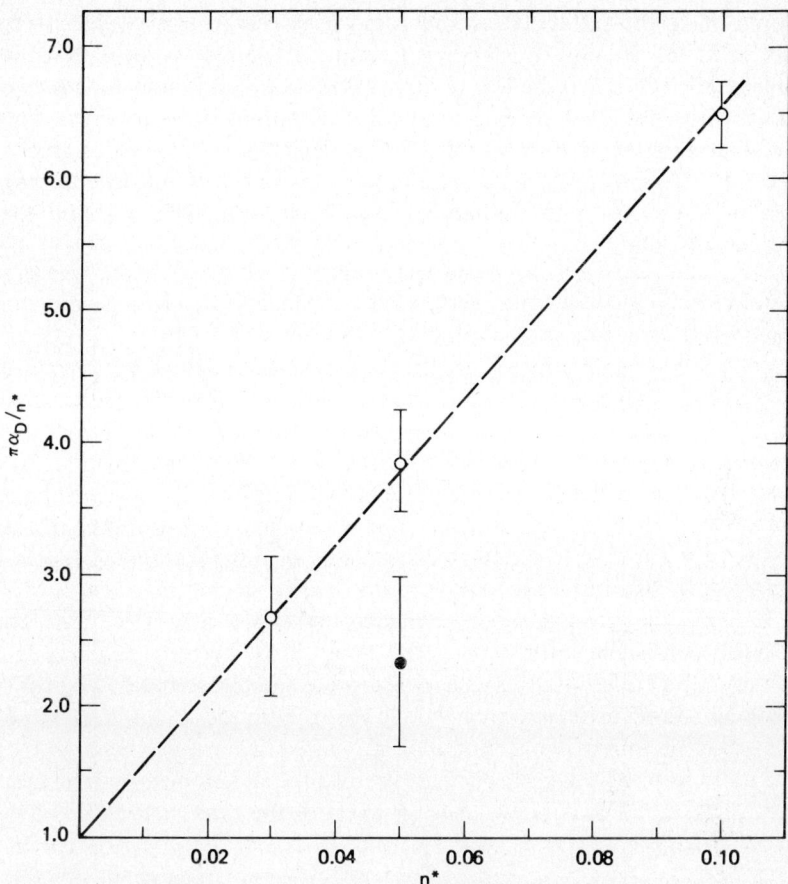

Fig. 1. The coefficient, α_D, of the inverse-square power-law tail of the velocity autocorrelation function at three low densities. The open circles represent the overlapping Lorentz gas and the closed circle represents the non-overlapping case [29]. The least-squares dashed line through the three data points extrapolates to the theoretical result of unity at zero density.

extrapolate to the low-density prediction of $\alpha_D = n^*/\pi$. Considering the large error bars and the large extrapolation by more than a factor of two from the lowest density result, only consistency with the graph theoretical prediction can be claimed. There is perhaps a theoretical suggestion that the next higher density correction to the low density asymptotic behavior of the velocity autocorrelation is also of the form $1/t^2$ and perhaps the class of graphs to obtain that result can be found. This would make the extrapolation much more reliable. It is also clear from the result for the non-overlapping Lorentz gas shown in fig. 1 that the density corrections to the low-density result will be different for the two cases. Only in the low-density limit will overlapping have no effect. The fact that an effect is observed at the low densities investigated, namely, that the diffusion coefficient is significantly larger in the non-overlapping case, is consistent with the large extrapolation required to obtain the low-density limit.

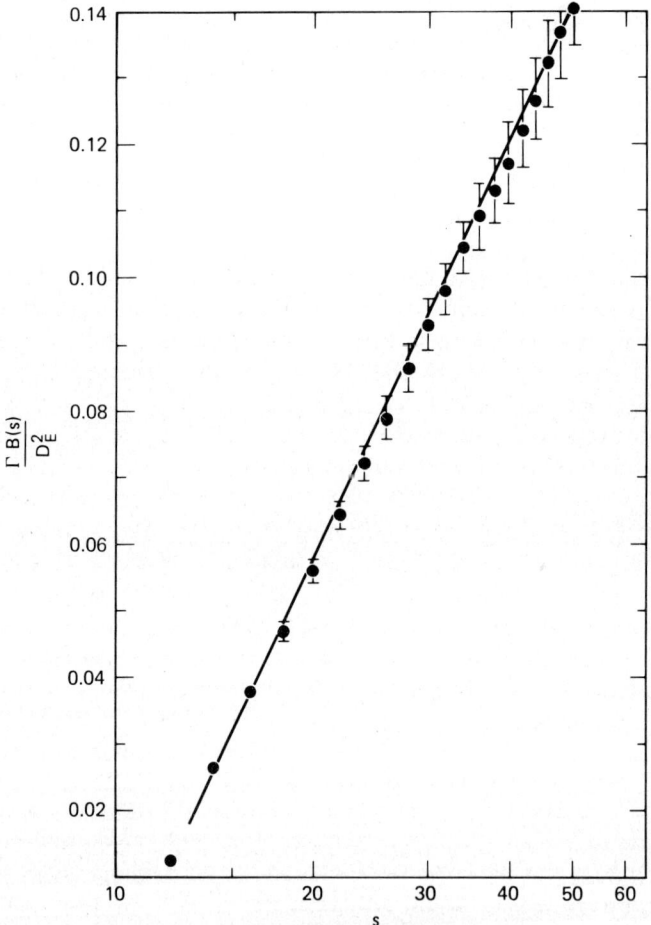

Fig. 2. The behavior of the dimensionally reduced Burnett coefficient at long times at $n^* = 0.05$ as deduced from a semi-logarithmic plot. The solid line merely indicates the validity of a linear relation.

The first non-vanishing Burnett coefficient, B, was shown at the three low densities investigated to have the predicted long-time decay of its autocorrelation function of α_B/t. This is illustrated in fig. 2, where the Burnett coefficient, made non-dimensional by multiplying by the collision rate, Γ, and by dividing by the Enskog diffusion coefficient squared, is plotted against the logarithm of the time, s, made non-dimensional in terms of the collision time. The numerical values of the coefficients, α_B, are given in table 1 and their low-density extrapolation has yet to be compared to the graph theoretical results once it becomes available. But the important fact that each higher cumulant depends on time, at large times, with one higher power appears to be verified. In the present case it means logarithmic divergence of even the first Burnett coefficient as vividly demonstrated by fig. 2.

5. Computer results at intermediate densities

Because the low density computer studies suffer from lack of accuracy, we have extended the studies in the overlapping Lorentz gas to higher density [28]. Even though graph theoretical results are unavailable anywhere but at low density, the hope is that some empirical rules can be established from the data at intermediate density that will suggest further theoretical development. Our principal aim in taking on the Lorentz gas was to understand the cause of the divergence of the Burnett coefficient for this simplest conceivable transport problem. Thus, our first objective was to establish that the velocity autocorrelation at all densities has a power law decay at long times of the form $-\alpha_D/t^{\beta_D}$, and to determine how β_D differs from its low density value of 2. We thought of this behavior as being determined by the complex geometry of the overlapping Lorentz gas at any given density, which must eventually be established by some topological considerations which were outside the scope of the question of the Burnett divergence. The only necessary fact to be established for present purposes was that the velocity autocorrelation decays non-exponentially. Then the question was the nature of the long-time behavior of the Burnett autocorrelation function. Specifically, does this autocorrelation function decay as α_B/t^{β_B} and can the low-density result be generalized, namely, $\beta_B = \beta_D - 1$? In other words, can we at least, given the value of the second moment or what is equivalent the velocity autocorrelation function, predict the behavior of the higher moments or cumulants at long times? If that were true, then we would maintain the simplifying feature of a Gaussian

Table 2.[a] Long-time behavior, at intermediate density, of the autocorrelation functions leading to the diffusion and Burnett coefficients, as well as the diffusion coefficients and the generalized Burnett coefficients themselves.

n^*	D/D_E^b	$-\alpha_D$	β_D	β_B	α_B	α_B^M	$D_4/D\lambda^2$
0.300	0.11_3	0.26_3	1.47_5	0.53_2	0.06_2	0.04_1	
0.240	0.27_2	0.20_2	1.48_6	0.50_5	0.07_1	0.08_1	
0.200	0.34_2	0.23_1	1.59_5	0.60_1	0.09_1	0.12_1	0.26_4
0.180	0.41_2	0.20_5	1.60_5	0.62_4	0.11_2	0.11_3	
0.143	0.52_2	0.22_5	1.80_5	0.87_3	0.15_2	0.16_3	0.22_3

[a] The last small number in each entry signifies the uncertainty in the last digit given.
[b] A tail correction has been applied.

stationary Markov process in which all higher moments can be expressed in terms of the second moment. Table 2 demonstrates that, within the accuracy of the data, the Burnett coefficients diverge with unit difference in the power-law tail from the diffusion coefficients.

Besides the relation between the tails of the moments, we thought we better establish (at the most favorable density from the numerical point of view) the nature of the distribution of distances at long times. The strong suspicion is that it is non-Gaussian because of the non-Markovian nature of the second moment and the divergence of the higher cumulants. Figure 3 gives the evidence. The distribution f is plotted relative to that of a Gaussian f_G, of the same half width $\langle x^2 \rangle^{1/2}$, at a given density, at three different times, as a function of distance measured in terms of the half width. The graph shows the large deviation from Gaussian behavior, even at the longest possible time that could be investigated. From the three different time curves it is clear that the approach to Gaussian behavior is extremely slow and it is an act of faith from this data to believe that the limiting distribution is Gaussian. The positive deviation of this distribution from Gaussian at short distances indicates particles that are trapped. As time proceeds that peak decreases indicating the reduced importance of such particles relative to the diffusing particles whose distance traveled continues to spread.

We must distinguish between two kinds of trapped particles. There are those that are located within enclosed regions of the overlapping scatterers and hence are truly trapped with zero diffusion coefficients. Since the distance these particles can travel is bounded, at long times their contribution to the diffusion coefficient vanishes. Nevertheless, they too have power law decays as demonstrated for the Ehrenfest model earlier and for the Lorentz model above the percolation density

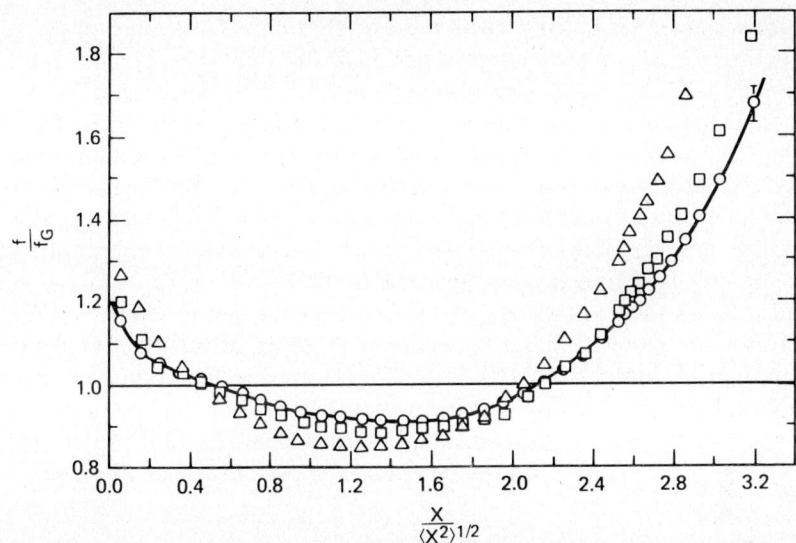

Fig. 3. The distribution function of distance traveled relative to a Gaussian distribution of the same half width, $\langle x^2 \rangle^{1/2}$, as a function of distance at $n^* = 0.2$ and at three different times: 8 collision times ($\triangle$); 48 collision times ($\square$); 100 collision times ($\bigcirc$). The solid curve is drawn as a visual aid through the largest time points.

in this study. Other particles are temporarily trapped in that they cannot escape from a region of space until they find a rarely occurring escape route. These particles must wait a very long time before they can diffuse out of the nearly trapped state, and that too causes higher than random return probabilities and the non-exponential tail even at infinite times. Evidence for the nearly trapped particle leading to power law decay comes from the non-overlapping Lorentz gas, where such behavior is found even though there can not be any permanently trapped regions except in the limit of close-packing.

6. Theory

Given the observations that the distribution approaches its long-time Gaussian limit in a non-Gaussian manner, that the higher moments are related to the second moment, and that long trapping times cause the power-law decays, it seemed physically plausible that the results could be explained by a generalization of the usual random walk [30–32].

Such a generalization has been successfully utilized in similar physical situations that involve jump times [33]. In this generalization, the particle takes a random step of length and direction determined from a spatial distribution, but between random steps the particle waits for a time as sampled from a waiting-time distribution. This waiting-time distribution corresponds to the different times it takes for the particle to escape from the various nearly-trapping regions it finds itself in. This waiting-time distribution must also have a long-time tail, corresponding to the very long time it takes for some particles to find their way out of the trap. In fact, knowledge of the waiting-time distribution is all the information required about the complex topology of the Lorentz gas to determine its transport coefficients. We circumvent this complex topology problem by requiring that the waiting-distribution $\psi(t)$ be such that the second moment of the distribution of distances as a function of time $\langle x^2(t) \rangle$, as determined on the computer, be reproduced at all times.

$$\psi(t) = \frac{2}{\langle l^2 \rangle} \int_0^t D(t-t')[\delta(t') - \psi(t')] \, dt',$$

where

$$D(t) = d\langle x^2(t) \rangle / 2 \, dt,$$

$\langle l^n \rangle$ is the nth moment of the step-size distribution, and $\delta(t)$ is the Dirac delta function. The above integral equation has been numerically solved for a given $D(t)$, with the result shown in fig. 4. It is seen that negative (and thus physically unacceptable) waiting times are obtained at intermediate times. At long times the waiting distribution is positive and does have the expected power-law decay. It is only at long times, when the higher order correlations have decayed or when the higher moments decouple that these are expressible in terms of the two point correlation or the second moment and only then does the random walk concept apply. Hence, in the analysis to follow, only long times will be considered.

At long times the fourth moment of the generalized random walk can be written in terms of the second moment in the following way:

$$\langle x^4(t) \rangle - 12 \int_0^t D(t-t') \langle x^2(t') \rangle \, dt' = \langle l^4 \rangle \langle x^2(t) \rangle / \langle l^2 \rangle.$$

If the long time expansion of $D(t-t')$,

$$D(t-t') = D(t) - t' \frac{dD}{dt} + \ldots = D(t) + t' \alpha_D / t^{\beta_D} + \ldots$$

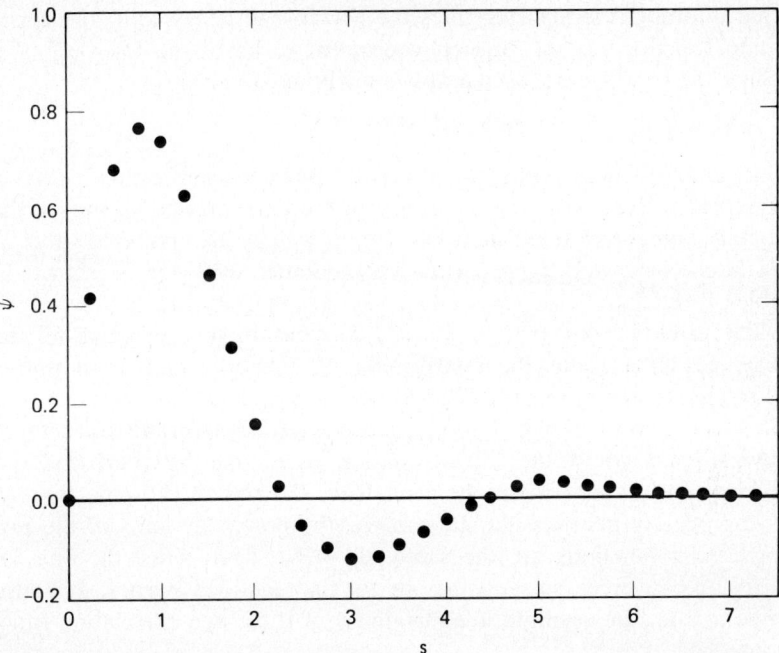

Fig. 4. The waiting-time distribution at $n^* = 0.20$ as a function of mean collision time s.

is substituted into the fourth moment, it is easy to show that, if the second moment had no long-time power-law decay, the resulting Burnett coefficient would be convergent. This is evident from substituting for $D(t-t')$ its long-time limit, namely the constant plateau value, D, upon which the left hand side becomes

$$\langle x^4(t) \rangle - 3 \langle x^2(t) \rangle^2$$

which, when differentiated with respect to time and divided by 4!, in the limit of long times, becomes the definition of the Burnett coefficient. It is well behaved because the right hand side of the above equation exists, by assumption, of a well behaved second moment.

If the second moment, on the other hand, has a long-time tail, then the next term, dD/dt, cannot be ignored. That term, which is just the velocity autocorrelation function and which, therefore, at long times from empirical evidence behaves as $-\alpha_D/t^{\beta_D}$, when substituted in the above equation, leads to an additional term, beyond the Burnett coefficient, on the left hand side that dominates the asymptotic behavior.

That additional term determines the asymptotic behavior of the correlation function for the Burnett coefficient to be of the form $\alpha_B^M/t^{\beta_B^M}$, where the coefficients, by the above analysis, are given by

$$\alpha_B^M = 4D\alpha_D/3D_E \quad \text{and} \quad \beta_B^M = \beta_D - 1.$$

This analysis thus leads to divergent Burnett coefficients with an asymptotic decay which is one power of the time larger than that of the velocity autocorrelation function. This, as well as the predicted value of the coefficient, α_B^M, agrees with the computer evidence as shown in tables 1 and 2.

The random walk with a waiting distribution thus explains all the observed facts about the distribution of distances and furthermore, allows us, by redefining the Burnett coefficients, to obtain well behaved moments. The redefined Burnett coefficient D_4 is naturally taken to be the left hand side of the above equation (after time differentiation and dividing by 4!) and it can be seen from the right hand side that D_4 asymptotically behaves like D, namely, the power-law tails of the two correlation functions are the same, $\beta_{D_4} = \beta_D$. Thus, since the one for diffusion converges, so must the one for the redefined Burnett. It is also evident from the asymptotic equivalence of these two correlation functions, that

$$\alpha_{D_4} = -\frac{1}{16}\frac{\Gamma}{D_E}\frac{\langle l^4 \rangle}{\langle l^2 \rangle}\alpha_D = -\frac{3}{16}\frac{\Gamma}{D_E}\lambda^2 \alpha_D,$$

where Γ is the collision rate. The last equality comes from identifying the step-size distribution as that of the free-path distribution, which is Gaussian, with an average which is the mean-free path, λ. Thus, a further consequence of this redefined Burnett coefficient and its asymptotic behavior is that

$$D_4/D\lambda^2 = \tfrac{1}{4},$$

also shown to hold within numerical uncertainty in tables 1 and 2 and graphically demonstrated in fig. 5. Hence, not only do we obtain a convergent fourth moment, but a quantitative evaluation of it as well.

The next higher order Burnett coefficient for the random walk with waiting distribution is predicted to be

$$D_6 = \frac{1}{6!}\frac{d}{dt}\left[\langle x^6(t)\rangle - 60\int_0^t D(t-t')\langle x^4(t')\rangle \, dt' \right.$$
$$\left. + 360\int_0^t D(t-t')\,dt'\int_0^{t'} D(t'-t'')\langle x^2(t'')\rangle \, dt''\right]$$
$$= 2D\langle l^6\rangle/6!\langle l^2\rangle = D\lambda^4/24.$$

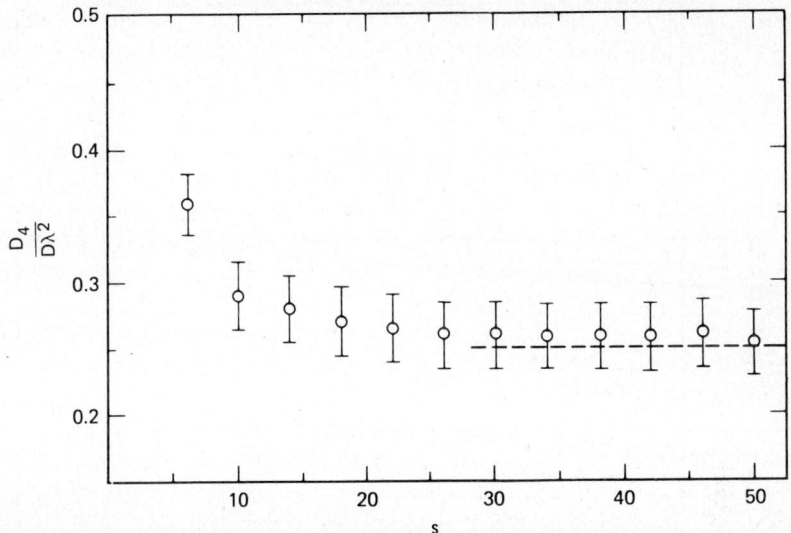

Fig. 5. The newly defined Burnett coefficient D_4 as a function of mean particle collision time at a density of $n^* = 0.20$. The vertical bars indicate the numerical uncertainty and the horizontal dashed line is the theoretical prediction.

The latter equality again assumes a Gaussian step-size distribution whose second moment is the mean-free path squared. Figure 6 shows that the marginal data that is available for this higher moment from the computer simulation is at least consistent with the prediction of a convergent coefficient.

It is clear that if this random walk with a waiting-time distribution derived from the second moment can reproduce all the information available about the long time behavior of the higher moments, it can also predict the asymptotic behavior of the distribution of distances itself. That distribution can be shown to obey the following integral differential equation:

$$\frac{\partial f(x,t)}{\partial t} = \sum_{n=1}^{\infty} \frac{D_{2n}}{D} \int_0^t \frac{dD(t-t')}{dt} \frac{\partial^{2n} f(x,t')}{\partial x^{2n}} dt',$$

where D_{2n} are the newly defined Burnett coefficients which have been determined to be

$$D_{2n} = 2D \langle l^{2n} \rangle / \langle l^2 \rangle (2n)!.$$

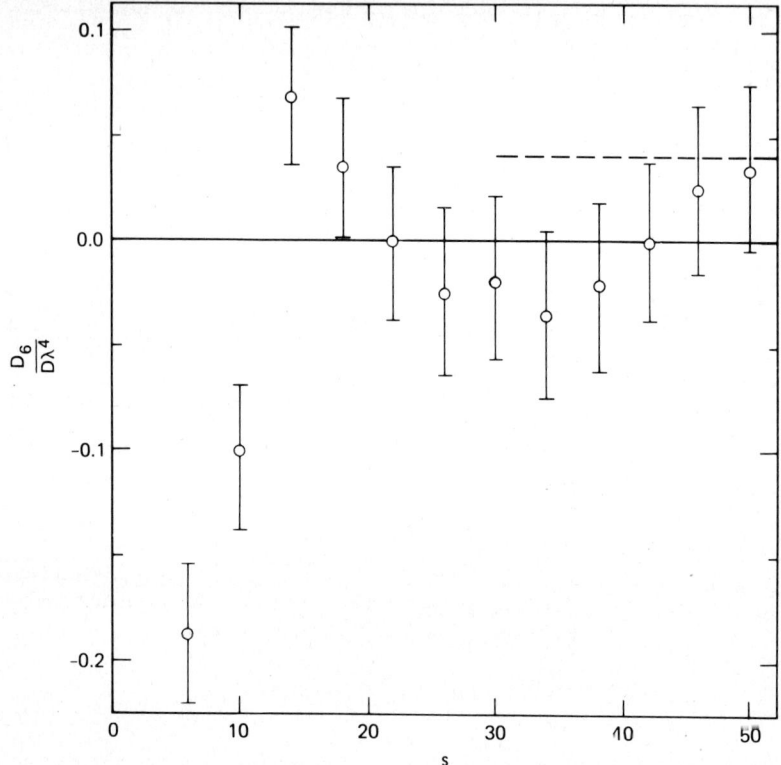

Fig. 6. The newly defined sixth-order Burnett coefficient D_6 as a function of mean-particle collision times. The vertical bars indicate the numerical uncertainty and the horizontal dashed line is the theoretical prediction.

Solution of the above equation has successfully reproduced the distribution at the largest time (100 collision times) of fig. 3 [32].

By the simple modification of the random walk with a waiting-time distribution we have thus been led to a rather significant modification of transport theory, namely, that Fick's law is not valid whenever the physical situation contains long-time correlations. It seems that the simplest systems have this characteristic. Under these circumstances a local transport equation is not adequate, instead a non-local (in time) theory is required. In the diffusion case the memory function characterizing the non-locality has been shown to be the velocity autocorrelation function (dD/dt). This non-local theory gets around the difficulty of Fick's law which predicts that the distribution of distances is not only Gaussian in the long-time limit, but at all long times, contradictary to computer generated evidence. Allowing for the non-Gaussian approach of the distribution to the Gaussian limit circumvents divergent Burnett coefficients and leads to a convergent Chapman-Enskog like expansion. For the transport theory of fluids a similar modification must be expected [32]. Thus, a significant new development was made possible through the correlation function approach to transport theory first given by Mel Green.

References

[1] H. A. Lorentz, Proc. Amst. Acad. **7** (1905) 438, 585, 684.
[2] E. H. Hauge, Lecture Notes in Physics, eds. G. Kirzenow and J. Marrow (Springer-Verlag, New York) **31** (1974) 337.
[3] Chapman and Cowling, The Mathematical Theory of Non-Uniform Gases (Cambridge University Press, 1970).
[4] M. S. Green, J. Chem. Phys. **20** (1952) 1281.
[5] M. S. Green, J. Chem. Phys. **22** (1954) 398.
[6] B. J. Alder and T. E. Wainwright, Phys. Rev. **A1** (1970) 18.
[7] B. J. Alder, D. M. Gass and T. E. Wainwright, J. Chem. Phys. **53** (1970) 3813.
[8] W. T. Ashurst and W. G. Hoover, Phys. Rev. Lett. **31** (1973) 206.
[9] W. T. Ashurst and W. G. Hoover, Phys. Rev. **11A** (1975) 658.
[10] J. M. J. van Leeuwen and A. Weyland, Physica **36** (1967) 457.
[11] A. Weyland and J. M. J. van Leeuwen, Physica **38** (1968) 35.
[12] J. L. Lebowitz and H. Spohn, J. Stat. Phys. **19** (1978) 633.
[13] L. A. Bunimovich, Th. Prob. Appl. **19** (1974) 65.
[14] Y. Sinai, Russ. Math. Surv. **25** (1970) 137.
[15] Y. Sinai, The Boltzmann Equation, eds. E. G. D. Cohen and W. Thirring (Springer-Verlag, New York, 1973), also Acta Physica Austriaca, Supp. X, p. 575.
[16] J. R. Dorfman and E. G. D. Cohen, Phys. Rev. Lett. **25** (1970) 1257.
[17] M. H. Ernst and A. Weyland, Phys. Lett. **34A** (1971) 39.
[18] E. H. Hauge and E. G. D. Cohen, Phys. Lett. **25A** (1967) 78.
[19] E. H. Hauge and E. G. D. Cohen, J. Math. Phys. **10** (1969) 379.
[20] S. Yip (private communication).
[21] M. H. Ernst and J. R. Dorfman, Phys. Lett. **38A** (1972) 269.
[22] Y. Pomeau and P. Resibois, Phys. Lett. **19C** (1975) 63.
[23] J. A. McLennan, Phys. Rev. **8A** (1973) 1479.
[24] I. M. de Schepper, H. van Beyeren and M. H. Ernst, Physica **75** (1974) 1.
[25] W. W. Wood and F. Lado, J. Comp. Phys. **7** (1971) 528.
[26] S. W. Haan and R. Zwanzig, J. Phys. A: Math. Gen. **10** (1977) 1547.
[27] C. Bruin, Physica **72** (1974) 261.
[28] B. J. Alder and W. E. Alley, J. Stat. Phys. **19** (1978) 341.
[29] J. C. Lewis and J. A. Tjon, Phys. Lett. **66A** (1978) 349.
[30] E. W. Montroll and G. H. Weiss, J. Math. Phys. **6** (1965) 167.
[31] B. J. Alder and W. E. Alley, Stochastic Processes in Non-equilibrium Systems, Lecture Notes in Physics (Springer-Verlag) **84** (1978) 184.
[32] W. E. Alley and B. J. Alder, Phys. Rev. Lett. **43**, (1979) 653.
[33] H. Scher and E. W. Montroll, Phys. Rev. **12B** (1975) 2455.

CHAPTER 2

Some Recent Developments in the Kinetic Theory of Gases

J. R. DORFMAN

Institute for Physical Science and Technology, and
Department of Physics and Astronomy
University of Maryland, College Park, MD 20742
U.S.A.

© *North-Holland Publishing Company 1981*

Perspectives in Statistical Physics
Ed. H. J. Raveché

Contents

1. Introduction   25
2. The generalized Boltzmann equation   25
3. The divergence difficulties   30
4. The time correlation function method   36
5. Density expansions with logarithmic terms   38
6. Long time tails and anomalous transport in fluids   45
7. Conclusion   54
References   54

1. Introduction

The year 1963, in which the fundamental paper by Green and Piccirelli [1] appeared, represents something of a turning point in the development of the kinetic theory of gases. At that time theorists were confident that one of the major problems of kinetic theory – the generalization of the Boltzmann transport equation to higher densities – had been solved, mainly due to the work of Bogoliubov [2], Uhlenbeck, Choh [3], Cohen [4] and Green [1, 5]. Then, all that remained to be done was to work out the consequences of the generalized Boltzmann equation using methods similar to those developed to treat the Boltzmann equation itself.

Now it has turned out that almost none of the anticipated consequences of the generalized Boltzmann equation has proved to be correct, and the theoretical situation as we understand it today is quite different from what we expected it to be in 1963. The purpose of this article is to give an outline of the situation – to contrast the expectations of 1963 with the present understanding and to explain briefly how all of this has come about. I will restrict the discussion to two topics where the most dramatic developments have occurred, namely (1) the properties of the density expansions for the transport coefficients of dense gases and (2) the properties of the gradient expansions of the hydrodynamic fluxes for systems in which some kind of transport is taking place.

I have chosen these topics because they are of central interest to the theory, and for that reason they were of central interest to Mel Green. His contributions to these topics were of fundamental importance, and one of them, the development of the time correlation function method [6], has of course, had a profound influence on our understanding of irreversible processes in general.

2. The generalized Boltzmann equation

For many years one of the most interesting and perplexing problems in the kinetic theory of gases was to find a generalization of the Boltzmann transport equation that would extend the range of kinetic theory from dilute gases to those of moderate and high densities. Interest in this problem stemmed not only from practical concerns for predicting the transport properties of dense gases, but also from the hope that a generalization of the Boltzmann equation would lead to a further clarification and understanding of the conceptual problems such as the recurrence and time reversal paradoxes that arise in Boltzmann's derivation of the transport equation [7]. In 1922 Enskog [8] proposed a transport equation for dense gases, but his equation had three shortcomings that sharply reduced its range of applicability: it applied only to gases composed of hard-sphere molecules, it took only binary collisions into account, and the gas had to be sufficiently close to equilibrium that an equilibrium-like pair distribution could be used to determine the frequency of binary collisions in the gas. Moreover, Enskog derived his equation using intuitive arguments similar to those used by Boltzmann, so his derivation was subject to the same fundamental difficulties. Of course, it was realized early that a satisfactory derivation of the Boltzmann equation, as well as of its generalization to higher densities, would have to be based on the Liouville equation; that is, the various elements in Boltzmann's derivation, such as the *Stosszahlansatz*, would have to be understood as describing the most probable behavior of a member of an ensemble of systems rather than the mechanically determined behavior of any individual system.

It was not until the work of Bogoliubov [2], published in 1947, that real progress was made in the derivation of the generalized Boltzmann equation from the Liouville equation. Bogoliubov was able to show that the Boltzmann equation, as well as its generalization to higher densities, could be derived from the Liouville equation provided one assumed that, as far as their time dependence is concerned, the two- and higher-particle distribution functions become functionally dependent on the one-particle function shortly after the gas starts to develop from some initial state. In addition, Bogoliubov needed a boundary condition on the s-particle distribution function for $s>1$; he had to assume that the s-particle distribution function would factorize to a product of s single-particle functions if one traced all the particle trajectories sufficiently far into the past, following the dynamics of the s-particle system treated in isolation from the rest of the particles in the gas.

Even though Bogoliubov's functional assumption and factorization boundary condition appeared, perhaps, to be less intuitive and more mysterious than the *Stosszahlansatz* they replaced, these conditions had decided advantages. First, they enabled Bogoliubov to derive the Boltzmann equation, and later Choh and Uhlenbeck [3] to derive the first density correction to it, directly from the Liouville equation. Second, these assumptions could be stated as properties of s-particle distribution functions, which made them more directly verifiable by the standard methods of statistical mechanics than the *Stosszahlansatz* had been.

It was this latter point that was taken up separately by Green [1, 5] and by Cohen [4, 9] and their co-workers. They realized that Bogoliubov's method could be greatly simplified and improved by adapting the cluster expansion techniques of equilibrium statistical mechanics to the non-equilibrium case. Moreover, the cluster expansion work suggested a more natural factorization assumption – that the *initial* state of the gas was such that the *initial* s-particle distribution function factorizes to a product of s single-particle distribution functions for all phases of the particles where the s particles are far apart. Using the cluster expansion method and the initial factorization assumption, Green, Cohen and their co-workers were able to derive a generalized Boltzmann equation where the general collision operator has an expansion in powers of the density in which the successive terms are determined by the dynamics of isolated groups of two, three,... particles, respectively, and which differs from Bogoliubov's expansion by a "correction term" that itself has a density expansion. All these expansions are very similar in form to the virial expansions of equilibrium thermodynamic properties. More specifically, Green and Cohen considered the single-particle distribution function $F_1(x_1, t)$, where $x_1 = (r_1, p_1)$ with r_1 and p_1 the position and momentum of a typical particle in the gas, labeled here by the subscript "1" and normalized so that $F_1(x_1, t)dx_1$ is the average number of particles in the gas in region dr_1 about r_1 and dp_1 about p_1 at time t.

An equation, the so-called first hierarchy equation, follows directly from the Liouville equation and the definition of F_1 as [7]

$$\frac{\partial F_1}{\partial t} + \frac{p_1}{m} \cdot \nabla_{r_1} F_1 = \int dx_2 \theta_{12} F_2(x_1, x_2, t), \qquad (2.1)$$

where F_2 is the two-particle distribution function, obtained by integrating the N-particle distribution appearing in the Liouville equation over all but two particles, and θ_{12} is given by:

$$\theta_{12} = \frac{\partial \phi(r_{12})}{\partial r_1} \cdot \left(\frac{\partial}{\partial p_1} - \frac{\partial}{\partial p_2} \right), \qquad (2.2)$$

where $\phi(r_{12})$ is the intermolecular pair potential, assumed to be a function of the relative distance $r_{12} = |r_1 - r_2|$. Using the first hierarchy equation and a cluster expansion method for evaluating F_2, Green and

Cohen derived an equation for F_1 of the form:

$$\frac{\partial F_1}{\partial t} + \frac{p_1}{m}\cdot\nabla_{r_1}F_1 = J_t(F_1, F_1) + K_t(F_1, F_1, F_1) + L_t(F_1, F_1, F_1, F_1) + \ldots$$
$$+ \varepsilon_2(F_1, F_1) + \varepsilon_3(F_1, F_1, F_1) + \varepsilon_4(F_1, F_1, F_1, F_1) + \ldots \quad (2.3)$$

Here J_t, K_t, L_t, ... are, respectively two, three, four, ... body collision operators that depend on the dynamical events that can take place between isolated systems of two, three, four, ... bodies in an interval of time of duration t, and ε_2, ε_3, ε_4, ... are initial condition correction terms that depend explicitly on the properties of the initial values of the two, three, four, ... particle distribution functions $F_2(x_1, x_2, 0)$, $F_3(x_1, x_2, x_3, 0)$, etc. The derivation of eq. (2.3) required only that one makes cluster expansion decomposition of the N-particle distribution function $F_N(x_1,\ldots x_N, 0)$ and of the N-particle time displacement operator $S_{-t}^{(N)}(x_1,\ldots x_N)$, an operator that replaces the phases of the N particles $x_1\ldots x_N$ by their values $x_1(-t)\ldots x_N(-t)$ at time t earlier. A typical cluster expansion of $S_{-t}^{(N)}(x_1,\ldots, x_N)$ is given in terms of lower-order time displacement operators $S_{-t}^{(s)}(x_1,\ldots, x_s)$ that involve only the dynamics of s particles interacting with each other and not with the other $N-s$ particles in the gas [4], as, for example,

$$S_{-t}^{(N)}(x_1,\ldots x_N) = S_{-t}^{(1)}(x_1)S_{-t}^{(N-1)}(x_2,\ldots x_N)$$
$$+ \sum_{j=2}^{N}\left[S_{-t}^{(2)}(x_1, x_j) - S_{-t}^{(1)}(x_1)S_{-t}^{(1)}(x_j)\right]$$
$$\times S_{-t}^{(N-2)}(x_2,\ldots x_{j-1}, x_{j+1},\ldots x_N) + \ldots \quad (2.4)$$

Similarly, $F_N(x_1\ldots x_N, 0)$ has the expansion

$$F_N(x_1\ldots x_N, 0) = F_1(x_1, 0)F_{N-1}(x_2\ldots x_N, 0)$$
$$+ \sum_{j=2}^{N}\left(F_2(x_1, x_j, 0) - F_1(x_1, 0)F_1(x_j, 0)\right)$$
$$\times F_{N-2}(x_2\ldots x_{j-1}, x_{j+1}\ldots x_N, 0) + \ldots \quad (2.5)$$

The cluster expansions given here have the form of Ursell expansions, familiar from equilibrium virial expansions, and are, for finite N at least, simply identities. Thus one could consider eq. (2.3) to be a formal identity that is written in a very suggestive form.

Bogoliubov's generalized Boltzmann equation follows from eq. (2.3) if it can be shown that (1) the initial condition correction terms $\varepsilon_2, \varepsilon_3, \ldots$ vanish rapidly for $t \gg t_d$, where t_d is the average duration of a binary collision, and (2) the collision integrals $J_t, K_t, L_t, \ldots$ reach asymptotic values $J_\infty, K_\infty, L_\infty \ldots$, again for $t \gg t_d$. This is one of the real achievements of the Green–Cohen approach; it allows one to derive the Boltzmann equation together with its generalization to higher densities in the form of a virial expansion, provided one can prove that the initial state of the gas is forgotten and that the dynamics of small groups of particles are such that the collision integrals rapidly reach asymptotic forms in their time behavior. If these various properties can be proved, then the one basic assumption on which the generalized Boltzmann equation depends is simply that the initial state of the gas is such that particles far from one another are statistically independent. This seems to be a clearer and more natural assumption than the *Stosszahlansatz* that it replaces.

If these various properties of the initial condition terms and collision integral could be proved, then the generalized Boltzmann equation would take the form, for $t \gg t_d$,

$$\partial F_1(x_1,t)/\partial t + v_1 \cdot \nabla_{r_1} F_1(x_1,t) = J_\infty\big(F_1(x_1,t)F_1(x_2,t)\big)$$
$$+ K_\infty\big(F_1(x_1,t)F_1(x_2,t)F_1(x_3,t)\big) + \ldots \quad (2.6)$$

Then by using the Chapman–Enskog procedure, one could derive the Navier–Stokes hydrodynamic equations from eq. (2.6), together with expressions for transport coefficients in terms of the intermolecular potential. These transport coefficients would also be obtained in the form of virial expansions such as

$$\mu/\mu_0 = 1 + a_1^{(\mu)}(n\sigma^3) + a_2^{(\mu)}(n\sigma^3)^2 + \ldots \quad (2.7)$$

where μ is a transport coefficient, such as the coefficient of shear viscosity or thermal conductivity or diffusion; μ_0 is the low-density, Boltzmann equation value; n is the number density of the gas; and σ is a length characterizing the range of the intermolecular forces.

In the thesis of Choh [3], this program was carried out far enough to obtain general expressions for $a_1^{(\mu)}$ in terms of the dynamics of three bodies, but no evaluations were carried out. It took several more years before enough was known about the dynamics of three bodies that $a_1^{(\mu)}$ could be calculated even for a simple model. After the work of Choh and Uhlenbeck, Green [10] was able to formulate the calculation of $a_1^{(\mu)}$

as a set of collision integrals related to, but considerably more complicated than those for μ_0. At Green's suggestion, this work was continued by Sengers [11], who was eventually able to compute $a_1^{(\mu)}$ for hard-sphere gases.

Moreover, it was expected that the Chapman–Enskog procedure applied to the generalized Boltzmann equation would also lead to Burnett, super-Burnett,... hydrodynamic equations, with density expansions for the associated transport coefficients. It is worth mentioning this point since we will refer to it later – that the Burnett and higher-order hydrodynamic equations result naturally from gradient expansions of the hydrodynamic fluxes, such as the energy flux or momentum flux, that appear in the conservation laws for mass, momentum and energy in a fluid [7]. The terms in the fluxes that are linear in the temperature or velocity gradients lead to the Navier–Stokes equations, while the second-order gradients lead to the Burnett equations and so on.

Finally, it was expected that the transport coefficients obtained from the generalized Boltzmann equation would agree with those obtained by the time correlation function method when these expressions are applied to dilute or moderately dense gases.

3. The divergence difficulties

The first important success of the cluster expansion approach discussed in the last section was the proof that $J_t(F_1, F_1)$ and $\varepsilon_2(F_1, F_1)$ have the properties that are required for a satisfactory derivation of the Boltzmann equation [1, 2, 4, 5, 9]. The proof relies on the fact that to compute the right hand side of the first hierarchy eq. (2.1) one needs only determine the two-particle distribution function for configurations where the two particles are interacting, due to the θ_{12} operator in the integral. One could prove that $\varepsilon_2 \to 0$ and that $J_t \to J_\infty$ for $t \gg t_d$ if one could show that if two particles are colliding at time t, they must have been far apart at much earlier times. Since ε_2 and J_2 are determined entirely by the properties of two particles considered in isolation from the rest of the particles in the gas, this separation requirement follows simply from the dynamics of a binary collision, at least if the potential is short range and purely repulsive. Extension of these arguments to potentials that have a long range and/or attractive parts requires a more detailed treatment, and in fact many aspects of the theory have not been as extensively developed for such potentials as they have for short-range repulsive ones [12, 13]. An exception is, of course, the Coulomb potential [14].

The first crack in the development of the theory began to appear when ε_3 and K_t were considered. When the dynamical events that contribute to K_t were examined, it was discovered that in addition to genuine triple collisions, there were contributions from sequences of three or more correlated binary collisions among the three particles. Some typical dynamical events that contribute to K_t are illustrated in fig. 1. In 1963, Green and Piccirelli [1], as well as Ono and Shizumi [15], discovered that the three-body collision operator K_t does not approach its asymptotic value K_∞ very rapidly, but approaches it as

$$K_\infty(F_1, F_1, F_1) - K_t(F_1, F_1, F_1) \approx O(t_d/t); \quad t \gg t_d \tag{3.1}$$

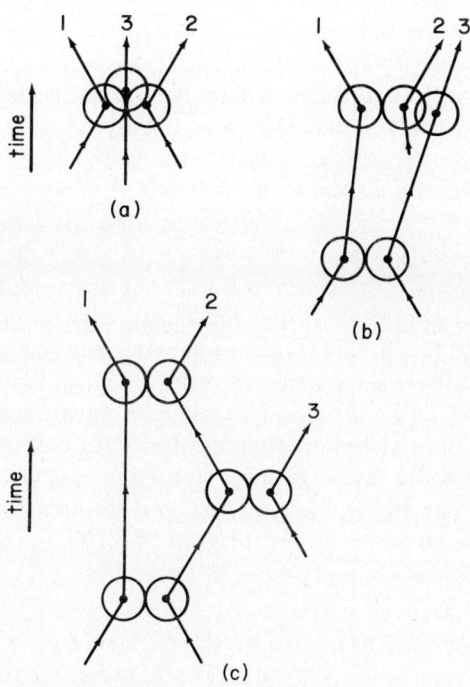

Fig. 1. Typical three-body events that contribute to K_t, and to the coefficient $a_1^{(\mu)}$ in the density expansion of the transport coefficients. We illustrate these events for the case of hard sphere molecules. Figure 1(a) represents a "triple collision". For this case it is an excluded volume correction included in the Enskog theory. Figures 1(b) and 1(c) represent sequences of collisions that take place over an interval of time.

for gases composed of particles with short-range repulsive forces in three dimensions. This slow approach is due to the contributions of extended sequences of three binary collisions that can take place over an interval of duration greater than t, and as a result contribute to K_∞ but not to K_t, no matter how long t is. Similarly, one can show that [9]

$$\varepsilon_3(t) \approx O\big((t_d/t)^2\big); \quad t \gg t_d, \tag{3.2}$$

so that the initial state is only slowly forgotten. These slow power decays can easily be understood in terms of the geometric features of the collision events that contribute to K_t and $\varepsilon_3(t)$ [7d]. For instance, consider the contribution of the collision sequence (12)(23)(12) illustrated in fig. 1c. Here we have a (12) collision followed by a (23) and then another (12) collision.

The events of this type that contribute to $\varepsilon_3(t)$ are such that the time interval between the first and last (12) collisions is exactly t. That is, if the final (12) collision is taking place at time t, then contributions to $\varepsilon_3(t)$ arise from initial states where the three particles are not all separated. The events of this type that contribute to $K_\infty - K_t$ are such that the time interval between the first and last (12) collisions is greater than t. In fig. 2 we illustrate the collision geometry in a coordinate system centered on particle 2 after the first (12) collision. In this coordinate system, particle 1 is traveling in the z direction with the relative velocity v_{12}. Then particle 3 must come and knock 2 back to 1. If we are considering $\varepsilon_3(t)$, 2 must hit 1 again after a time t has elapsed from the time of the initial (12) collision. This requirement means that 3 must knock 2 into a solid angle of order of magnitude $((\sigma/v_{12}t)^2)$ in three dimensions and must transfer enough momentum to 2 to enable it to catch up with 1 at time t after the first (12) collision. The crucial point turns out to be the solid angle restriction. $\varepsilon_3(t)$ is proportional to the solid angle and hence is of order $(t_d/t)^2$ in three dimensions, given by eq. (3.2). To get $K_\infty - K_t$, we integrate this solid angle factor from t to ∞, as

$$K_\infty - K_t \sim \int_t^\infty \mathrm{d}\tau (t_d/\tau)^2 \sim (t_d/t). \tag{3.3}$$

Mel Green recognized that the unexpectedly slow decay of K_t to K_∞ would be troublesome for the theory, and he speculated in early 1964 [16] that there might be terms of order $n^2 \ln n$ in the density expansion of the transport coefficients, although he was not able to substantiate this. It was not long, though, before theorists were able to show that terms of order $n^2 \ln n$ must be present in the density expansions of the transport coefficients, and even that virial expansions in general cannot be used to adequately represent properties of a gas that is not in equilibrium. Before we turn our attention to this point, however, we should dwell for a moment on some other practical consequences of the results given by eqs. (3.1) and (3.2).

In spite of the fact that K_t approaches its asymptotic value K_∞ slowly or that ε_3 decays to zero as $(t_d/t)^2$ for large t, it is still possible for large times and not too dense gases to approximate the generalized Boltzmann equation by:

$$\frac{\partial F_1}{\partial t} + \frac{\boldsymbol{p}_1}{m} \cdot \boldsymbol{\nabla}_{r_1} F_1 = J_\infty(F_1, F_1) + K_\infty(F_1, F_1, F_1). \tag{3.4}$$

To describe hydrodynamic processes in the gas, one is interested in times of order L/c, where L is some macroscopic length and c is the

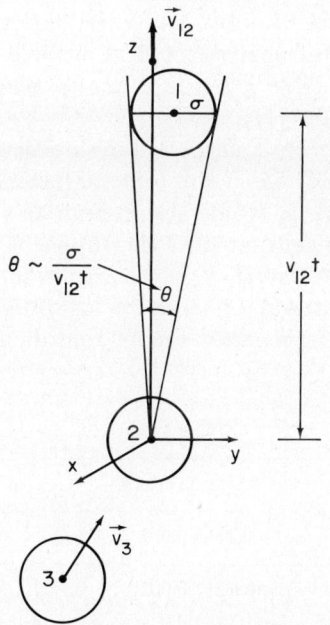

Fig. 2. Geometric configuration of the three particles participating in the re-collision event illustrated in fig. 1(c). Particles 1 and 2 have collided once already. Particle 3 will collide with particle 2 in such a way that particle 2 will collide again with particle 1.

velocity of sound in the gas. Since $L/c \gg t_d$, one would expect that eq. (3.4) could be used to describe hydrodynamic flows and then to obtain the first density corrections to the Boltzmann equation values for the transport coefficients, i.e. the $a_1^{(\mu)}$ in eq. (2.7). Further, eq. (3.2) essentially guarantees that the $a_1^{(\mu)}$ will be finite. These coefficients have now been computed by Sengers and co-workers for gases composed of hard spheres. Table 1 shows the results for $a_1^{(\mu)}$ obtained by Sengers et al. [11] for $\mu = D$, the coefficient of self-diffusion; $\mu = \eta$, the coefficient of shear viscosity; and $\mu = \lambda$, the coefficient of thermal conductivity [11]. These values are compared with the results obtained from the theory of Enskog mentioned earlier in which only the overlapping collision event illustrated in fig. 1a is taken into account. One can see that taking into account the contributions of dynamical events of the types illustrated in figs. 1b and c, that take into account dynamical correlations among the three particles amounts to a correction of up to 10% to Enskog's theory. It should be pointed out that Sengers et al. had to carry out some very difficult numerical integrations in order to obtain these results. It does not seem easy to extend their calculations to more general potentials, and even for hard spheres it took a few years to develop the techniques that made such a careful analysis of $a_1^{(\mu)}$ possible. Their calculation illustrates what a formidable job it is to try to extend the Boltzmann equation to higher densities [17].

Now we return to the consequence of the slow decay of $K_\infty - K_t$ and ε_3 to zero. First, we note that if the same arguments were applied to two-dimensional systems, ε_3 would decay only as (t_d/t) and $K_T - K_t$ would diverge logarithmically as $\log(T/t)$, so that K_∞ does not exist in two dimensions [9, 18]. Although K_∞ does exist in three dimensions, the four-particle collision integral L_∞ diverges logarithmically in three dimensions. Some of the dynamical events contributing to L_t that are

Table 1

μ	$a_1^{(\mu)}$ (Sengers et al.)	$a_1^{(\mu)}$ Enskog theory
D (self-diffusion)	-1.201 ± 0.002	$-5\pi/12 \simeq -1.309$
η (shear viscosity)	$+0.403 \pm 0.002$	$+7\pi/60 \simeq +0.367$
λ (thermal conductivity)	$+1.252 \pm 0.001$	$+23\pi/60 \approx +1.204$

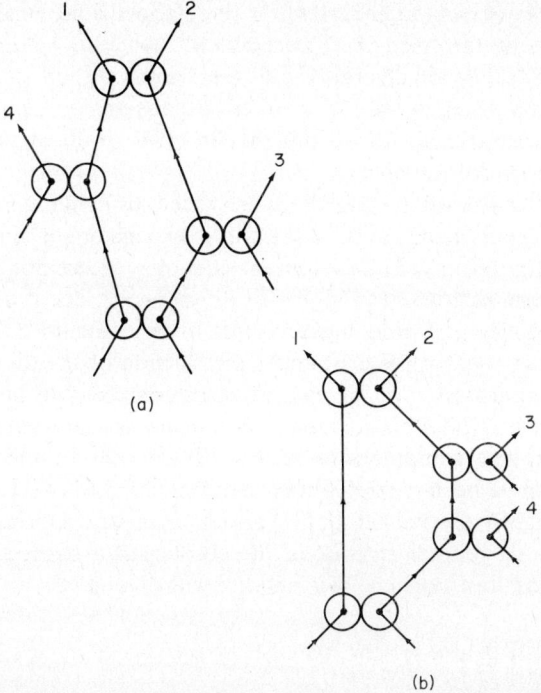

Fig. 3. Some typical four-body collision events that give diverging contributions to the collision operator L_t and to $a_2^{(\mu)}$ in the virial expansion of the transport coefficients.

responsible for the divergence in L_∞ are illustrated in fig. 3. They are sequences of four binary collisions among four particles, and the contributions of these sequences to ε_4 and L_t can be analyzed in a way similar to that for ε_3 and K_t. The fourth particle is responsible for an additional power of t in the phase space estimate because only one of the particles, 3 or 4, has to be aimed properly for the (12) collision to take place, while the other particle, 4 or 3, can be anywhere in a collision cylinder whose volume is proportional to t. Then in d dimensions one finds that

$$\varepsilon_4(t) \approx (t_d/t)^{d-2}, \quad t \gg t_d \tag{3.5}$$

and

$$L_\infty - L_t \approx \int_t^\infty dt (t_d/t)^{d-2}, \quad t \gg t_d. \tag{3.6}$$

Thus L_∞ does not exist in either two or three dimensions, and, as similar arguments show, the five, six... particle collision integrals in the virial expansion eq. (2.6), all diverge. Thus the representation of the generalized Boltzmann equation as a virial expansion in eq. (2.6) must certainly be abandoned, for all but the first (for $d=2$) or the first two (for $d=3$) terms do not exist.

Further, if one tried to use the generalized Boltzmann equation to compute the coefficients $a_i^{(\mu)}$ in the virial expansion of the transport coefficients (ignoring for the moment that $L_\infty \ldots$ do not exist), one would find that $a_2^{(\mu)}$ and the higher-order transport coefficient diverge for $d=3$, and that $a_1^{(\mu)}$ and higher terms diverge for $d=2$. The divergences of these coefficients are simply a reflection of the divergences in the collision operators. For the case of gases composed of hard disks or hard spheres, it is possible to replace the phase space arguments given above by explicit computations of the divergences in the $a_i^{(\mu)}$. Such calculations have been carried out by Sengers [19] and Haines et al. [20] for hard disks, by Gervois et al. [21] and by Kan and Dorfman [22] for hard spheres. In general, in spite of the absence of a rigorous proof, we expect that the transport coefficients $a_i^{(\mu)}$ will diverge as:

$$a_i^{(\mu)} \sim \int_{T_0}^{\infty} \left(\frac{t}{t_d}\right)^{i-d} dt, \tag{3.7}$$

where T_0 is a time on the order of a few t_d.

Before we discuss the resolution of these divergence difficulties, we should turn our attention to another, more general, method for computing transport coefficients, and discuss its relation to the Boltzmann equation method.

4. The time correlation function method

At roughly the same time as the structure of the generalized Boltzmann equation was being worked out, another, more general approach to the theory of irreversible processes was being developed. This was the time correlation function method. This method has its origin in Einstein's work on Brownian motion [23], but its modern version is due to Green [6], Kubo, Mori, McLennan and Zwanzig [24]. Basically, the time correlation function method provides a means of deriving hydrodynamic equations directly from the Liouville equation without the intermediate step of deriving a kinetic equation, and it leads to general expressions for transport coefficients in terms of the microscopic properties of the fluid. For example, the transport coefficients in the linearized Navier–Stokes equations, such as the coefficient of shear viscosity (η), thermal conductivity (λ), or self-diffusion (D), can be expressed in terms of time correlation functions as:

$$\mu = \lim_{t\to\infty} \lim_{\substack{V\to\infty \\ N\to\infty \\ N/V=n}} C_\mu \int_0^t d\tau \langle J_\mu(\tau) J_\mu(0) \rangle. \tag{4.1}$$

Here $\mu = \eta$, λ or D; C_μ is a factor depending on N, density and temperature; J_μ is a microscopic current of momentum, energy, or particles; $J_\mu(0)$ is the value of this current at some initial time $t=0$; $J_\mu(\tau)$ is its value at time τ later; and the angular brackets denote an equilibrium ensemble average. For a classical N-particle system with pairwise, central forces, the microscopic currents for η, μ, and D are given by:

$$J_\eta = \sum_{i=1}^N \frac{p_{ix} p_{iy}}{m} - \sum_{i<j} r_{ij,x} \frac{\partial \phi(r_{ij})}{\partial r_{ij,y}}, \tag{4.2a}$$

$$J_\lambda = \sum_{i=1}^N \frac{p_{ix}}{m} \left(\frac{p_i^2}{2m} + \frac{1}{2} \sum_{j\neq i} \phi(r_{ij}) - h \right) - \frac{1}{2} \sum_{i\neq j} r_{ij,x} \frac{p_i}{m} \cdot \frac{\partial \phi(r_{ij})}{\partial r_{ij}} \tag{4.2b}$$

and

$$J_D = p_{1x}/m. \tag{4.2c}$$

Here p_i is the momentum of the ith particle, and its components with respect to some laboratory system are p_{ix}, p_{iy}, p_{iz}; $r_{ij} = r_i - r_j$, $\phi(r_{ij})$ is the pair potential; and h is the equilibrium enthalpy per particle. For the case of self-diffusion we follow the motion of a tagged particle which we label 1.

Although there are several ways to derive the hydrodynamic equations and the time correlation function formulas from the Liouville equation, probably the most interesting from our point of view here is based on the Chapman–Enskog method [25]. In the usual Chapman–Enskog procedure [7] one assumes that shortly after some initial state, the gas is close to a state of local equilibrium, and the Boltzmann equation is then used to compute the deviation from local equilibrium as an expansion in the gradients of the hydrodynamic variables. This procedure leads to expansions of the momentum and energy fluxes in powers of the gradients of the hydrodynamic variables. The coefficients in these expressions are the transport coefficients, which appear as integrals involving the Boltzmann collision operator.

Exactly the same procedure can be applied to the Liouville equation. One assumes that the system is close to local equilibrium, the N-particle distribution function is expressed as a local equilibrium distribution function plus a correction, and the Liouville equation is used to compute the deviation from the local equilibrium distribution function. The hydrodynamic fluxes can then be computed as expansions in terms of the gradients of the hydrodynamic variables with coefficients that are the time correlation function expressions for the transport coefficients. As in the case of the Boltzmann equation, this procedure gives rise to a hierarchy of hydrodynamic equations: the Euler, Navier–Stokes, Burnett,... equations, where, in the hydrodynamic fluxes, one takes into account terms of the zeroth, first, second, ... order, respectively, in the gradients of the hydrodynamic variables. One now has general expressions in the form of time correlation functions for the Navier–Stokes, Burnett and higher-order transport coefficients of a fluid.

Thus we have on the one hand expressions for transport coefficients for a moderately dense gas from the generalized Boltzmann equation, and on the other, general expressions for transport coefficients for fluids from the Liouville equation. Since a moderately dense gas is a special case of a fluid, we would expect that the time correlation functions, when evaluated for a gas, would yield results identical with those obtained from the generalized Boltzmann equation. This turned out to be the case. When cluster expansions similar to those used to derive the generalized Boltzmann equation from the Liouville equation are used to evaluate the time correlation functions, one obtains virial expansions of the transport coefficients that are term by term identical with those obtained from the generalized Boltzmann equation [26]. The two methods are perfectly consistent and both methods lead to virial expansions of transport coefficients where all but the first few terms diverge.

5. Density expansions with logarithmic terms

The difficulties in the virial expansions of the transport coefficients can be traced to the fact that they do not properly describe an important collective effect, the mean-free-path "damping" of particle trajectories in a fluid. The divergence in the coefficient $a_i^{(\mu)}$ in eq. (2.7) results from sequences of binary collisions in which the particles travel freely over large distances (compared to a molecular size) between collisions. In actuality such large free trajectories are not possible, for, on the average, a particle cannot travel more than a mean free path or so without colliding with another particle in the gas. Now the mean-free-path damping is contained in the virial expansions, but broken up in a very unnatural way. We would expect collision integrals that involve free particle trajectories of length x, say, to include functions such as $\exp[-x/l]$ where l is the mean-free-path length, to take into account the fact that free trajectories much longer than l are unlikely. In making virial expansions of the transport coefficients, we are forced to expand these exponential factors as power series in x/l, or equivalently in $n\sigma^2 x$, since $l \approx (n\sigma^2)^{-1}$. The divergences in the transport coefficients, as well as in the generalized Boltzmann equation, indicate that the virial expansion should not have been carried out, and that at least a partial summation of the virial expansion must be made before the actual physical situation in the gas is properly described. Mathematically, one has to sum up the most divergent terms in the virial expansion before there is any hope of obtaining a well-behaved expression.

The summation of the most divergent terms in the virial expansions of the transport coefficients was carried out first by Oppenheim and Kawasaki [27] and by Weinstock [28] and later, for the generalized Boltzmann equation, by Cohen and Dorfman [29] and by Frieman and Goldman [18]. In the case of the Navier–Stokes transport coefficients for three-dimensional systems, this resummation leads to an expansion of the form [22]:

$$\mu/\mu_0 = 1 + a_1^{(\mu)}(n\sigma^3) + \tilde{a}_2^{(\mu)}(n\sigma^3)^2 \ln(n\sigma^3) + a_2^{\prime(\mu)}(n\sigma^3)^2 + \ldots \quad (5.1)$$

Thus, Green's intuition that the slow decay of K_t in eq. (2.3) to its asymptotic form, K_∞ indicated the presence of an $n^2 \ln n$ term in the density expansion of the transport coefficient was correct. It is not hard to see how the resummation of the most divergent term in eq. (2.7) would lead to an expansion of the form (5.1). The resummation leads to a "renormalized" four-body contribution to the transport coefficient, roughly of the form of eq. (3.7) for $i=2$ ($d=3$), but with exponential damping factors that limit the distance particles travel between collisions. That is, $a_2^{(\mu)}$ is replaced by:

$$\tilde{a}_2^{(\mu)} \sim \int_{T_0}^\infty dt \, \frac{e^{-t/t_{\mathrm{mfp}}}}{(t/t_d)} \sim \ln \frac{t_d}{t_{\mathrm{mfp}}} \approx \ln n\sigma^3, \quad (5.2)$$

where t_{mfp} is the mean free time for a particle in the gas. The renormalization produces a series, eq. (5.1), where $\tilde{a}_2^{(\mu)}$ is determined by contributions from collision sequences involving four particles. However, the coefficient $a_2'^{(\mu)}$ is no longer determined only by four-particle dynamics, as it would be if the virial expansion were correct. Instead, it is determined by contributions from collision sequences involving four or more particles.

There have been several attempts to verify eq. (5.1) by fitting experimental data on the transport coefficients and showing that the fit is better with an $n^2 \ln n$ term than without it [30]. It seems fair to say that the results of such studies do not indicate that an $n^2 \ln n$ term is necessary for representing the data as a density expansion. In many cases, the experimental data are better represented if a small $n^2 \ln n$ term is included, but the error bars on the coefficient usually include the value zero.

The situation as to the theoretical prediction for the coefficients $\tilde{a}_2^{(\mu)}$ in eq. (5.1) is not yet clear. So far, two groups have attempted to calculate these coefficients. Pomeau, Gervois and Normand-Alle [21] numerically computed $\tilde{a}_2^{(\mu)}$ for a gas of hard spheres for $\mu = \eta$, λ and D, while Kan and Dorfman [22] were able to estimate these coefficients by approximating some collision integrals that appear in expressions for $\tilde{a}_2^{(\mu)}$. The results of Pomeau et al. are consistently larger than the estimates of Kan and Dorfman. Although the hard sphere model does not apply to real systems, one can suppose that it does and compare the theoretical predictions with the experimental results for $\tilde{a}_2^{(\mu)}$; all that is necessary is to provide an effective hard sphere diameter for the particles of the real gas one is considering. Such a diameter is usually obtained by fitting low-density data at the temperature of interest to the Boltzmann equation result, μ_0, for a hard sphere system.

When such a comparison is made, the predictions of Pomeau et al. are outside the possible range of $\tilde{a}_2^{(\mu)}$, while the Kan and Dorfman's values are within this range [30]. It is even more interesting to compare the various theoretical predictions with "experimental" values for the case of the coefficient of self-diffusion, D, for a hard sphere system. These experimental values were obtained by computer-simulated molecular dynamics by Alder, Wainwright and Gass [31] and by Wood and Erpenbeck [32]. In fig. 4 we plot three curves,

$$D/D_0 = 1 - 1.201 n\sigma^3 \tag{5.3a}$$

$$D/D_0 = 1 - 1.201 n\sigma^3 - 6.418(n\sigma^3)^2 \ln(n\sigma^3) \tag{5.3b}$$

$$D/D_0 = 1 - 1.201 n\sigma^3 - 0.9065(n\sigma^3)^2 \ln(n\sigma^3) \tag{5.3c}$$

as functions of the reduced density $n\sigma^3$ and compare them with the

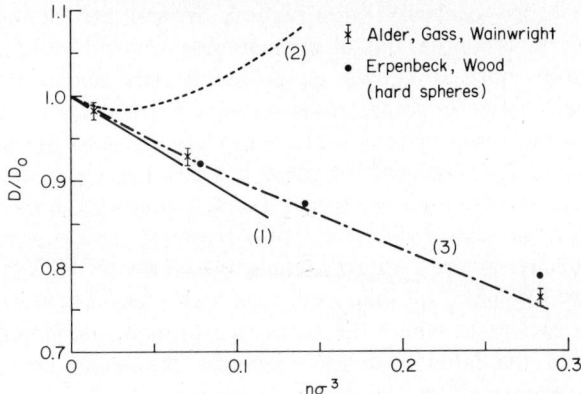

Fig. 4. A plot of D/D_0 as a function of $n\sigma^3$ for hard sphere molecules, where n is the number density of the hard spheres and σ is their diameter. The results of Alder, Gass and Wainwright and Erpenbeck and Wood are obtained by computer simulated molecular dynamics. Curve (1) is a plot of eq. (5.3a) obtained by Sengers et al. for the first correction to the low density value of D; curve (2) is a plot of eq. (5.3b) with the coefficient of the $n^2 \ln n$ term obtained by Gervois et al. and curve (3) is a plot of eq. (5.3c) with the coefficient of the $n^2 \ln n$ term given by Kan's estimate.

computer data. Equation (5.3a) is simply the Boltzmann equation result plus the first density correction as computed by Sengers et al. [11] while eq. (5.3b) has the $n^2 \ln n$ term with the coefficient computed by Pomeau et al. and eq. (5.3c) uses Kan's [22] estimate for $\tilde{a}_2^{(D)}$. As can be seen from fig. 4, Kan's value leads to better agreement with the computer data than does the value of Pomeau et al. Such a comparison may be misleading, however. First, it is not consistent to have a term proportional to $n^2 \ln n$ in the density expansion of D without including a term of order n^2, since a change in units of density will affect the coefficient of the n^2 term. Therefore, we must add a parabola to each of the curves in fig. 4 whose magnitudes are presently unknown. All we can do here is point out that if one improves the fit of the curve with Pomeau's coefficient by adding an appropriate n^2 term, the curve deviates from the experimental data at a somewhat higher density than eq. (5.3b) does, while a small parabolic correction to eq. (5.3c) improves the fit over the density range covered in fig. 4. Nevertheless, in spite of the nice fit obtained with eq. (5.2c), eq. (5.2b) may still be correct; one could only conclude then that the density expansion, eq. (5.1), is not a very rapidly converging series and many terms must be taken into account before an adequate representation of the data is obtained, even at moderate densities. In order to clarify this situation, Sengers and Kamgar-Parsi are repeating the computations of Pomeau, Gervois and Normand-Alle.

Considerably more is known about the density expansion of the transport coefficients for some model systems, particularly the two- and three-dimensional hard sphere Lorentz models [33] and the Ehrenfest wind-tree model [34]. In each of these models there is a set of fixed scatterers placed at random in space, and one studies the diffusion of light particles that can collide only with scatterers and not with each other. In the hard sphere Lorentz models, the scatterers are spheres and the light particles make specular collisions with them. Further, one can consider the two cases where the scatterers do or do not interact with each other. In the latter case the scatterers can really be placed at random in space; in the former, their placement is affected by the requirement that they cannot overlap. The wind-tree model is similar. In this two-dimensional model the scatterers (trees) are squares with diagonals along the x and y axes and the light particles can travel in only one of four directions, the $\pm x$ and $\pm y$ directions. Here, too, one can consider that the trees can overlap or that they cannot. Typical dynamical features of the Lorentz and wind-tree models are illustrated in figs. 5 and 6, respectively.

The theoretical properties of the Lorentz model were studied by Van Leeuwen and Weijland [33], and Bruin [35] made computer simulations to test the theoretical predictions. Because of the relative simplicity of the model, Van Leeuwen and Weijland were able to compute several terms in the density expansion of the diffusion coefficient of the light particle. For example, for a two-dimensional Lorentz gas with non-overlapping scatterers they find:

$$D/D_0 = 1 + \tfrac{4}{3}n\sigma^2 \ln n\sigma^2 - 0.8775 n\sigma^2$$
$$+ 4.519(n\sigma^2)^2 (\ln n\sigma^2)^2 + \ldots \tag{5.4}$$

where n is the number density of the scatterers, σ is their diameter, and the light particles are treated as points. Figure 7 shows a comparison of the computer results of Bruin with eq. (5.4), and with a simpler form where the $O(n\sigma^2 \ln n\sigma^2)$ term is included but the $O(n\sigma^2)$ and higher terms are not. The two curves are fairly close to each other and are in reasonably good agreement with the computer data for a density up to $n\sigma^2 \approx 0.1$. Then, for the Lorentz model at least, one does not make a serious mistake by truncating the series after the $n \ln n$ term, but the series does not converge rapidly for $n\sigma^2 > 0.1$.

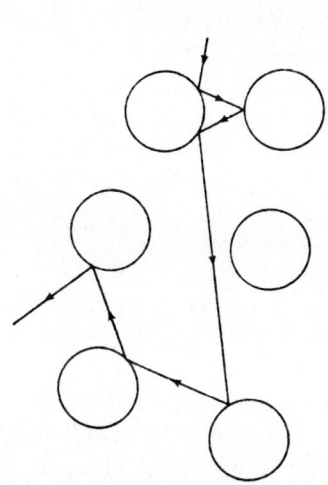 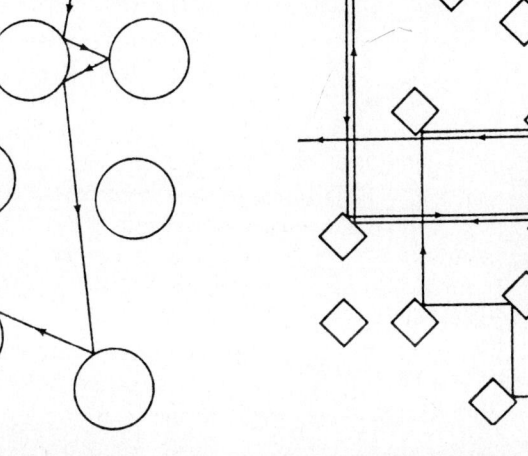

Fig. 5. Some typical dynamical events in the Lorentz model. A light particle is scattered by fixed spheres or disks placed at random in space.

Fig. 6. Some typical dynamical events in the wind-tree model. The "trees" are fixed squares with diagonals oriented along the horizontal and vertical directions. The "wind" particles can travel only in the horizontal or vertical directions with constant speed. Their velocity changes direction only on collision with a tree. Here, the case of overlapping trees is illustrated, and a "reflector" formed from two overlapping trees is shown.

The wind-tree model is dynamically even simpler than the Lorentz model; it has been studied theoretically in great detail by Cohen and Hauge [34], and computer simulations have been made by Wood and Lado [36]. There are no logarithmic terms in the density expansion of the diffusion coefficient of the wind particle, due to the special nature of the wind-tree collision. However, in the case where the trees can overlap, something very spectacular happens – the diffusion coefficient vanishes for all densities. The vanishing of the diffusion coefficient means, of course, that the mean square displacement of a typical wind particle does not grow linearly with time, t, for large times, but rather with some lower power. The cause of this anomalous diffusion is the presence of "reflectors" formed from overlapping trees that reverse the trajectories of the particles that encounter them. One such reflector is

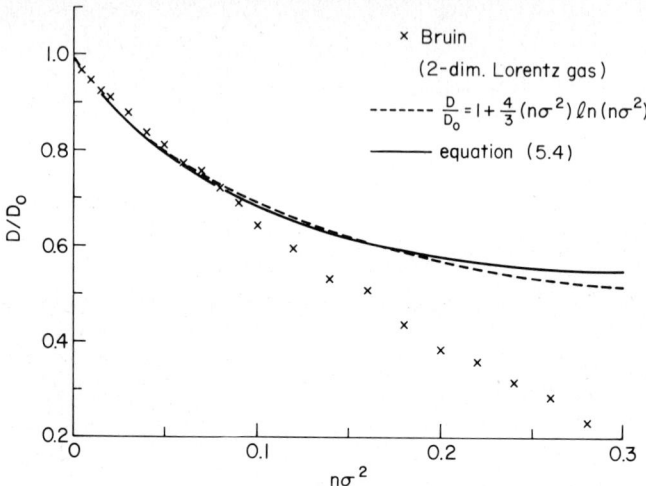

Fig. 7. A plot of D/D_0 as a function of the reduced density $n\sigma^2$ for the two dimensional Lorentz model studied by van Leeuwen, Weijland and Bruin. The crosses represent the results obtained by computer simulation of the model. Here n is the number density and σ the radius of the circular scatterers.

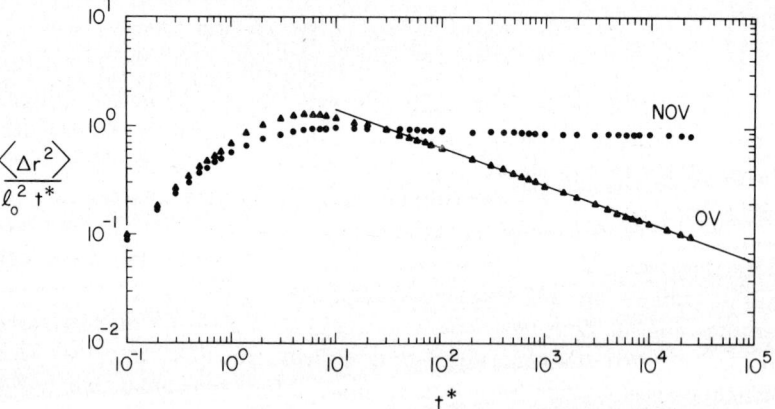

Fig. 8. The reduced mean square displacement function for the wind-tree model as a function of time. Here l_0 is the Boltzmann mean free path length and $t^* = t/t_B$, where t_B is the Boltzmann mean free time. The triangles denote the computer results for the case of overlapping trees and the circles, results for the non-overlapping trees.

illustrated in fig. 6. In order to see the anomalous diffusion on a computer, Wood and Lado had to follow the dynamics of the wind particles for thousands of collision times, and their results are shown in fig. 8. Here we plot the mean square displacement of a wind particle divided by t, as a function of t. The limit of this function as $t \to \infty$ is the self-diffusion coefficient, and one can see that for the case of overlapping trees, D is heading for zero as $t \to \infty$. Hauge and Van Beijeren [27] showed that for low densities of the trees, the mean-square displacement is proportional to $t^{1-(3/4)n\sigma^2}$ for large times, where 2σ is the length of a diagonal of a tree. This result is in good agreement with the computer data.

Now it has turned out that cases of anomalous transport are more numerous than we once suspected. In fact, as discussed in the next section, transport for ordinary fluids in two and three dimensions has turned out to be anomalous too, but in a different sense than in the wind-tree model.

6. Long time tails and anomalous transport in fluids

Having discussed the density expansion of the transport coefficients, we now turn to the second topic mentioned in the introduction, the properties of the gradient expansions of the hydrodynamic fluxes.

As discussed earlier, the Chapman–Enskog method, applied either to the Boltzmann equation or to the Liouville equation, leads to an expansion of hydrodynamic fluxes in terms of the gradients of the hydrodynamic variables, with coefficients that we identify as transport coefficients. Here we will be interested in the general theory as follows from the Liouville equation and, as a result, in the time correlation function expressions for the various transport coefficients that appear. To be more specific, let's consider a simple case, the diffusion of a tagged particle in a fluid of mechanically identical particles. This is the phenomenon of self-diffusion [38].

If $P(\mathbf{r}, t)$ is the probability of finding the tagged particle at point $\mathbf{r}$ at time t and $\mathbf{J}_p(\mathbf{r}, t)$ is the probability flux of the tagged particle, then the conservation of probability requires that $P(\mathbf{r}, t)$ and $\mathbf{J}_p(\mathbf{r}, t)$ be related by:

$$\frac{\partial P(\mathbf{r}, t)}{\partial t} + \nabla \cdot \mathbf{J}_p(\mathbf{r}, t) = 0. \tag{6.1}$$

The time correlation function method can be applied to the process of self-diffusion and it leads to an additional relation between J_p and $P(r, t)$ of the form:

$$J_p(r, t) = -D(t)\nabla P(r, t) - D_2(t)\nabla(\nabla^2 P(r, t)) + \ldots \quad (6.2)$$

Here one can identify $D(t)$ as the (time dependent) coefficient of self diffusion given by:

$$D(t) = \int_0^t d\tau \langle v_x(0) v_x(\tau) \rangle, \quad (6.3)$$

where $\langle v_x(0) v_x(t) \rangle$ is the time auto-correlation function of the x component of the velocity of the tagged particle. If the velocity auto-correlation function $\langle v_x(0) v_x(t) \rangle$ decays to zero after some microscopic time, then $D(t)$ reaches a constant value, D, after this time, and we can identify D with the macroscopic self-diffusion coefficient. Physically, the decay of $\langle v_x(0) v_x(t) \rangle$ to zero means that the velocity of the tagged particle at time t is no longer correlated with its initial value. Presumably the collisions of the tagged particle with the other particles in the fluid should lead to such a randomization of its velocity, after a few mean free times. Further, $D_2(t)$ is the super-Burnett self-diffusion coefficient given by:

$$\begin{aligned} D_2(t) = \int_0^t dt_1 \int_{t_1}^t dt_2 \int_{t_2}^t dt_3 [&\langle v_x(0) v_x(t_1) v_x(t_2) v_x(t_3) \rangle \\ &- \langle v_x(0) v_x(t_1) \rangle \langle v_x(t_2) v_x(t_3) \rangle \\ &- \langle v_x(0) v_x(t_2) \rangle \langle v_x(t_1) v_x(t_3) \rangle \quad (6.4)\\ &- \langle v_x(0) v_x(t_3) \rangle \langle v_x(t_1) v_x(t_2) \rangle]. \end{aligned}$$

In principle one could continue the series in eq. (6.2), including fifth and higher order derivatives of $P(r, t)$. Now when eq. (6.2) is inserted in the conservation law, eq. (6.1), one obtains a hydrodynamic equation for $P(r, t)$,

$$\frac{\partial P(r, t)}{\partial t} = D(t) \nabla^2 P(r, t) + D_2(t) \nabla^2 \nabla^2 P(r, t) + \ldots \quad (6.5)$$

The terms on the right-hand side of eq. (6.5) represent the Navier–Stokes, the super-Burnett, and higher-order terms, respectively in the hydrodynamic description of self-diffusion, provided that $D(t), D_2(t), \ldots$ reach constant values for times larger than some microscopic time.

Almost all diffusion processes are well described by the Navier–Stokes equation but the higher-order equations are of some theoretical interest, even though there are rather severe questions about how and when they can be applied [39].

In the 1960's Alder and Wainwright began a systematic study of the velocity autocorrelation function, $\langle v_x(0)v_x(t)\rangle$, using computer-simulated molecular dynamics for systems of hard spheres or hard discs [31, 40, 41]. Later this study was also taken up and extended by Wood and Erpenbeck [32]. The general features of their results are illustrated in fig. 9 where we plot $\rho_D^{(d)}(t) = \langle v_x(0)v_x(t)\rangle / \langle v_x^2(0)\rangle$ as a function of $t/t_{\mathrm{mfp}} = t^*$ where d is the number of dimensions of the system and t_{mfp} is the mean free time. We show two curves, one typical for low densities

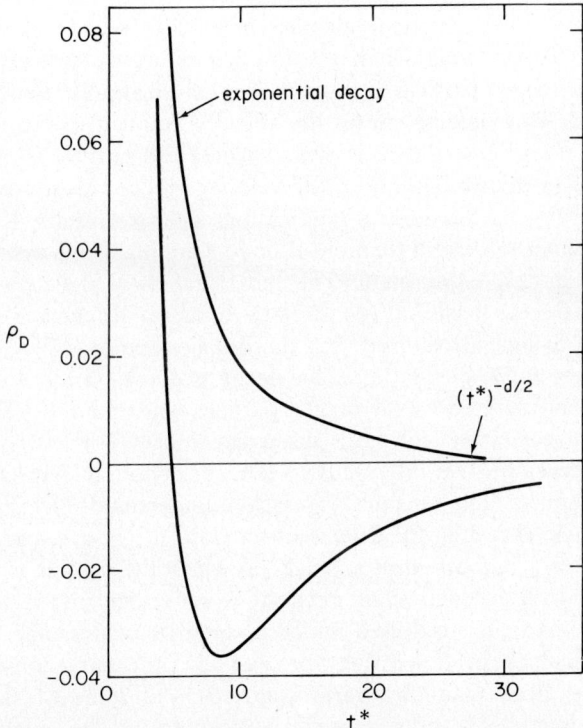

Fig. 9. A sketch of the main features of the computer results for the normalized velocity autocorrelation function $\rho_D^{(d)}$ as a function of time for hard sphere or hard disk systems. The upper curve shows the main features for systems at low densities; the lower curve, for high densities. Here $t^* = t/t_{\mathrm{mfp}}$, where t_{mfp} is the mean free time between collisions.

($V/V_0 \geq 1.8$ where V_0 is the volume ($d=3$) or area ($d=2$) at close packing) and one for high densities ($V/V_0 \leq 1.8$). For times t less than a few t_{mfp}, $\rho_D^{(d)}(t)$ decays exponentially. After several t_{mfp}, the decay is no longer exponential. If the density is not too high $\rho_D^{(d)}(t)$ remains positive and appears to decay to zero as an inverse power of t over the time intervals studied on the computer, $0 \leq t^* \leq 80-100$. In d dimensions $\rho_D^{(d)}(t)$ decays as $(t/t_{\text{mfp}})^{-d/2}$ for large times. At high densities $\rho_D^{(d)}(t)$ becomes negative for $t > 5 t_{\text{mfp}}$, and at longer times ($t > 30 t_{\text{mfp}}$) becomes positive again and appears to decay to zero as $(t/t_{\text{mfp}})^{-d/2}$. Similar behavior has been noted by other authors who carried out computer studies on particles that interact according to other potentials [42]. Alder and Wainwright's molecular dynamics calculation of $\rho_D^{(2)}(t)$ revealed a vortex type of velocity correlation between the tagged particle and the surrounding molecules similar to the hydrodynamic flow field surrounding a moving volume element in a fluid initially at rest [41]. This vortex pattern suggests that a fraction of the momentum transferred by the tagged particle to those in front is eventually returned to it from behind. This process causes the velocity autocorrelation function to be larger than it would be if these vortices did not occur, and it is connected with the slow decay of the velocity autocorrelation function.

Before discussing the general implications of these results, I want to describe briefly the current theoretical understanding of the microscopic processes responsible for them. The initial exponential decay can be understood on the basis of the Enskog equation. In fact, Lebowitz, Percus, and Sykes [43] showed that the initial slope of $\rho_D^{(d)}(t)$ is given exactly by the Enskog equation. This decay is due to random uncorrelated collisions suffered by the tagged particle, since for short times the sequences of correlated collisions discussed in sect. 3 have not had a chance to develop. The intermediate time behavior, in particular the negative region at high densities, is less well understood. Physically, the tagged particle is feeling the effects of its close neighbors, and at high densities these neighbors form a "cage" in which the particle is trapped for a while, with a concomitant reversal of its velocity from its initial value, as the particle is reflected by the boundaries of this cage. In spite of the difficulties in describing the various processes, considerable progress has been made by various groups and some of the main theoretical ingredients of a correct description of the intermediate region are becoming clear [44–46]. The long time $t^{-d/2}$ decays, or "long time tails", are clearly understood [47]. They are due to the same types of dynamical events that are responsible for the divergences in the virial expansions of the transport coefficients, namely sequences of correlated binary collisions of the type illustrated in figs. 1 and 3. Like the logarithmic terms that appear in the density expansion of the transport coefficients (for $d=3$), the long time tails are due to collective effects produced by all of the correlated collision sequences taken together. Thus the resummation of the (diverging) virial expansion produces two collective effects – the $n^2 \ln n$ terms (for $d=3$) and the long time tails. To see how this can come about, we recall that the $n^2 \ln n$ terms result from the mean-free-path damping of the trajectories of the particles taking part in a particular collision sequence. However, another effect is possible. This is the apparent "random walk" of a particle between the first and last collisions of a correlated collision sequence. For example, if we consider all correlated collision sequences beginning with a (12) collision and ending with a (12) collision, then both particles 1 and 2 can make any number of collisions between the first and last (12) collisions. Further, these intermediate collisions can occur randomly. The sequence is correlated by the fact that the same pair participate in the first and last collisions in the sequence. The average distance traveled by particle 1 or particle 2 between collisions will, of course, be a mean free path, but if there is a sufficiently long time interval between the first and last (12) collisions particles 1 and 2 appear to be carrying out a random walk between these collisions, as illustrated in fig. 10. Although this particular process is only one of many similar processes taking place in the gas that contribute to the long-time behavior of $\rho_D^{(d)}(t)$, we can use this one to get at least a qualitative feeling for the dynamical origin of the $t^{-d/2}$ decays. For if we consider the motions of particles 1 and 2 as random walks, or diffusion processes, then we can easily estimate the probability that these particles will collide at time t, if they were known to have collided at time $t=0$. To do this we suppose that the initial (1,2) collision takes place at the origin of some coordinate system. Then, for another (1,2) collision to occur at time t later both particles must be at the same point at time t, if we ignore the slight difference in position required by the finite size of the particles. Now we use the fact that $P(r, t)$, the probability that a particle starting from the origin at $t=0$ will be at r at time t, is given by $P(r,t) = (4\pi Dt)^{-d/2}[\exp -(r^2/4Dt)]$, where D is the diffusion coefficient. Then the probability that particles 1 and 2 will both be at r at time t is proportional to $P^2(r, t)$. Since the location of the final (1,2) collision is not important for the time correlation function, $\rho_D^{(d)}(t)$, the relevant probability for a (1,2) collision at time t is obtained by integrating $P^2(r, t)$ over all r, that is, by computing

$$\int d\boldsymbol{r} P^2(\boldsymbol{r}, t) = \frac{1}{(4\pi Dt)^d} \int d\boldsymbol{r} \exp[-r^2/2Dt] = \frac{1}{(8\pi Dt)^{d/2}}. \quad (6.6)$$

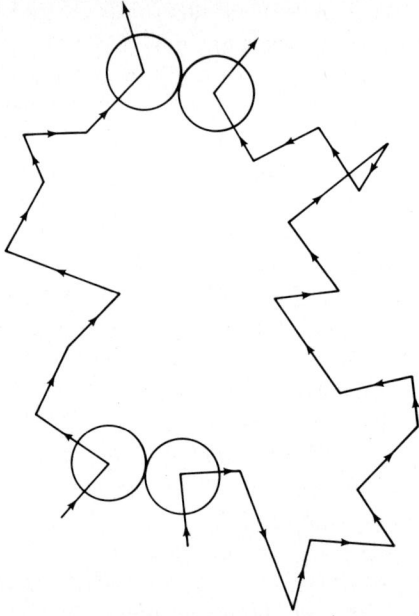

Fig. 10. A "recollision" event with several intermediate collisions. Two particles collide with each other, then they each undergo a random walk, produced by the intermediate collisions, before they collide with each other again.

Thus, the total probability of a recollision is proportional to $t^{-d/2}$. The $t^{-d/2}$ decay of this process is then reflected in the time dependence of $\rho_D^{(d)}(t)$, since the long time behavior of $\rho_D^{(d)}(t)$ is governed by recollision and similar events with random intermediate collisions.

This qualitative description of the long time tail has been substantiated by more careful theoretical calculations. To be specific, the computer experiments lead to a result of the form [32, 41]:

$$\rho_D^{(d)}(t) \approx \alpha_D^{(d)}(n)(t^*)^{-d/2} \quad \text{for} \quad t^* \gg 1, \tag{6.7}$$

where $\alpha_D^{(d)}(n)$ is a coefficient that for hard spheres or disks depends only on the density n. The resummation of the most divergent terms in the virial expansion of the time correlation function for D leads to this behavior, and gives an expression for $\alpha_D^{(d)}$ valid at low densities. To obtain $\alpha_D^{(d)}$ at higher densities one has to sum up less divergent contributions to the virial expansion. This was carried out by Dorfman and

Cohen, who resummed excluded volume corrections to the most divergent terms and obtained an expression for $\alpha_D^{(d)}(n)$ that is in excellent agreement with the computer data for a wide range of densities [47]. A comparison is shown in fig. 11. Finally, it should be mentioned that kinetic theory is by no means the only approach to the long time tails. There are now a large number of ways to derive them and the literature is extensive [47].

The reason for the great interest in the long time tails is that they have remarkable consequences for the general theory of transport in fluids. By referring to eq. (4.1), one can see that the self-diffusion coefficient is simply proportional to the time integral of $\rho_D^{(d)}(t)$. Since this correlation function measures how fast a particle forgets its initial velocity, we might have expected that $\rho_D^{(d)}(t)$ approaches zero after a few mean free times, since the collisions a particle suffers would tend to randomize its velocity and make it forget its initial value. However, this is not what actually happens. Instead, the velocity autocorrelation function decays to zero very slowly and is measurably nonzero even

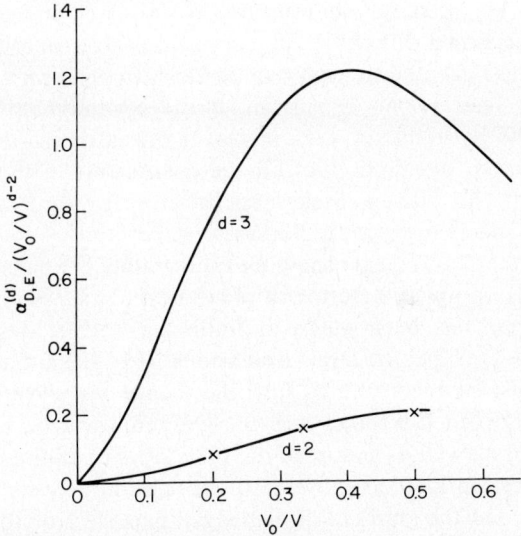

Fig. 11. $\alpha_D^{(d)}/(V_0/V)^{d-2}$ plotted as a function of the reduced density V_0/V, where V_0 is the volume at close packing for hard disks ($d=2$), or for hard spheres ($d=3$). The crosses indicate the results of Alder and Wainwright [41] for $d=2$. The curve is taken from Dorfman and Cohen [47]. The subscript E denotes the fact that Enskog theory values are used for the transport coefficients that appear in the theoretical expression for $\alpha_D^{(d)}$.

after many collision times. Further, both theoretical [47] and computer studies [48] indicate that the time correlation functions that determine η and λ also decay as $(t^*)^{-d/2}$.

Probably the most spectacular consequence of the long time tails is the divergence of the Navier–Stokes transport coefficients for two-dimensional systems due to the t^{-1} decay of the time correlation functions. If this decay persists for asymptotically long times, then the time integral that determines D, as well as those that determine the other transport coefficients, will diverge for large t. This divergence is not due to an unphysical treatment of a collective effect, as was the case with the divergences in the virial expansion of the transport coefficients, rather it is the result of a proper treatment of a collective effect, as discussed above. Therefore the divergence difficulties associated with the long time tails must be considered a real physical effect.

In three dimensions, the asymptotic $t^{-3/2}$ decay of the time correlation function is sufficiently rapid to guarantee the existence of the Navier–Stokes transport coefficients. However, it appears that the transport coefficients associated with the Burnett and higher-order hydrodynamic equations diverge too [32, 47, 49, 50].

Despite the fact that no rigorous proofs exist yet, it would appear from these divergence difficulties that even in such a simple fluid as a moderately dense gas, the transport of particles, momentum and energy is anomalous. There is no expansion of the hydrodynamic fluxes in gradients of the hydrodynamic variables. If one assumes that such an expansion is valid, one finds that all the coefficients in the expansion diverge, except the Navier–Stokes coefficients for three-dimensional systems. This case is the most important one, certainly, but the difficulties with the other coefficients show that the gradient expansions do not give a useful or complete description of transport processes in fluids.

Some progress has been made in finding a correct description of transport in gases in two or three dimensions. It appears that three new ingredients must be incorporated into the theory now. We expect that the relation between the fluxes and the gradients will be non-local in space and time as well as non-analytic [49–51]. For example, the time correlation function method shows that the relation between the tagged particle flux J_p and the gradient of the tagged particle probability $P(r, t)$ can also be written in the form:

$$J_p(r, t) = - \int dr' \int_{-\infty}^{t} dt' \, D(r-r'; t-t') \nabla_{r'} P(r', t'), \qquad (6.8)$$

as well as in the form of eq. (6.2). Here $D(r-r'; t-t')$ is a self-diffusion kernel. The divergent coefficients in eq. (6.2) imply that one cannot assume that the kernel is short range in $r-r'$ or in $t-t'$; rather it appears that $D(r-r'; t-t')$ has a complicated analytic structure [49, 50].

Viscous shear flows in two- and three-dimensional systems have been studied also [51]. One case has been treated in some detail, namely steady plane Couette flow. For this case one finds that the stress tensor depends in a non-analytic way on the gradient of the velocity field $\boldsymbol{u}$. For example, when $\boldsymbol{u} = \hat{x} u_x(y)$ the x-y component of the pressure tensor, $\boldsymbol{P}$ depends on the gradient of the local velocity as:

$$P_{xy} = +\tilde{\eta}_0 \frac{\partial u_x}{\partial y} \ln\left[a\left|\frac{\partial u_x}{\partial y}\right|\right] + \dots \tag{6.9a}$$

for two dimensional systems, and as

$$P_{xy} = -\eta_0 \frac{\partial u_x}{\partial y} + \eta_1 \frac{\partial u_x}{\partial y}\left|\frac{\partial u_x}{\partial y}\right|^{1/2} + \dots \tag{6.9b}$$

for three-dimensional systems. Here u_x is the x component of the fluid velocity, y is a coordinate in a perpendicular direction and a is a constant with the dimension of time. These results were obtained in recent studies designed to find the correct form of the transport equations for a simple situation [51]. Much more has to be done before the general structure is worked out, though. It is worth pointing out that one can see from eqs. (6.9a and b) why one obtains divergent results for the Navier–Stokes viscosity in two dimensions and for the Burnett coefficients in three. For example the form of eq. (6.9a) shows that $[-P_{xy}/(\partial u_x/\partial y)]$ does not exist as the velocity gradient goes to zero. If this limit were to exist, it would be identified with the coefficient of shear viscosity. Consequently the divergence indicates that the correct description requires a non-analytic form as $\partial u_x/\partial y \to 0$. A similar discussion applies to the Burnett coefficients in three dimensions. Finally, we should mention some recent attempts by Naitoh and Ono [52] to detect the presence of the $(\partial u_x/\partial y)^{3/2}$ terms in P_{xy} for hard sphere systems using computer simulated molecular dynamics. However, the coefficient η_1 in eq. (6.9b) is very small and Zwanzig [53] has pointed out that their results are better explained by more ordinary effects predicted by the non-linear Boltzmann equation, when its solution is expanded as a power series in $(\partial u_x/\partial y)$.

7. Conclusion

In the years since the paper of Green and Piccerelli, we have had to revise many of the ideas that were current when that paper was written. Because of the various divergence difficulties discussed in the preceding sections, we no longer believe that the generalized Boltzmann equation can be expanded in powers of the density, that the transport coefficients have a virial expansion in powers of the density, or that the hydrodynamic fluxes can be expanded in powers of the gradients of density, velocity, or temperature, or in higher-order gradients of those quantities. As part of the revised theoretical description of non-equilibrium process that was necessitated by these difficulties, we were able to obtain a number of quantitative results, all of which are either in striking agreement with or consistent with the best available experimental or computer data. There is every reason to believe that we are on the right track. If we are, then it was in large part due to the many important contributions and ideas of Mel Green. I hope I have been able to convey some idea of just how much we are indebted to him.

Acknowledgement

The author would like to thank Professor E. G. D. Cohen and Mr. T. Kirkpatrick for their valuable comments on this paper as it was being prepared.

References

[1] M. S. Green and R. A. Piccirelli, Phys. Rev. **132** (1963) 1388.
[2] N. N. Bogoliubov, in: Studies in Statistical Mechanics, eds. G. E. Uhlenbeck and J. de Boer, Vol. I (North-Holland, Amsterdam, 1962);
E. G. D. Cohen and T. H. Berlin, Physica **26** (1960) 717.
[3] S. T. Choh and G. E. Uhlenbeck, The Kinetic Theory of Dense Gases, University of Michigan Report, Ann Arbor (1958);
G. E. Uhlenbeck and G. W. Ford, Lectures on Statistical Mechanics, American Mathematical Society, Providence (1963).
[4] E. G. D. Cohen, Physica **28** (1962) 1025, 1045, 1060; J. Math. Phys. **4** (1963) 143.
[5] M. S. Green, J. Chem. Phys. **25** (1956) 836; Physica **24** (1958) 393.
[6] M. S. Green, J. Chem. Phys. **22** (1954) 398.
[7] (a) S. T. Chapman and T. G. Cowling, The Mathematical Theory of Non-Uniform Gases, 3rd ed. (Cambridge University Press, London, 1970);

(b) P. Resibois and M. DeLeener, Classical Kinetic Theory of Fluids (John Wiley, New York, 1977);
(c) E. G. D. Cohen in Transport Phenomena in Fluids, ed. H. J. M. Hanley (Marcel Dekker, New York, 1970);
(d) J. R. Dorfman and H. Van Beijeren, in: Statistical Mechanics B, Time Dependent Processes, ed. B. Berne (Plenum Press, New York, 1977);
(e) J. H. Ferziger and H. G. Kaper, Mathematical Theory of Transport Processes in Gases (North Holland, Amsterdam 1972);
(f) J. O. Hirschfelder, C. F. Curtiss and R. B. Bird, Molecular Theory of Gases and Liquids (John Wiley, New York, 1954);
(g) C. Cerciganani, Theory and Application of the Boltzmann Equation (Elsevier, New York, 1975).
[8] D. Enskog, in: Kinetic Theory, Vol. 3, ed. S. Brush (Pergamon Press, New York, 1972).
[9] J. R. Dorfman and E. G. D. Cohen, J. Math. Phys. **8** (1967) 282.
[10] M. S. Green, Phys. Rev. **136A** (1964) 905.
[11] J. V. Sengers, in: The Boltzman Equation, eds. E. G. D. Cohen and W. Thirring (Springer, Vienna,1973); J. V. Sengers, D. T. Gillespie and J. J. Perez-Esandi, Physica **90A** (1978) 365 and references contained therein.
[12] K. Kawasaki and I. Oppenheim, Phys. Rev. **139A** (1965) 649 and J. W. Dufty and K. E. Gubbins, Chem. Phys. Lett. **64** (1979) 142.
[13] J. T. Lowrey and R. F. Snider, J. Chem. Phys. **61** (1974) 2320; R. D. Olmsted and C. F. Curtiss, J. Chem. Phys. **62** (1975) 903,3979; and **63** (1975) 1966.
[14] D. A. Montgomery and D. Tidman, Plasma Kinetic Theory (McGraw Hill, New York, 1964).
[15] S. Ono and T. J. Shizume, Phys. Soc. Japan **18** (1965) 29.
[16] Proceedings of the International Seminar on the Transport Properties of Gases, eds. J. Kestin and J. Ross (Brown University, Providence, Rhode Island, 1964).
[17] H. J. M. Hanley, R. D. McCarty and E. G. D. Cohen, Physica **60** (1972) 322; **83A** (1976) 215 and references contained therein;
R. F. Snider and C. F. Curtiss, Phys. Fluids **1** (1958) 122;
D. K. Hoffman and C. F. Curtiss, Phys. Fluids **8** (1965) 890;
V. M. Kuznetsov, High Temperature Research **16** (1979) 1005;
R. D. Olmstead and C. F. Curtiss in ref. [13].
[18] J. Weinstock, Phys. Rev. **132** (1963) 454; **140** (1965) 460;
R. Goldman and E. A. Frieman, Bull. Amer. Phys. Soc. **10** (1965) 531; J. Math. Phys. **7** (1966) 2153; **8** (1967) 1410;
J. R. Dorfman and E. G. D. Cohen, Phys. Lett. **16** (1965) 124 and J. Math. Phys. **8** (1967) 282;
S. Brush, Kinetic Theory, Vol. 3 (Pergamon Press, New York, 1972).
[19] J. V. Sengers, Phys. Fluids **9** (1966) 1685.
[20] L. K. Haines, J. R. Dorfman and M. H. Ernst, Phys. Rev. **144** (1966) 207.
[21] A. Gervois, C. Normand-Alle and Y. Pomeau, Phys. Rev. **A12** (1975) 1570;
A. Gervois and Y. Pomeau, Phys. Rev. **A9** (1975) 2196.
[22] Y. Kan and J. R. Dorfman, Phys. Rev. **A16** (1977) 2447
Y. Kan, Physica **93A** (1978) 191.
[23] A. Einstein, Ann. Phys. **17** (1905) 549.

[24] R. Zwanzig, Ann. Rev. Phys. Chem. **16** (1965) 67;
W. A. Steele, in: Transport Phenomena in Fluids, ed. H. J. M. Hanley, (Marcel Dekker, New York, 1970);
J. R. Dorfman and H. Van Beijeren, ref. [7d];
S. Brush, Kinetic Theory, Vol. 3, op cit.;
C. K. Wong, J. A. McLennan, M. Lindenfeld and J. W. Dufty, J. Chem. Phys. **68** (1978) 1563;
R. Zwanzig, Prog. Theor. Phys. Suppl. **64** (1978) 74.
[25] M. H. Ernst, Ph.D. Thesis, University of Amsterdam (1965);
M. H. Ernst, E. H. Hauge and J. M. J. Van Leeuwen, J. Stat. Phys. **15** (1976) 23.
[26] M. H. Ernst, J. R. Dorfman and E. G. D. Cohen, Physica **31** (1965) 493;
M. H. Ernst, Physica **32** (1966) 209;
J. Sharma, Ph.D. Thesis, University of Maryland (1977).
[27] K. Kawasaki and I. Oppenheim **139A** (1965) 1763 and in: Statistical Mechanics, ed. T. Bak, (Benjamin, New York, 1967) p. 313.
[28] J. Weinstock, Phys. Rev. Lett. **17** (1966) 130.
[29] E. G. D. Cohen, in: Statistical Mechanics at the Turn of the Decade, ed. E. G. D. Cohen (Marcel Dekker, New York, 1971).
[30] Y. Kan, J. R. Dorfman and J. V. Sengers, in: Proc. of the Seventh Symposium on Thermophysical Properties, ed. A. Cezairliyan (ASME, New York, 1977) p. 652;
H. R. Van den Berg and N. J. Trappeniers, Chem. Phys. Lett. **58** (1978) 12;
J. Kestin, O. Korfali and J. V. Sengers, Physica **100A** (1980) 335.
[31] B. J. Alder, T. E. Wainwright and D. M. Gass, J. Chem. Phys. **53** (1970) 3813.
[32] W. W. Wood, in: The Boltzmann Equation, eds. E. G. D. Cohen and W. Thirring, (Springer, Vienna, 1973);
W. W. Wood, in: Fundamental Problems in Statistical Mechanics, Vol. III, ed. E. G. D. Cohen, (North-Holland, Amsterdam, 1975);
J. J. Erpenbeck and W. W. Wood; Ann. Rev. Phys. Chem. **27** (1976) 319.
[33] J. M. J. Van Leeuwen and A. Weijland, Physica, **36** (1967) 457; **38** (1968) 35.
[34] E. H. Hauge and E. G. D. Cohen, Phys. Lett. **25A** (1967) 78; J. Math. Phys. **10** (1969) 347.
[35] C. Bruin, Physica, **72** (1974) 261.
[36] W. W. Wood and F. Lado, J. Comp. Phys. **7** (1971) 528.
[37] H. Van Beijeren and E. H. Hauge, Phys. Lett. **39A** (1972) 397.
[38] J. R. Dorfman, in: Fundamental Problems in Statistical Mechanics, Vol. III, ed. E. G. D. Cohen (North-Holland, Amsterdam, 1975).
[39] There are a number of problems. Since the super-Burnett equation is a fourth-order equation many boundary conditions are needed to completely specify the solutions. However, no way is available to determine boundary conditions. Further, the super-Burnett equation can have solutions where $P(r, t)$ becomes negative, c.f. C. Cercignani ref. [7g] for a discussion of some of these problems as they apply to the Kramers slip flow problem.
[40] S. G. Brush, T. E. Wainwright and B. J. Alder, Bull. Am. Phys. Soc. (II) **5** (1960) 377;
B. J. Alder and T. E. Wainwright, Phys. Rev. Lett **18** (1967) 988; J. Phys. Soc. Japan, Suppl. **26** (1968) 267;
J. H. Dymond and B. J. Alder, J. Chem. Phys. **48** (1968) 343.
[41] B. J. Alder and T. Wainwright, Phys. Rev. **A1** (1970) 18.
[42] D. Levesque and W. T. Ashurst, Phys. Rev. Lett. **33** (1974) 277;
S. Toxvaerd, Phys. Rev. Lett **43** (1979) 529.

[43] J. Lebowitz, J. Percus and J. Sykes, Phys. Rev. **188** (1969) 487.
[44] I. M. de Schepper and E. G. D. Cohen, Phys. Lett. **55A** (1976) 385; **68A** (1978) 308;
D. Lieberworth and E. G. D. Cohen, Phys. Lett. **58A** (1976) 209; and in: Fundamental Problems in Statistical Mechanics, Vol. IV, eds. E. G. D. Cohen and W. Fiszdon (Polish Acad. of Science, Warsaw, 1978).
[45] P. Resibois and J. L. Lebowitz, J. Stat. Phys. **12** (1975) 483;
P. Resibois, J. Stat. Phys. **13** (1975) 393.
[46] P. M. Furtado, G. F. Mazenko and S. Yip, Phys. Rev. **A14** (1976) 869;
J. T. Hynes, Ann. Rev. Phys. Chem. **28** (1977) 301;
S. Yip, Ann. Rev. Phys. Chem. **30** (1979) 547.
[47] J. R. Dorfman and E. G. D. Cohen, Phys. Rev. Lett. **25** (1970) 1257; Phys. Rev. **A6** (1972) 776, **A12** (1975) 292;
Y. Pomeau, Phys. Lett. **27A** (1968) 602, Phys. Rev. **A3** (1971) 1174;
J. Dufty **A5** (1972) 2247;
R. Goldman, Phys. Rev. Lett. **17** (1966) 910;
Y. Pomeau and P. Resibois, Phys. Reports **19C** (1975) 63.
[48] B. J. Alder, D. M. Gass and T. E. Wainwright, Phys. Rev. **A4** (1971) 233;
W. W. Wood and J. Erpenbeck, Molecular Dynamics Calculation of Shear Viscosity Time Correlation Function for Hard Spheres (preprint);
D. Evans, Mol. Phys. 37 (1979) 1747; J. Stat. Phys. 22 (1980) 81.
[49] Cohen and Dorfman, ref. [47].
M. H. Ernst and J. R. Dorfman, Physica **61** (1972) 157 and J. Stat. Phys. **12** (1975) 311;
I. M. De Schepper, H. Van Beijeren and M. H. Ernst, Physica **75** (1974) 1;
I. M. De Schepper and M. H. Ernst, Physica **87A** (1977) 35;
T. Keyes and I. Oppenheim **70** (1973) 100;
I. A. Michaels and I. Oppenheim, Physica **81A** (1975) 522;
H. H. H. Yuan and I. Oppenheim, Physica **90A** (1978) 1, 21, 561.
[50] W. E. Alley and B. J. Alder, Phys. Rev. Lett. **43** (1979) 653.
[51] M. H. Ernst, B. Cichocki, J. R. Dorfman, J. Sharma and H. Van Beijeren, J. Stat. Phys. **18** (1978) 237;
A. Onuki, Phys. Lett. **70A** (1979) 31;
K. Kawasaki and J. Gunton, Phys. Rev. **A8** (1973) 2048;
T. Yamada and K. Kawasaki, Prog. Theor. Phys. **53** (1975) 111.
[52] T. Naitoh and S. Ono, J. Chem. Phys. **70** (1979) 4515.
[53] R. Zwanzig, J. Chem. Phys. **71** (1979) 4416.

CHAPTER 3

Non-equilibrium Fluctuations and the Hierarchy

M. H. ERNST

Instituut voor Theoretische Fysica
Rijksuniversiteit Utrecht
Princetonplein 5, P.B. 80.006
3508 TA Utrecht, The Netherlands

E. G. D. COHEN

The Rockefeller University
New York, NY 10021
U.S.A.

© *North-Holland Publishing Company 1981* *Perspectives in Statistical Physics*
Ed. H. J. Raveché

Contents

1. Introduction   61
2. Hierarchies of distribution and correlation functions   62
3. Low density calculations   67
Appendix   71
References   71

1. Introduction

Many years ago M. S. Green [1] initiated a method to derive and generalize the Boltzmann equation for a gas not in thermal equilibrium starting from the Liouville equation. This method used a generalization of the cluster expansions employed in equilibrium statistical mechanics to obtain the activity and virial expansions of the thermodynamic properties as well as of the equilibrium pair and higher order distribution functions of a gas. As a consequence, Green was led to consider non-equilibrium distribution functions, in particular the non-equilibrium pair distribution function. This work by Green was later modified [2] and generalized to the study of equilibrium time correlation functions such as those that occur in the Green–Kubo formulae for the transport coefficients [3]. While the distribution functions depend on one time t, these time correlation functions in principle depended on two times t_1 and t_2 although in equilibrium they actually only involve a single time $\tau = t_2 - t_1$ because of the invariance of the equilibrium average for a time translation. In this paper we further generalize Green's method to non-equilibrium multiple time correlation functions, in particular to the non-equilibrium two time correlation function. Now the average behavior and the fluctuations around this average of a many body system, whether in equilibrium or not, can be completely described by the set of distribution and time correlation functions. In this paper we will outline how the above mentioned method allows one at least in principle to obtain a closed set of equations for these functions. We will illustrate the method in particular for the case of a dilute gas. We also note that our approach is similar in spirit to that of van Kampen [4] in that we derive equations that describe the average – van Kampen's macroscopic behavior – as well as the fluctuations around this average from one point of view without having to make the kind of ad hoc assumptions about fluctuations and their correlations as in the case of a Langevin-type of approach. Although we neglect certain contributions in our low density theory, their effect can in principle – and in some cases in practice – be included and their consequences be studied. We realize that our equations are similar to those derived by many others [5–7].

We believe, however, that our approach unifies and simplifies in certain respects that of previous authors and allows a treatment of the behavior of fluctuations in dense gases not in equilibrium, using the same methods as have been used before in the kinetic theory for distribution and equilibrium time correlation functions of such systems.

2. Hierarchies of distribution and correlation functions

The hierarchy for the one and two-time distribution functions can be obtained most conveniently as the averaged equations of motion of the microscopic densities $\psi(1t), \psi(12t)\ldots$ We define:

$$\psi(1t) = \psi(x_1 t) = \sum_i^N \delta(X_i(t) - x_1),$$

$$\psi(12t) = \psi(x_1 x_2 t) = \sum\sum_{i \neq j}^N \delta(X_i(t) - x_1) \delta(X_j(t) - x_2) \tag{2.1}$$

and similarly for $\psi(1't') = \psi(x_1' t'),\ldots$. The lower case phase variables $x_i = (\mathbf{r}_i, \mathbf{v}_i)$ with $i = 1, 2, 3\ldots$ are field variables; the capital variables $X_i(t) = (\mathbf{R}_i(t), \mathbf{V}_i(t))$ represent the phase of the ith particle at time t in Γ-space. In the ψ-functions only their dependence on the field variables and the time are explicitly indicated.

Hamilton's equations of motion determine the time evolution of the microscopic densities, i.e.,

$$\partial_t \psi = \{\psi, H(X^N)\} = L(X^N)\psi. \tag{2.2}$$

The Liouville operator $L(X^N)$ is defined through the Poisson brackets with the N-particle Hamiltonian $H(X^N) = \sum_i^N \frac{1}{2} m V_i^2 + \sum\sum_{i<j}^N \phi(R_{ij})$, where m is the mass of the particles and $\phi(R_{ij})$ is the interparticle pair potential, which depends only on the relative distance $R_{ij} = |\mathbf{R}_i - \mathbf{R}_j|$. It reads explicitly

$$L(X^N) = \sum_i^N L_0(X_i) - \sum\sum_{i<j}^N \theta(X_i X_j) \tag{2.3}$$

with

$$L_0(X_i) = \mathbf{V}_i \cdot \frac{\partial}{\partial \mathbf{R}_i},$$

$$\theta(X_i X_j) = m^{-1} \frac{\partial \phi(R_{ij})}{\partial \mathbf{R}_{ij}} \cdot \left(\frac{\partial}{\partial \mathbf{V}_i} - \frac{\partial}{\partial \mathbf{V}_j}\right). \tag{2.4}$$

Due to the presence of δ-functions in the definition of the microscopic densities ψ, the operators L_0 and θ can be made to act on the field variables $x_1 x_2 \ldots$, instead of on the Γ-space variables $X_1 X_2 \ldots$. The result is a hierarchy of coupled equations,

$$[\partial_t + L_0(1)]\psi(1, t) = \int dx_2\, \theta(12)\psi(12, t)$$

$$[\partial_t + L_0(12) - \theta(12)]\psi(12, t) = \int dx_3 (1 + P_{12})\theta(13)\psi(123, t). \quad (2.5)$$

We will follow the convention that the numbers 12... are short for the field variables $x_1 x_2 \ldots$, so that $L_0(1) = L_0(x_1)$ and $\theta(12) = \theta(x_1 x_2)$. The permutation operator P_{ij} interchanges the labels of the field variables x_i and x_j; and $L_0(12) = L_0(1) + L_0(2)$.

This set of equations has the same structure as the BBGKY hierarchy for the non-equilibrium distribution functions, but the microscopic densities ψ still depend on all Γ-space variables. In order to obtain the distribution functions, one has to average the microscopic densities or products of densities over some initial non-equilibrium ensemble $D(X^N, 0)$, i.e.

$$\langle A \rangle = \int dx^N D(X^N, 0) A(X^N), \quad (2.6)$$

where $dX^N = dX_1 dX_2 \ldots dX_N$ and $A(X^N)$ is a function of dynamical variables. The ordinary (one time) distribution functions $f(1t)$, $f(12t)$ are defined as:

$$f(1t) = f(x_1 t) = \langle \psi(1t) \rangle,$$
$$f(12t) = f(x_1 x_2) = \langle \psi(12t) \rangle. \quad (2.7a)$$

Using Liouville's theorem, this is easily seen to agree with the usual definitions:

$$f(X_1 t) = N \int dX_2 \ldots \int dX_N D(X^N, t),$$

$$f(X_1 X_2 t) = N(N-1) \int dX_3 \ldots \int dX_N D(X^N, t). \quad (2.7b)$$

We note that $f(1t)$ is the probability density to find a particle at time t with phase x_1, and similarly for $f(12t)$. The two-time distribution functions $F(1t, 1't')$, $F(12t, 1't') \ldots$ are averages over products of pairs of microscopic densities taken at two different times:

$$F(1t, 1't') = F(x_1 t, x'_1 t') = \langle \psi(1t)\psi(1't') \rangle,$$
$$F(12t, 1't') = F(x_1 x_2 t, x'_1 t') = \langle \psi(12t)\psi(1't') \rangle. \quad (2.8)$$

Thus $F(1t, 1't')$ is the joint probability density of finding a particle at time t with phase x_1 and at time t' a particle (the same or a different one) with phase x'_1. From eqs. (2.8), (2.7) and (2.1) one easily verifies that:

$$F(1t, 2t) = f(12t) + \delta(x_1 - x_2) f(1t). \quad (2.9)$$

The equations of motion for the one-time distribution functions are the usual BBGKY-hierarchy, which can be obtained directly by averaging eq. (2.5) over some initial ensemble, using eq. (2.7). In doing so we interchange the averaging over Γ-space variables with differentiation with respect to field variables present in $L_0(i)$ and $\theta(ij)$. The result is:

$$[\partial_t + L_0(1)] f(1t) = \int dx_2 \, \theta(12) f(12t),$$
$$[\partial_t + L_0(12) - \theta(12)] f(12t) = \int dx_3 (1 + P_{12}) \theta(13) f(123t). \quad (2.10)$$

To obtain the equations of motion for the two-time distribution functions, we multiply eq. (2.5) with $\psi(1't')$, and perform again an average over an initial ensemble. Notice that $\psi(1't')$ can be interchanged with the operators $L_0(i)$ and $\theta(ij)$. In this way we find a hierarchy of equations, first obtained by Tolmachev [8]:

$$[\partial_t + L_0(1)] F(1t, 1't') = \int dx_2 \, \theta(12) F(12t, 1't'),$$
$$[\partial_t + L_0(12) - \theta(12)] F(12t, 1't') = \int dx_3 (1 + P_{12}) \theta(13) F(123, 1't').$$
$$(2.11)$$

The next step is the introduction of cluster functions, based on the idea of weakening of correlations for spatially separated groups of particles, although these non-equilibrium correlations can decay rather slowly with increasing relative distance and with increasing time. They are defined through the recursion relations:

$$f(12t) = f(1t)f(2t) + g(12t)$$
$$f(123t) = f(1t)f(2t)f(3t) + f(1t)g(23t) + f(2t)g(13t)$$
$$+ f(3t)g(12t) + g(123t) \quad (2.12)$$

and

$$F(1t,1't') = f(1t)f(1't') + C(1t,1't')$$
$$F(12t,1't') = f(1t)f(2t)f(1't') + f(1t)C(2t,1't')$$
$$+ f(2t)C(1t,1't') + f(1't)g(12t) + C(12t,1't'). \quad (2.13)$$

These cluster or correlation functions have also a very direct meaning in terms of fluctuation formulae. If we introduce $\delta\psi(1t) = \psi(1t) - \langle\psi(1t)\rangle = \psi(1t) - f(1t)$ as the density fluctuation at the point x_1 in μ-space, then its correlation function at two different times is given by:

$$\langle\delta\psi(1t)\delta\psi(1't')\rangle = F(1t,1't') - f(1t)f(1't') = C(1t,1't'). \quad (2.14)$$

Density fluctuations at equal times are essentially measured by the non-equilibrium pair correlation function $g(12t)$, since it follows directly from eq. (2.9) that:

$$\langle\delta\psi(1t)\delta\psi(2t)\rangle = C(1t,2t) = \delta(x_1 - x_2)f(1t) + g(12t). \quad (2.15)$$

The equations of motion for the correlation functions can be derived directly from the hierarchy eqs. (2.10) and (2.11). The net result is that after inserting eqs. (2.12) and (2.13) into eqs. (2.10) and (2.11) all disconnected terms disappear, where we consider the θ-operators and the correlation functions g and C as connecting links between particles (i.e. $\theta(12)$, $g(12)$ or $C(12)$ as a connecting link between the particles 1 and 2 etc.). We only quote the first few equations, which will be needed later on. For the equal (one) time correlation functions they are:

$$[\partial_t + L_0(1)]f(1t) = \int dx_2 \theta(12)[f(1t)f(2t) + g(12t)]$$
$$[\partial_t + L_0(12) - \theta(12)]g(12t) = \theta(12)f(1t)f(2t)$$
$$+ \int dx_3(1+P_{12})\theta(13)(1+P_{13})f(3t)g(12t)$$
$$+ \int dx_3(1+P_{12})\theta(13)g(123t). \quad (2.16)$$

For the two-time correlation functions we need:

$$[\partial_t + L_0(1)]C(1t,1't') = \int dx_2 \theta(12)(1+P_{12})f(2t)C(1t,1't')$$
$$+ \int dx_2 \theta(12)C(12t,1't'). \quad (2.17)$$

We are interested in the application of these equations to fluids, in which the particles interact through a pair potential which is in general short ranged and has a steep repulsive part. This implies that the solution of these equations through a perturbation expansion in powers of the potential or its derivatives, i.e. in terms of $\theta(ij)$, is divergent, and a resummation of this series to one in terms of binary collision operators $T(ij, t)$ rather than $\theta(ij)$ is required. These binary collision operators $T(ij, t)$ are non-vanishing only for times $t < t_s$ where t_s is the time required to traverse the soft part of the potential. In the special case of hard spheres the binary collision operators are instantaneous, i.e. $T(ij, t) = T(ij)\delta(t)$, and the binary collision expansion can be simply performed by introducing pseudo-streaming operators as shown by Ernst et al. [9]. The net result for hard sphere systems is that the θ-operators in all previous equations have to be replaced by binary collision operators $T(ij)$. For this reason we restrict ourselves in the subsequent part of this paper to hard sphere interactions. However, we expect that our results are also valid for more general potentials as long as we are interested in times $t > t_s$. A derivation of the above results, as well as the explicit expressions for the binary collision operators are given in an appendix. Here we only quote the results. For equal time correlation functions and $t > 0$, eq. (2.16) becomes:

$$[\partial_t + L_0(1)] f(1t) = \int dx_2 \overline{T}_-(12)[f(1t)f(2t) + g(12t)] \quad (2.18\text{a})$$

$$[\partial_t + L_0(12) - \overline{T}_-(12)] g(12t)$$
$$= \overline{T}_-(12) f(1t) f(2t)$$
$$+ \int dx_3 (1 + P_{12}) \overline{T}_-(13)(1 + P_{13}) f(3t) g(12t)$$
$$+ \int dx_3 (1 + P_{12}) \overline{T}_-(13) g(123t). \quad (2.18\text{b})$$

For unequal time correlation functions and $t - t' > 0$ eq. (2.17) yields:

$$[\partial_t + L_0(1)] C(1t, 1't') = \int dx_2 \overline{T}_-(12)(1 + P_{12}) f(2t) C(1t, 1't')$$
$$+ \int dx_2 \overline{T}_-(12) C(12t, 1't'). \quad (2.19)$$

The binary collision operators $\overline{T}_-$ are defined in the appendix. Of course, all these hierarchy equations form sets of coupled equations since the right-hand side of each equation contains a correlation function that involves one more particle than the correlation function that appears on the left-hand side of the equation.

3. Low density calculations

In this section we will show that the general method developed for obtaining kinetic equations for non-equilibrium distribution functions and for equilibrium time correlation functions can be applied directly to non-equilibrium time correlation functions. This will be illustrated here by developing in close parallel the kinetic equations for equal and unequal time correlation functions in the special case of hard spheres in the limit of low densities. A convenient way of doing so is to use the method of Ernst and Dorfman [10], in which successive higher density approximations to the kinetic equations are obtained by neglecting higher and higher correlation functions on the basis that they are of higher and higher order in the density. In this way the hierarchy equations can be decoupled and solved.

We apply the method first to eq. (2.18a) for $f(1, t)$ and obtain then the low density approximation by neglecting the two particle correlation function $g(12t)$. This leads to the non-linear Boltzmann equation for the single particle distribution function of a dilute gas of hard spheres, i.e.

$$(\partial_t + L_0(1))f(1t) = \int dx_2 \, \hat{T}(12)f(1t)f(2t). \tag{3.1}$$

Here we have introduced the point-T-operator,

$$\hat{T}(12) = \delta(\boldsymbol{r}_{12})T_0(12) = \delta(\boldsymbol{r}_{12})\sigma^2 \int_{\boldsymbol{v}_{12}\cdot\hat{\boldsymbol{\sigma}}<0} d\hat{\boldsymbol{\sigma}} |\boldsymbol{v}_{12}\cdot\hat{\boldsymbol{\sigma}}|(b_{\hat{\sigma}}-1). \tag{3.2}$$

The diameter of the spheres is σ, and $\boldsymbol{v}_{12} = \boldsymbol{v}_1 - \boldsymbol{v}_2$. We integrate over a hemisphere $\boldsymbol{v}_{12}\cdot\hat{\boldsymbol{\sigma}} < 0$, where $\hat{\boldsymbol{\sigma}}$ is a unit vector. The operator $b_{\hat{\sigma}}$ transforms the precollision velocities $\boldsymbol{v}_i$ $(i = 1, 2)$ into postcollision velocities $\boldsymbol{v}_i'$,

$$\begin{aligned}b_{\hat{\sigma}}\boldsymbol{v}_1 &= \boldsymbol{v}_1' = \boldsymbol{v}_1 - \hat{\boldsymbol{\sigma}}(\hat{\boldsymbol{\sigma}}\cdot\boldsymbol{v}_{12}) \\ b_{\hat{\sigma}}\boldsymbol{v}_2 &= \boldsymbol{v}_2' = \boldsymbol{v}_2 + \hat{\boldsymbol{\sigma}}(\hat{\boldsymbol{\sigma}}\cdot\boldsymbol{v}_{12}),\end{aligned} \tag{3.3}$$

where $\sigma\hat{\boldsymbol{\sigma}}$ defines the point of closest approach in the binary collision. In writing down eq. (3.1) we have also replaced the $\overline{T}_-$-operator in eq.

(2.18) by $\hat{T}$-operators. While the former operators take into account the difference in position of the centers of the spheres at collision, the $\hat{T}$-operators neglect this difference. Therefore, as an additional approximation we have neglected spatial variations in $f(1t)$ over the range of a sphere diameter σ, as is customary in the Boltzmann equation for a dilute gas.

Next, we consider the two particle correlation function $g(12t)$, describing the equal time fluctuations, eq. (2.15), in non-equilibrium. To obtain $g(12t)$ in lowest approximation in the density we neglect the three particle correlation function $g(123t)$ on the right hand side of eq. (2.18b) and the term $\bar{T}_-(12)g(12t)$ on the left hand side of eq. (2.18b). The latter term represents collision sequences in which the spheres 1 and 2 collide more than once as can be seen from an iterated solution of the equation. At low densities such collision sequences can be neglected since they require the consideration of at least a third particle and, therefore, lead to higher density effects. Also, as in eq. (3.1), we neglect spatial variations over distances smaller than σ by replacing $\bar{T}_-$ by $\hat{T}$. The resulting equation for the non-equilibrium pair correlation function is then:

$$\{\partial_t + L(1t) + L(2t)\}g(12t) = \hat{T}(12)f(1t)f(2t), \tag{3.4}$$

where the operator L is defined as:

$$L(1t) = L_0(1) - \Lambda(1t) \tag{3.5a}$$

$$\Lambda(1t) = \int dx_3 \hat{T}(13)(1 + P_{13})f(3t). \tag{3.5b}$$

The function $f(it)$ with $i = 1, 2, 3$ is the solution of a non-linear Boltzmann equation (3.1). Equation (3.4) is our final result for the pair correlation function in non-equilibrium. To obtain some insight in the nature of this equation, we can argue as follows. If we replace $L(it)$ with $i = 1, 2$ by free streaming operators $L_0(i)$, we would obtain spatial correlations – i.e. non-zero values of $g(12t)$ for vectors $r_{12} = r_1 - r_2$ parallel to the relative velocity $v_1 - v_2$ – that grow with time and are due to the "memory" of a binary collision between the particles 1 and 2 at time $t = 0$. In phase space these correlations define "zones of memory", as Green showed for the first time. Our operators $L(i, t)$ contain, however, in addition to $L_0(i)$ another operator that leads to a damping through collisions, which will restrict the spatial correlations. Green had conjectured that these correlations would be on the order of the mean free path. In general, however, the spatial correlations in a dilute gas not in equilibrium are very long ranged $\sim 1/r_{12}$ [11].

For the equal time correlation function $C(1t,2t)$ we obtain with eqs. (2.15), (3.1) and (3.4) the equation:

$$\{\partial_t + L(1t) + L(2t)\} C(1t,2t) =$$

$$+ \hat{T}(12)f(1t)f(2t) + \delta(x_1 - x_2) \int dx_3 \hat{T}(13)f(1t)f(3t)$$

$$- [\Lambda(1t) + \Lambda(2t)] \delta(x_1 - x_2) f(1t). \tag{3.6}$$

After treating the equal time correlation functions, we proceed to the correlation functions at different times. The lowest density approximation to $C(1t, 1't')$ is obtained by neglecting the three particle correlation function $C(12t, 1't')$ in eq. (2.19). In addition, we replace again $\bar{T}$ by $\hat{T}$, thereby neglecting spatial variations over distances of order σ. With eq. (3.5) our results for non-equilibrium fluctuations in a dilute gas become finally for $t > t'$,

$$(\partial_t + L(1t)) C(1t, 1't') = 0, \tag{3.7}$$

where the initial value of $C(1t, 1't)$ follows from eq. (2.15):

$$C(1t, 2t) = \delta(x_1 - x_2) f(1t) + g(12t). \tag{3.8}$$

Here the functions $f(1t)$ and $g(12t)$ are solutions of the eqs. (3.1) and (3.4) respectively. These results, eqs. (3.7) and (3.8), have been obtained also by Kirkpatrick et al. [7] from the Langevin approach. Using the hierarchy method Hinton [5] also obtains the equation of motion (3.7), but instead of eq. (3.8) he gives as an initial condition for a dilute gas:

$$C_{\text{app}}(1t, 2t) = \delta(x_1 - x_2) f(1t). \tag{3.9}$$

An argument against dropping $g(12t)$ in eq. (3.8) as a higher density correction is the following. Consider the fluctuations $\langle \delta A_\alpha(t) \delta A_\beta(t) \rangle$ in the total number of particles N, total momentum P and total energy H, for short $A_\alpha = \int dx_1 a_\alpha(v_1) \psi(1t)$ with $a_\alpha(v) = (1, v, v^2)$ and $\delta A_\alpha = A_\alpha - \langle A_\alpha \rangle$. This fluctuation is clearly constant in time. However, the incorrect approximation (3.9) would yield spurious fluctuations,

$$\langle \delta A_\alpha \delta A_\beta \rangle_{\text{app}} = \int dx_1 a_\alpha(v_1) a_\beta(v_1) f(1t), \tag{3.10}$$

which are time-dependent. In fact, using the Chapman–Enskog solution for $f(1, t)$ in eq. (3.4) reveals that $g(12t)$ is of the same order in the density as the first term on the right hand side of eq. (3.9). Spurious fluctuations in the densities of conserved quantities caused by neglecting $g(12t)$, were also noted by Kirkpatrick et al.

As argued above, although the equations for the distribution and correlation functions in this low density theory were derived for the special case of hard spheres, they are also expected to hold for more general interparticle interactions. In fact, this can be achieved by replacing the binary collision operators $\hat{T}(12) = \delta(r_{12}) T_0(12)$ by $\hat{T}(12) = \delta(r_{12}) T_0(v_1 v_2)$ with:

$$T_0(v_1 v_2) = \int d\Omega \, g I(g, \chi)(b_\Omega - 1). \tag{3.11}$$

Here $g = v_1 - v_2$ and $I(g, \chi)$ is the differential scattering cross section, $d\Omega = \sin \chi \, d\chi \, d\phi$, where χ is the scattering angle in the binary collision of the particles 1 and 2, with $g^2 \cos \chi = g \cdot g'$ and ϕ the azimuthal angle which determines the plane of scattering. The operator b_Ω acts only on the velocities v_i ($i = 1, 2$) with the result $b_\Omega v_i = v'_i$, and $g' = v'_1 - v'_2$, where the primed velocities are velocities after the binary collision.

With this binary collision operator the kinetic equation (3.1) becomes the non-linear Boltzmann equation for a general short range interparticle potential. Our main results for the equal and unequal time correlation function, the eqs. (3.4) and (3.7) respectively, can then be similarly interpreted as valid for general interparticle potentials.

To summarize the restrictions of our equations: they only apply (i) to lowest order in the density because only terms that lead to contributions of lowest order in the density have been kept; (ii) to spatial variations on a length scale larger than the range of the forces σ, because $\hat{T}$-operators we have used neglect the difference in position of two colliding particles; (iii) to times t and τ larger than the time needed to traverse the soft part of the potential as discussed below eq. (2.17).

It is not the aim of this paper to work out systematically higher approximations to these equations, such as those that contain the contributions of three body collisions, rings or repeated rings to the equal or unequal time correlation functions in non-equilibrium, although this could be done in a manner analogous to that used for distribution functions before. Our main goal was to illustrate that the methods used so far to derive generalized kinetic equations for the non-equilibrium distribution and for the equilibrium time correlation functions can be extended directly to two or multiple time correlation functions in non-equilibrium.

As an illustration of the simplicity and usefulness of our results we calculate the two-time correlation formula in a stationary state described by a single particle distribution functions $f_s(1) = f_s(x_1)$. The solution of eq. (3.4) for the stationary pair correlation function $g_s(x_1 x_2)$ ignoring the homogeneous solution of eq. (3.4), which is associated with correlations over a range of the order σ, is:

$$g_s(x_1 x_2) = \{L_s(1) + L_s(2)\}^{-1} \hat{T}(12) f_s(x_1) f_s(x_2) \tag{3.12}$$

where $L_s(1)$ is given by eq. (3.5) after replacing $f(3t)$ by $f_s(3)$. The unequal time correlation functions $C(1t+\tau, 1't) = C_s(x_1 x_1', \tau)$ in the stationary state is obtained by taking the Laplace transform of (3.7) and using the initial condition (3.8). The result is:

$$\begin{aligned}\hat{C}_s(x_1 x_2, z) &= \{z + L_s(1)\}^{-1} \{\delta(x_1 - x_2) f_s(x_1) + g_s(x_1 x_2)\} \\ &= \{z + L_s(1)\}^{-1} \{\delta(x_1 - x_2) f_s(x_1) + [L_s(1) + L_s(2)]^{-1} \\ &\quad \times \hat{T}(12) f_s(x_1) f_s(x_2)\}. \end{aligned} \tag{3.13}$$

Equation (3.13) is identical to one obtained by Kirkpatrick, Cohen and Dorfman, using a fluctuating Boltzmann equation with a random force added. Their bilinear terms are contained in the operator $L_s(1)$, and account for the deviation of $f_s(x_1)$ from its equilibrium value. As already discussed, for a gas not in equilibrium the term $g_s(x_1 x_2)$ in eq. (3.13) is not a higher density correction to $\delta(x_1 - x_2) f_s(x)$ and should be taken into account. For a gas in equilibrium, however, where $f_s(x_1)$ is the Maxwellian, the operator $L_s(1)$ reduces to the linearized Boltzmann collision operator and $g_s(x_1 x_2)$ can be neglected.

Acknowledgement

One of us (M.H.E.) would like to thank the Rockefeller University for its hospitality during January 1979, when this work was started.

Appendix

For the special case of hard sphere interactions the resummation of the perturbation expansion, described at the end of sect. 2 can be performed very elegantly by means of pseudo-streaming operators S_t^{ps}, introduced by Ernst et al. [9]. These authors have shown that the time dependence of the ψ-functions (2.1) can be represented as:

$$\psi(12\ldots,t) = S_t^{ps}(X^N)\psi(12\ldots,0). \tag{A.1}$$

The operator S_t^{ps} generates the physical trajectories in Γ-space starting from the physical initial positions in which hard spheres are not overlapping. The streaming operators S_t^{ps} generate also unphysical trajectories for unphysical (overlapping) initial positions. However, in the averages to be considered here, i.e.

$$\langle \psi(12\ldots,t) \rangle = \int dX^N D(X^N,0) S_t^{ps}(X^N) \psi(12\ldots,0) \tag{A.2}$$

$$\langle \psi(12\ldots,t)\psi(1't') \rangle$$

$$= \int dX^N D(X^N,0) S_t^{ps}(X^N) \psi(1',0) S_{t-t'}^{ps}(X^N) \psi(12\ldots,0), \tag{A.3}$$

the overlapping initial configurations have vanishing weight $D(X^N,0)$, so that the pseudo-streaming operators generate the proper dynamics inside averages of the form (A.2) and (A.3). It is important to notice that the above mentioned arguments apply equally well to products of streaming operators as occurring in eq. (A.3).

The explicit form of S_t^{ps} depends on the sign of t. In eq. (A.2) we are clearly interested in $t>0$. Furthermore, in eq. (A.3) we finally need only $\langle \psi(1t)\psi(1't') \rangle$ with $t>t'$; the case $t<t'$ can be obtained by interchanging primed and unprimed variables. Hence, we use only forward streaming operators ($t>0$), defined as:

$$S_t^{ps}(X^N) = \exp tL_+^{ps}(X^N). \tag{A.4}$$

The pseudo-Liouville operator is:

$$L_+^{ps}(X^N) = \sum_i^N L_0(X_i) + \tfrac{1}{2} \sum_{i \neq j}^N T_+(X_i X_j), \tag{A.5}$$

and the binary collision operator:

$$T_+(X_1 X_2) = \sigma^2 \int_{V_{12}\cdot\hat{\sigma}<0} d\hat{\sigma} |V_{12}\cdot\hat{\sigma}| \delta(R_{12} - \sigma\hat{\sigma})(b_{\hat{\sigma}} - 1), \tag{A.6}$$

where the symbols are defined in eq. (3.3). We briefly retrace the steps of sect. 2 for forward pseudo-streaming operators. The derivative of eq.

(A.1) can be cast again into a form, analogous to eq. (2.5) by using the relation:

$$T_+(X_iX_j)\delta(X_i-x_1)\delta(X_j-x_2) = \bar{T}_-(x_1x_2)\delta(X_i-x_1)\delta(X_j-x_2).$$
(A.7)

This is nothing but a statement about the hermitean adjoint, i.e. $(T_+)^\dagger = \bar{T}_-$, as demonstrated in ref. [9], where

$$\bar{T}_-(X_1X_2) = \sigma^2 \int_{V_{12}\cdot\hat{\sigma}>0} d\hat{\sigma} |V_{12}\cdot\hat{\sigma}| [\delta(R_{12}-\sigma\hat{\sigma})b_{\hat{\sigma}} - \delta(R_{12}+\sigma\hat{\sigma})].$$
(A.8)

The resulting hierarchy has the form:

$$(\partial_t + L_0(1))\psi(1t) = \int dx_2 \bar{T}_-(12)\psi(12t),$$

$$[\partial_t + L_0(12) - \bar{T}_-(12)]\psi(12t) = \int dx_3(1+P_{12})\bar{T}_-(13)\psi(123t).$$
(A.9)

From this follows that the one-time distribution functions defined in eq. (A.2), obey the BBGKY-hierarchy for hard sphere systems,

$$(\partial_t + L_0(1))f(1t) = \int dx_2 \bar{T}_-(12)f(12t),$$

$$[\partial_t + L_0(12) - \bar{T}_-(12)]f(12t) = \int dx_3(1+P_{12})\bar{T}_-(13)f(123t).$$
(A.10)

Consider next the t-derivative of the two-time distribution functions $F(12\ldots t, 1't')$, for which the representation (A.3) is most convenient. The t-derivative of $S_{t-t'}^{ps}\psi(1,0)$ can be expressed as in (A.9), and with the help of (A.3) we arrive at the hierarchy of the two-time distribution functions,

$$(\partial_t + L_0(1))F(1t, 1't') = \int dx_2 \bar{T}_-(12)F(12t, 1't'). \quad (A.11)$$

The subsequent steps are completely identical to those in sect. 2, and lead directly to eqs. (2.18) and (2.19).

References

[1] M. S. Green, Physica, **24** (1958) 393.
[2] E. G. D. Cohen, Physica **28** (1962) 1045.
[3] E. G. D. Cohen, Fundamental Problems in Statistical Mechanics II (North-Holland, Amsterdam, 1968) p. 228.
[4] N. G. van Kampen, in: Topics in Statistical Mechanics and Biophysics, ed. R. A. Piccirelli, AIP-Conference Proceeding **27** (American Institute of Physics, New York, 1976) p. 153.
[5] F. L. Hinton, Physics of Fluids **13** (1970) 857;
J. Keizer, J. Chem. Phys. **63** (1975) 398;
A. Onuki, J. Stat. Phys. **18** (1978) 475;
J. Dufty, Phys. Rev. **A13** (1976) 2299.
[6] G. Ludwig, Physica **28** (1962) 841;
O. Seeberg, J. Stat. Physics **4** (1972) 83;
N. G. van Kampen, Phys. Lett. **50A** (1974) 237.
[7] T. Kirkpatrick, E. G. D. Cohen and J. R. Dorfman, Phys. Rev. Lett. **42** (1979) 862.
[8] V. V. Tolmachev, Sov. Phys. Doklady **2** (1957) 85.
[9] M. H. Ernst, J. R. Dorfman, W. R. Hoegy and J. M. J. van Leeuwen, Physica **45** (1969) 127.
[10] M. H. Ernst and J. R. Dorfman, Physica **61** (1972) 157.
[11] Y. Pomeau, J. Math. Phys. **12** (1971) 2286;
A. Onuki in ref. [5] and ref. [7] plus references therein.

CHAPTER 4

Green's Contributions to Non-equilibrium Statistical Mechanics Revisited

L. S. GARCIA-COLIN* and J. L. Del RIO**

Department of Physics, UAM-Iztapalapa
Mexico 13, D.F.

---

*Also at the Facultad de Ciencias, UNAM México 20, D.F. and Miembro del Colegio Nacional.
**Also at the Escuela Superior de Física y Matemáticas del IPN, México 14, D.F.

Contents

1. Introduction 77
2. Principles of non-equilibrium statistical mechanics 77
3. Non-equilibrium averages and coarse grained variables 79
4. The Fokker–Planck equation and the Green–Kubo formulae 80
5. Concluding remarks 86
References 87

1. Introduction

With the publication of Melville Green's paper "Markoff random processes and the statistical mechanics of time dependent phenomena" [1] in 1952 a whole set of phenomena in the field of non-equilibrium statistical mechanics began to show their understanding from the microscopic equations which govern the motion of classical particles. Twenty-eight years after the time it was published it may be considered one of the classical papers on the subject. Its ideas and results as well as those contained in its companion paper [2] are frequently used in the study of several non-linear phenomena in transport theory. Typical examples of these are Zwanzig's derivation of exact kinetic equations from first principles [3], Kawasaki's mode–mode coupling equation [4], the non-linear theory of dynamic fluctuations [5–9] and the theory of non-equilibrium stationary states [10]. As a modest way of paying homage to Mel Green's significant contributions to time dependent statistical mechanics we would like to stress the important role that his ideas have played in some modern developments on the subject. In particular, concepts such as the hypercell in phase space [7, 9], the concept of coarse grained variables (mesoscopic dynamics) [11] and the results which one can obtain from them following a systematic scheme, lean heavily on the ideas introduced by him. Here we shall point out how his results are connected with some recent developments in this field.

In sect. 2 we briefly review the basic principles underlying Green's work. In sect. 3 the concept of an hypercell and averages taken over it in non-equilibrium situations are related to equilibrium averages and this is used to derive the formulae for transport coefficient without invoking the "regression of fluctuations" assumption. In sect. 4 the derivation of Green's Fokker–Planck equation and the Kubo–Green formulae is presented in a more modern language and sect. 5 contains some pertinent concluding remarks.

2. Principles of non-equilibrium statistical mechanics

The three basic principles invoked by Green [1] are stated below together with some clarifying comments and a brief discussion of the main quantities which will be of further use.

First principle: "The appropriate objects of study of a statistical mechanics of time dependent phenomena are the random processes $A_i(\Gamma, t)$ with initial distribution of Γ, $\rho_E(\Gamma)$ (the microcanonical distribution), for all energies of interest and for all gross variables of interest."

Second principle: "The random processes associated with a complete set of gross variables are stationary Markoff random processes."

Third principle: "The time rate of change of the space functions $A_i(\Gamma, 0)$ are controlled by a slowness parameter δ which implies $\dot{A}_i(\Gamma, 0) \approx \delta$ with $\delta < 1$, the relevant terms are those of order δ and δ^2, and these characterize the slow behavior of the space functions."

Random processes are used because every time an experiment is repeated the numerical values a_i of the phase-space functions $A_i(\Gamma)$ behave as stochastic variables. The linear regression of fluctuations hypothesis is implicitly made because when choosing the initial distribution function to be a microcanonical distribution, one is interested in fluctuations which occur in an aged system.

The description of a random process is made using the set of distribution functions

$$g_n(a_1, t_1, \ldots, a_n, t_n) = \int \rho(\Gamma, 0) \prod_i \delta[A(\Gamma, t_i) - a_i] \, d\Gamma \tag{1}$$

where $\rho(\Gamma, 0)$ is the initial distribution function, $[A(\Gamma, t)]$ is a complete set of space functions and $\{a_i\}$ a set of their numerical values. Notice then that if $\rho(\Gamma, 0) = \rho_E(\Gamma)$ the process is stationary, $g_1(a, t)$ is time independent and corresponds to the equilibrium distribution of the a' variables

$$g_1(a, t) = g_{eq}(a). \tag{2}$$

The restriction imposed by the second principle implies that the relevant dynamical quantity is the equilibrium conditional probability $P_{eq}(a, t|b)$; namely the probability that $a \leqslant A(\Gamma, t) \leqslant a + da$ if at time $t = 0$ the fluctuations are such that $A(\Gamma, 0) = b$. This quantity is assumed to obey the Chapman–Kolmogoroff equation. Thus,

$$P_{eq}(a_3, t_3|a_1, t_1) = \int da_2 \, P_{eq}(a_3, t_3|a_2, t_2)$$
$$\times P_{eq}(a_2, t_2|a_1, t_1). \tag{3}$$

Finally, the third principle restricts the process to a slow one [12] and although it is not referred to it as a slow Markoffian process, it is systematically used throughout this entire paper.

3. Non-equilibrium averages and coarse grained variables

In order to give a simple derivation of the Fokker–Planck equation obtained by Green and the formulae for the transport coefficients as well, we shall use a formalism developed recently [11] which is based on the concept of a phase space hypercell. Consider the time evolution of a many body system which is initially in a constrained equilibrium state. When some of the constraints are removed it evolves towards a final less constrained equilibrium state. Let $\alpha(t)$ denote the set of macroscopic observables associated to the space functions $A(\Gamma, 0)$ in the usual way, namely,

$$\alpha(t) = \int d\Gamma \rho(\Gamma, t) A(\Gamma, 0) = \int d\Gamma \rho(\Gamma, 0) A(\Gamma, t) \tag{4}$$

where $\rho(\Gamma, t)$ is the non-equilibrium probability distribution function at time t. The region in phase space characterized by the condition that $a \leq A(\Gamma, t) \leq a + da$ is the hypercell corresponding to the set $\{a\}$ in phase space. The average of any phase space function $f(\Gamma)$ over this cell is defined as,

$$\langle f(\Gamma); a \rangle = \left[\int_{hc} d\Gamma \rho_{eq}(\Gamma) \right]^{-1} \int_{hc} d\Gamma f(\Gamma) \rho_{eq}(\Gamma) \tag{5}$$

where $\int_{hc} = \int_{a < A < a + da}$ is the integral over the hypercell. Equation (5) may be rewritten in a more compact way as

$$\langle f(\Gamma, t); a \rangle = g_{eq}^{-1}(a)[f(\Gamma), G(a, 0)] \tag{6}$$

where use is made of the internal product of two arbitrary phase functions,

$$(A, B) = \int d\Gamma \rho_{eq}(\Gamma) A(\Gamma) B(\Gamma) \tag{7}$$

and $G(b, t) \equiv \delta[A(\Gamma, t) - b]$ is the hypercell's characteristic function.

The conditional probability $P_{eq}(a, t | b)$ may be expressed as an average over a hypercell after its definition is transformed using eq. (6).

Thus,

$$P_{eq}(at|b) = g_{eq}^{-1}(b)g_2(a,t;b,0) = \langle G(a,t); b \rangle. \tag{8}$$

Defining the coarse grained variables of the system $\overline{a(t)}^b$ as the averages of the numerical values of the phase functions A, taken with respect to $P_{eq}(a,t|b)$ [11] we have

$$\overline{a(t)}^b = \int da\, a P_{eq}(a,t|b) = \langle A(\Gamma,t); b \rangle \tag{9}$$

where the last equality follows from the identity,

$$A(\Gamma,t) = \int da\, a G(a,t). \tag{10}$$

We must now seek the relationship between $\alpha(t)$ and $\overline{a(t)}^b$. To do so we start from the assumption that

$$\rho(\Gamma,0) = \rho_{eq} \nu[A(\Gamma,0)] \tag{11}$$

where $\rho_{eq}(\Gamma)$ is the equilibrium distribution function of the final equilibrium state and $\nu[A(\Gamma,0)]$ is an arbitrary function of the phase functions, chosen such that

$$\nu[A(\Gamma,0)] \equiv \int db\, \nu(b) G(b,0). \tag{12}$$

Substitution of eqs. (9), (11) and (12) into eq. (4) leads immediately to the result that

$$\alpha(t) = \int db\, \nu(b) g_{eq}(b) \overline{a(t)}^b \tag{13}$$

which is the sought relationship between the macroscopic and the coarse grained variables. Equation (13) is the relevant equation in the derivation of formulae for the transport coefficients. The coarse grained variables were used by Onsager in the characterization of the average regression of fluctuations in a system in equilibrium [13].

4. The Fokker–Planck equation and the Green–Kubo formulae

The purpose of this section, which is the central part of this paper is to give a derivation of the most relevant results obtained by Green in his original papers using the language developed in the two previous sections. This we hope will emphasize on the elegance and importance of Green's contributions.

In order to derive the Fokker–Planck equation we shall take the following steps:

(a) We use the second principle of sect. 2 implying that $P_{eq}(a, t|b)$ obeys the Kramers–Moyal (KM) expansion and we express the derivative moments as averages over the hypercell.

(b) Using the third principle we reduce the KM expansion to a Fokker–Planck equation and the first two derivative moments are computed up to terms which are of second-order in the slowness parameter.

(c) The results obtained in (b) are further simplified introducing the assumption that the correlation of the velocities taken over a hypercell has the same behavior as the equilibrium correlations.

The KM expansion for the Markoffian process described by $P_{eq}(a, t|b)$ may be written in the following way [14], namely;

$$\frac{\partial P_{eq}(a, t|b)}{\partial t} = \sum_{n=1}^{\infty} \frac{1}{n!} \sum_{\alpha_1} \cdots \sum_{\alpha_n} \left(-\frac{\partial}{\partial \alpha_1}\right) \cdots \left(\frac{-\partial}{\partial \alpha_n}\right)$$

$$\times \mathbb{K}_{\alpha_1 \ldots \alpha_n}(a) P_{eq}(a, t|b) \tag{14}$$

where $\mathbb{K}_{\alpha_1 \ldots \alpha_n}(a)$ is the nth order derivative moment defined by

$$\mathbb{K}_{\alpha_1 \ldots \alpha_n}(a) = \lim_{T \to 0} \frac{1}{T} \int \prod_{i=1}^{n} (c_{\alpha_i} - a_{\alpha_i}) P_{eq}(c, T|a) \, dc. \tag{15}$$

Using eq. (8) we are allowed to write eq. (15) as an average over the hypercell, namely,

$$\mathbb{K}_{\alpha_1 \ldots \alpha_n}(a) = \lim_{T \to 0} \frac{1}{T} \int_0^T d\tau_1 \ldots \int_0^T d\tau_n \langle \dot{A}_{\alpha_1}(\Gamma, \tau_1) \ldots \dot{A}_{\alpha_n}(\Gamma, \tau_n); a \rangle \tag{16}$$

where use has been made of eq. (10) and of the identity,

$$\Delta A_i(\Gamma, t) = A_i(\Gamma, t) - A_i(\Gamma, 0) = \int_0^t \dot{A}_i(\Gamma, t) \, dt. \tag{17}$$

Now we can introduce Green's third principle into eq. (16) since according to it $\mathbb{K}_{\alpha_1 \ldots \alpha_n}(a)$ is $O(\delta^n)$. Keeping in eq. (16) terms which are at most of order δ^2 leads to the following expression for eq. (14), namely:

$$\frac{\partial P_{eq}(a,t|b)}{\partial t} = \sum_i \left(-\frac{\partial}{\partial a_i}\right) \mathbb{K}_i^{(2)}(a) P_{eq}(a,t|b)$$
$$+ \sum_i \sum_j \frac{1}{2} \frac{\partial}{\partial a_i} \frac{\partial}{\partial a_j} \mathbb{K}_{ij}^{(2)}(a) P_{eq}(a,t|b) \quad (18)$$

where the superindices indicate that we are keeping contributions up the second order in δ. Equation (18) is the equation that Green proposes as the equivalent to the Chapman–Kolmogoroff equation without explicitly mentioning the slow approximation, although it is implicitly used in the calculation of the derivative moments.

To calculate the second order contributions of the terms $\mathbb{K}_i(a)$ and $\mathbb{K}_{ij}(a)$ we notice that in eq. (16) the relevant hypercell averages are $\langle \dot{A}_i(\Gamma,\tau_1); a \rangle$ and $\langle \dot{A}_i(\Gamma,\tau_i)\dot{A}_j(\Gamma,\tau_2); a \rangle$. To extract the contributions in δ and δ^2 we expand the characteristic function $G(a,t)$ to terms in first order in δ, namely:

$$G(a,t) = \delta[A(\Gamma,0) - a + \Delta A(\Gamma,t)]$$
$$= G(a,0) - \sum_k \frac{\partial}{\partial a_k} \int_0^t ds\, \dot{A}_k(\Gamma,s) G(a,0) + O(\delta^2). \quad (19)$$

On the other hand, since $f(\Gamma,t) = \exp[iLt]f(\Gamma,0)$ and Liouville's operator L is hermitian, $(iLA, B) = -(A, iLB)$ eq. (6) may be rewritten as

$$\langle f(\Gamma,t); a \rangle = g_{eq}^{-1}(a)(f(\Gamma,0), G(a,t)) \quad (20)$$

Using now eqs. (19), (20) and the fact that $\dot{A}_i(\Gamma,\tau_1)\dot{A}_j(\Gamma,\tau_2) = \exp[iL\tau_1]A_i(\Gamma,0)\dot{A}_j(\Gamma,\tau_2-\tau_1)$ one obtains that,

$$\langle \dot{A}_i(\Gamma,\tau_1); a \rangle = V_i(a) + g_{eq}^{-1}(a) \sum_k \frac{\partial}{\partial a_k} g_{eq}(a) \int_0^{\tau_1} ds\, \sigma_{ik}(a,-s)$$
$$+ O(\delta^3) \quad (21)$$

where

$$V_i(a) = \langle \dot{A}_i(\Gamma,0); a \rangle \quad (22)$$

and

$$\sigma_{ik}(a,t) = \langle \dot{A}_i(\Gamma,0)\dot{A}_k(\Gamma,t); a \rangle. \quad (23)$$

Also,

$$\langle \dot{A}_i(\Gamma,\tau_1)\dot{A}_j(\Gamma,\tau_2);a\rangle = \langle \dot{A}_i(\Gamma,0)\dot{A}_\tau(\Gamma,\tau_2-\tau_1);a\rangle + O(\delta_2) \quad (24)$$

Using eq. (21) back in eq. (16) leads to the value for $\mathbb{K}_i^{(2)}(a)$ namely,

$$\mathbb{K}_i^{(2)}(a) = V_i(a) + g_{eq}^{(-1)}(a) \sum_k \frac{\partial}{\partial a_k} g_{eq}(a)$$

$$\times \lim_{T\to 0} \frac{1}{T}\int_0^T d\tau_1 \int_0^{\tau_1} ds\, \sigma_{ik}(a,-s) + O(\delta^3). \quad (25)$$

To find $\mathbb{K}_{ij}^{(2)}(a)$ one first rearranges the expression for $\mathbb{K}_{ij}(a)$ arising from eq. (16) in the following way [1]:

$$\mathbb{K}_{ij}(a) = \lim_{T\to 0} \frac{1}{T}\int_0^T d\tau_2 \int_0^{\tau_2} d\tau_1 \big[\langle \dot{A}_i(\Gamma,\tau_1)\dot{A}_j(\Gamma,\tau_2);a\rangle$$

$$+ \langle \dot{A}_j(\Gamma,\tau_1)\dot{A}_i(\Gamma,\tau_2);a\rangle\big] \quad (26)$$

which after use has been made of eqs. (23) and (24) leads to

$$\mathbb{K}_{ij}^{(2)}(a) = \lim_{T\to 0} \frac{1}{T}\int_0^T d\tau_2 \int_0^T d\tau_1 \big[\sigma_{ij}(a,\tau_2-\tau_1) + \sigma_{ji}(a,\tau_2-\tau_1)\big]. \quad (27)$$

The final step consists in carrying out the time integrals in eqs. (25) and (27) using the assumption stated in (c), namely that $\langle \dot{A}_i(\Gamma,0)\dot{A}_j(\Gamma,t);a\rangle$ has the same behavior as the average taken over a microcanonical ensemble $\langle \dot{A}_i(\Gamma,0)\dot{A}_j(\Gamma,t)\rangle_{\text{micro}}$. This asumption imposes the following conditions over the hypercell averages [1];

(i) $\quad \sigma_{ij}(a,t) = \sigma_{ji}(a,-t) \quad (28)$

(ii) $\quad \lim_{t\to\infty} \sigma_{ij}(a,t) = V_i(a)V_j(a) \quad (29)$

an ergodicity condition, and
(iii) if $T \gg t_{\text{micro}}$ where t_{micro} is the relaxation time of $\sigma_{ij}(a,t)$, then the integral

$$\int_0^T dt\big[\sigma_{ij}(a,t) - V_i(a)V_j(a)\big] = L_{ij}^{(2)}(a) \quad (30)$$

where $L_{ij}^{(2)}(a)$ is a Γ-independent function.

Using eq. (28) it is easy to see that

$$I = \int_0^T d\tau_1 \int_0^{\tau_1} ds\, \sigma_{ij}(a,s) = T\int_0^T ds(1-s/T)\sigma_{ij}(a,s).$$

If we add and subtract $V_i(a)V_j(a)$ in the integrand we find, using eqs. (29), (30) and the fact that $s/T \simeq 0$ for macroscopic times, that

$$I = TL_{ij}^{(2)}(a) + \tfrac{1}{2}T^2 V_i(a)V_j(a) \tag{31}$$

so that

$$\lim_{T\to 0}(1/T)\int_0^T d\tau_1 \int_0^{\tau_1} ds\, \sigma_{ij}(a,s) = L_{ij}^{(2)}(a). \tag{32}$$

Using eqs. (28) and (32) in eq. (25) we obtain finally that

$$\mathbb{K}_i^{(2)}(a) = V_i(a) + g_{eq}^{-1}(a)\sum_k \frac{\partial}{\partial a_k} g_{eq}(a) L_{ki}^{(2)}(a). \tag{33}$$

Letting $s = \tau_2 - \tau_1$ in eq. (27) and using once more eq. (32) we also find that

$$\mathbb{K}_{ij}^{(2)}(a) = L_{ij}^{(2)}(a) + L_{ji}^{(2)}(a). \tag{34}$$

If eqs. (33) and (34) are put back into eq. (18) for the time evolution of the equilibrium conditional probability $P_{eq}(a,t|b)$ we recover the Fokker–Planck equation first obtained by Green.

In order to derive the Green–Kubo formulae for the transport coefficients we take the following steps:

(a) Derive the equation of motion for the coarse grained variables;
(b) Linearize these equations in the a-variables;
(c) Use the relationship between non-equilibrium averages and coarse grained variables as implied by eqs. (4) and (13) to find the linear transport equations.

The first step is readily accomplished. Multiplying eq. (18) by a_i, using eq. (9) defining the coarse grained variables and integrating over da we get, after an integration by parts is performed assuming that all terms vanish at the boundaries of the a-space, that

$$\frac{\partial \overline{a_i(t)}^b}{\partial t} = \int da \left[V_i(a) + g_{eq}^{-1}(a) \sum_k \frac{\partial}{\partial a_k} g_{eq}(a) L_{ki}^{(2)}(a) \right]$$
$$\times P_{eq}(a,t|b) \tag{35}$$

where also, use was made of eq. (33).

Since in general $V_i(a)$ and $L_{ki}^{(2)}(a)$ are non-linear functions in the a's the regression of fluctuations equation that arises from eq. (35) is a non-linear one. The linearization procedure consists in the introduction of two assumptions in eq. (35), namely that (i) $V_i(a)=0$ and (ii) that $L_{ki}^{(2)}(a)$ may be replaced by its average value $\mathring{L}_{ki}$

$$\mathring{L}_{ki} = \int d\boldsymbol{a}\, g_{eq}(a) L_{ki}^{(2)}(a). \tag{36}$$

These assumptions have a strong bearing on the structure of the time evolution equations for the coarse grained variables. Indeed, since $V_i(a)$ is assumed to be neglible eq. (30) reduces to a simple form,

$$\int_0^\infty dt\, \sigma_{ij}(a,t) \cong L_{ij}^{(2)}(a) \tag{37}$$

which used in eq. (36) implies that

$$\mathring{L}_{ki} = \int_0^\infty dt \left(\dot{A}_k(\Gamma,0), \dot{A}_i(\Gamma,t) \right) \tag{38}$$

where use has been made of eq. (23) and of the fact that $\int d\Gamma \rho_{eq} f(\Gamma) = \int \langle f(\Gamma); a \rangle g_{eq}(a) d\boldsymbol{a}$.

Substituting eq. (38) back in eq. (35), recalling assumption (i) and assuming that the A's are even functions under time reversal, we find that

$$\frac{\partial \overline{a_i(t)}^b}{\partial t} = \sum_k \mathring{L}_{ik} \int d\boldsymbol{a}\, \frac{\partial \ln g_{eq}(a)}{\partial a_k} P_{eq}(a,t|b). \tag{39}$$

Invoking the central limit theorem one can take for the distribution function for the fluctuations in equilibrium a Gaussian function [15] so

$$g_{eq}(a) = c \exp\left[-(2k_B)^{-1} \sum_l \sum_m g_{lm} a_l a_m \right] \tag{40}$$

where k_B is Boltzmann's constant and $g_{lm} = |\partial^2 S/\partial a_l \partial a_m|$, S being the entropy. Substitution of eq. (40) into eq. (39) leads at once to the linear law for the regression of fluctuations, namely,

$$\frac{\partial \overline{a(t)}^b}{\partial t} = -M \overline{a(t)}^b \tag{41}$$

where $M = k_B^{-1} \mathring{L} \cdot g$. It is worth emphazising that eq. (41) is a direct consequence of the linearization assumptions (i) and (ii) applied to the slow Markoffian equation of motion for $P_{eq}(a,t|b)$.

Finally if we take the time derivative of eq. (13) and make use of eq. (41) we find that

$$\frac{\partial \alpha(t)}{\partial t} = -M \cdot \alpha(t) \tag{42}$$

where M is the matrix for the transport coefficients which is readily obtained from eq. (38) and the definition of M. Whence,

$$M = \frac{1}{k_B} \int_0^\infty \langle \dot{A}(\Gamma,0)\dot{A}(\Gamma,t)\rangle_{eq} \cdot g \tag{43}$$

which are the famous Green–Kubo formulae. These results where first explicitly derived by Kubo, Yokota and Nakajima [16] in both their classical and quantum versions using Onsager's linear law for the regression of fluctuations as the starting point. In Green's work eq. (43) is a consequence of a trivial calculation which is not given in his paper which could be the reason why it took longer to recognize that they were a direct consequence of his own theory.

We would like to stress here that in this derivation one does not use the Onsager's regression hypothesis. In this more contemporary language it is a direct consequence of the definition of coarse grained variables and its relation with non equilibrium averages.

5. Concluding remarks

The ideas developed by Mel Green have had a profound influence in the development of non-equilibrium statistical mechanics during the last two decades. Just to mention some of the most relevant results it is worth mentioning Zwanzig's derivation [3] of an exact kinetic equation for the distribution function $g(a,t)$ starting from Liouville's equation using a non-linear projection operator. For slow Markoffian processes his results reduce to those given here by eqs. (18), (33) and (34). He also showed that if $V_i(a)$ is a quadratic function in the a's eq. (18) gives rise to a renormalization of the transport coefficients [5, 6]. Furthermore he also showed how these ideas are related to the mode–mode coupling scheme developed by Kawasaki [4]. In this theory the starting point to construct the appropriate kinetic equations are also eqs. (18), (33) and (34).

The generalized Langevin equation approach developed by Mori in 1965 [17] leads to linear equations of motion for the phase functions $A_i(\Gamma)$. If the ideas behind the problem of Brownian motion are used in the hypercell in phase space [7] one can see, as Mori showed, how Zwanzig's exact kinetic equation is recovered and furthermore how one can obtain a new expansion in terms of the slowness parameter [8] which corresponds essentially to a KM expansion for non-Markovian processes. From these results, eq. (18) follows easily. The approaches of Zwanzig and Mori can be framed within a unified context [13, 18] that allows us to see how Green's results and others of similar nature are related to first principles.

Finally it is also important to mention that the non-linear Langevin type equations that have also been derived by Kawasaki [19] and Mori [7] reduce under different approximations to eq. (35) for the coarse grained variables [20]. These results do partial justice to the many original contribution of M. S. Green to non-equilibrium statistical mechanics.

References

[1] M. S. Green, J. Chem. Phys. **20** (1952) 1281.
[2] M. S. Green, J. Chem. Phys. **22** (1954) 398.
[3] R. W. Zwanzig, Phys. Rev. **124** (1961) 983.
[4] K. Kawasaki, Proc. of the Sixth IUPAP Conf. on Statistical Mechanics, eds. S. Rice, K. Freed and J. S. Light (The Univ. of Chicago Press, 1972) p. 259.
[5] R. W. Zwanzig, Proc. of the Sixth IUPAP Conf. on Statistical Mechanics, eds. S. Rice, K. Freed and J. S. Light (The Univ. of Chicago Press 1972) p. 241.
[6] R. Zwanzig, K. S. J. Nordholm and W. C. Mitchell, Phys. Rev. **5A** (1972) 2680.
[7] H. Mori and H. Fujisaka; Prog. Theor. Phys. (Kyoto) **49** (1973) 764.
[8] H. Mori, H. Fujisaka and H. Shigematsu, Prog. Theor. Phys. (Kyoto) **51** (1974) 109.
[9] L. S. Garcia-Colin and R. M. Velasco, Phys. Rev. **12A** (1975) 646.
[10] K. Kawasaki, Prog. Theor. Phys. (Kyoto) **51** (1974) 1064.
[11] L. S. Garcia-Colin and J. L. Del Rio, Physica **96A** (1979) 606.
[12] J. L. Del Rio and L. S. Garcia-Colin, J. Stat. Phys. **19** (1978) 109.
[13] L. Onsager, Phys. Rev. **37** (1931) 405; Phys. Rev. **38** (1931) 2265.
[14] R. L. Stratonovich, Topics in the Theory of Random Noise (Gordon and Breach Co. New York, 1963) Ch. 3.
[15] S. R. de Groot and P. Mazur "Non-equilibrium Thermodynamics" (North-Holland, Amsterdam, 1962) Chap. 7.
[16] R. Kubo, M. Yokota and S. Nakajima, J. Phys. Soc. Japan (Tokyo) **12** (1957) 1203.
[17] H. Mori, Prog. Theor. Phys. (Kyoto) **33** (1965) 423.
[18] L. S. Garcia-Colin and J. L. Del Rio, J. Stat. Phys. **16** (1977) 235.
[19] K. Kawasaki, J. Phys. A (London) **A6** (1973) 289.
[20] K. Kawasaki, in: Phase Transitions and Critical Phenomena, eds. M. S. Green and C. Domb (Academic Press, New York, 1976) Vol. **5A**.

CHAPTER 5

Stochastic Description of Many-body Systems

N. G. van KAMPEN

Institute for Theoretical Physics of the University of Utrecht, Netherlands

Before coming here I reread the two fundamental papers of Mel Green [1, 2]. It was a revelation to see how much of my present ideas about stochastic processes are clearly stated in them, although I had forgotten their source. The best start for my talk at this occasion is therefore to briefly review the physical ideas in his work and to emphasize a few points that seem to me particularly important. Subsequently I shall introduce a new feature and use it to obtain some results from the equations that he developed.

Statistical Mechanics deals with systems having an enormous number of microscopic variables obeying microscopic equations of motion, which for the present discussion are taken to be the classical Hamilton equations. Experience has taught us that a 'contracted' description is possible in much fewer macroscopic or gross variables, which miraculously again satisfy a set of self-contained differential equations. In the present literature this miracle is often taken for granted, but Green said explicitly: *'One of the facts to be explained by a statistical mechanics of time-dependent phenomena is that in certain systems the time-behavior of the gross variables as observed with large scale methods can be represented very well by differential equations which are of the first order in time.'*

These macroscopic equations are not exact: deviations from them show up in the form of fluctuations. They are normally very small. (That is not so around an *unstable* solution of macroscopic equations; this is a topic of much interest at present, but we shall not consider it here.) The question is: How do these macroscopic equations relate to the microscopic ones, and how does one describe the fluctuations? This question touches the very foundations of statistical mechanics, because it involves the transition from reversible to irreversible equations.

Let X stand for the 10^{23} variables that specify a point in phase space, and X_t for the point into which it is carried by the Hamilton equations [3]. Let $A_i(X)$ be a set of grossly observable variables. Green: *'One of the important problems of the theory is to define the conditions under which a system has a complete set of gross variables and to give a rule for determining them.*' In fact, this question is crucial for understanding how macroscopic equations can exist. Unfortunately, little attention has been paid to it. Most physicists and chemists (and all students) prefer to work with streamlined formalisms such as projection operator techniques and linear response theory, which tacitly *assume* the existence of gross variables endowed with all the properties desired by the authors.

Having chosen a set $A_i(X)$ one defines phase cells as domains in phase space delineated by

$$a_i < A_i(X) < a_i + \delta a_i \quad (i = 1, 2, \ldots)$$

and denotes their volume by

$$w(a)\,\delta a_1 \delta a_2 \ldots .$$

Saying that at $t°$ the system is in a certain macroscopic state given by the values $a_i°$ of the variables $A_i(X)$ merely tells that X lies in a certain phase cell, but gives no information about all the microscopic variables needed to specify the exact point in that cell. Non-fluctuating macroscopic theory now makes the *randomness assumption* (Stosszahlansatz, molecular chaos): The future values $A_i(X_t)$ do not depend on the precise position of X within the phase cell. *Whether this is true or not depends on the structure of the system* (i.e., the microscopic equations), *and on the choice of the $A_i(X)$.* For a dilute gas we are pretty sure that the local density, velocity and energy are a proper choice. Or, on a more refined level, the Boltzmann distribution function. But only a more profound investigation can tell whether this distribution function is still a proper choice for denser gases, as Bogolyubov surmised [4].

If the randomness assumption is correct it means that one might as well describe the state of the system by an ensemble that has a constant density in that particular phase cell (and zero density outside it). It is clear, however, that the streaming in phase space will not carry that whole ensemble into the next phase cell, but the ensemble spreads out over all neighboring cells. This is the cause of the fluctuations. Let the fraction of the ensemble that at $t > t°$ finds itself in the phase cell $a, \delta a$ be $T(a°, t°/a, t)\delta a$.

Of course, if one knew T for all $t > t°$ the whole evolution would be known. However, in practice it is only feasible to compute T for $t - t° = \Delta t$ small enough. One then knows the distribution at $t° + \Delta t$. Now again the randomness assumption is invoked to say that it does not matter how the fraction in each cell at $t° + \Delta t$ is situated: one may again smear it out evenly through each of the cells. This *repeated randomness assumption* permits to treat the evolution during the next interval Δt in the same manner as from $t°$ to $t° + \Delta t$. This is expressed by the equation

$$T(a°, t°/a, t+2\Delta t) = \int T(a°, t°/a', t°+\Delta t)\, da'$$
$$\times T(a', t°+\Delta t/a, t°+2\Delta t).$$

Thus the succession of values of $a_i(t)$ has become a Markov process with transition probability T, which obeys this Chapman–Kolmogorov equation. Since we expect the fluctuations to be small it will be a good approximation to assume that the transition probabilities are sharply peaked, so that the equation can be replaced with a Fokker–Planck equation [5]

$$\frac{\partial T}{\partial t} = -\sum_i \frac{\partial}{\partial a_i}\bar{v}_i(a)T + \sum_{i,j}\frac{\partial^2}{\partial a_i \partial a_j}\bar{\xi}_{ij}(a)T.$$

The coefficients are given in the well-known way by

$$\bar{v}_i(a) = \frac{\langle \Delta a_i \rangle_a}{\Delta t}, \quad \bar{\xi}_{ij}(a) = \frac{\langle \Delta a_i \Delta a_j \rangle_a}{\Delta t},$$

where $\langle\ \rangle_a$ denotes an average with fixed a. By some algebra Green obtains

$$\bar{v}_i(a) = \langle \dot{A}_i(X) \rangle_a + \frac{1}{w(a)}\sum_i \frac{\partial}{\partial a_k}w(a)\xi_{ki}(a),$$

$$\bar{\xi}_{ij}(a) = \tfrac{1}{2}\{\xi_{ij}(a) + \xi_{ji}(a)\},$$

where

$$\xi_{ij}(a) = \int_0^\infty \{\langle \dot{A}_i(X)\dot{A}_j(X_t) \rangle_a - \langle \dot{A}_i(X) \rangle_a \langle \dot{A}_j(X) \rangle_a\}\, dt.$$

This is the major result: *The coefficient of the diffusion term in the Fokker–Planck equation for a equals the time-integral over the autocorrelation function of the microscopic rate of change of a.* One example is the ordinary diffusion coefficient

$$D = \tfrac{1}{2}\int_0^\infty \langle \dot{x}(0)\dot{x}(t) \rangle\, dt,$$

but this special formula can be found in an elementary fashion.

This is the way in which Green introduced the stochastic process to describe the evolution of a non-equilibrium system. He did not mince the physical assumptions and approximations needed and did not just derive a smooth mathematical scheme that leaves to the reader the decision why and when it applies. In that respect he resembles Ehrenfest rather than Gibbs. It is also the reason why I prefer his work to the now common derivation of the Kubo relation; ostensibly that derivation has no need of these assumptions, but if one tries to justify its application to an actual situation one finds that the same repeated randomness assumption is again indispensable. *This assumption and the related problem of the proper choice for the gross variables cannot be avoided.* Unfortunately it may well be also the reason why his work has not received the attention it deserves.

The resulting Fokker–Planck equation is therefore

$$\frac{\partial T}{\partial t} = -\sum_i \frac{\partial}{\partial a_i}\left(v_i + \frac{1}{w}\sum_j \frac{\partial}{\partial a_j} w\xi_{ji}\right)T + \sum_{i,j} \frac{\partial^2}{\partial a_i \partial a_j} \bar{\xi}_{ij}T,$$

where $v_i(a) = \langle \dot{A}_i(X)\rangle_a$ and ξ_{ij} is expressed in the auto-correlation function of $\dot{A}$. A crucial consequence is that *the macroscopic evolution of a many-body system is a stochastic process rather than a deterministic differential equation*. The popular view in physics is that the evolution is given by the familiar macroscopic equations and that someone who is interested in fluctuations should tag them on somehow in the form of a Langevin term. I have often marvelled how tenaciously this view is applied even to cases in which it is manifestly wrong, e.g., radio-active decay and chemical reactions. Green emphatically states *that the stochastic process is the first stage, and that the next step is to extract from it the deterministic equations one is used to in macroscopic physics* [14].

This means that one has to eliminate the fluctuations. The way in which he did this is by first deducing from the Fokker–Planck equation rigorously

$$\frac{d}{dt}\langle a_k\rangle_{a°} = \langle v_k(a)\rangle_{a°} + \left\langle \frac{1}{w(a)}\sum_j \frac{\partial}{\partial a_j} w(a)\xi_{jk}(a)\right\rangle_{a°}.$$

Now suppose that the fluctuations are small so that $\langle v_k(a)\rangle_{a°} \approx v_k(\langle a\rangle_{a°})$, etc. Then the average $\langle a\rangle_{a°} \equiv \alpha$ approximately obeys

$$\frac{d\alpha_k}{dt} = v_k(\alpha) + \frac{1}{w(\alpha)}\sum_j \frac{\partial}{\partial \alpha_j} w(\alpha)\xi_{jk}(\alpha).$$

Thus the rate of change is not just $v_k = \langle \dot{A}_k \rangle$, but an additional term appears involving the auto-correlation function of $\dot{A}$ as well.

Unfortunately this result is not unique for the following reason. The Fokker–Planck equation describes a probability distribution $T(a°, t°/a, t)$ in a-space, which starts out at $t°$ as a delta peak at $a°$, and ends up as the equilibrium distribution. If the fluctuations are indeed small, that means that at all times this distribution is a sharp peak and as such moves bodily in a-space from $a°$ to a^{eq}. Its location in a-space as a function of time is the macroscopic state α, and the width determines the fluctuations. From this picture it is evident, however, that its 'location' is defined only with a margin of the size of the fluctuations. Hence also the macroscopic equation that purports to describe the location as a function of time is afflicted with that same margin of uncertainty. To put it differently, there is no compelling reason for identifying the macroscopic value α with the average; any other value within the margin set by the fluctuations is equally acceptable. This would lead to the conclusion that the additional term in Green's macroscopic equation is not realistic, and that one should be equally justified in writing the naive equation

$$\dot{\alpha}_k = v_k(\alpha).$$

The same dilemma confronted those authors who took a more phenomenological approach [6]. This consists in *postulating* a Fokker–Planck equation with as yet unknown coefficients and then imposing general conditions on them, viz., time invariance and detailed balance. Of course the connection with auto-correlation functions cannot be found in this way, nor does it solve our dilemma. Yet it was already known that the dilemma gave rise to practical difficulties for determining the noise in non-linear electronic systems [7]. A more recent version is the controversy between the Itô and Stratonovich interpretations of the non-linear Langevin equation, which as it stands is meaningless [8].

The only way to obtain a well-defined, unique, macroscopic, deterministic description is by finding a physical parameter which permits to scale down the fluctuations and studying the limit in which the fluctuations vanish. Such a parameter is the size of the system, the mass of a Brownian particle, or the capacity of a condenser. A systematic treatment based on an expansion in the reciprocal of this parameter leads to the following result [5, 9]. Normally the first approximation gives the macroscopic deterministic equation, the next one gives the fluctuations in linear Gaussian approximation, and the higher approximations describe the influence of the non-linearity on the fluctuations. No non-linear Fokker–Planck equation [10] ever appears in this 'normal' case. However, when the first term in this expansion happens to be zero the terms must be rearranged and the lowest approximation now does have the form of a non-linear Fokker–Planck equation. In particular this occurs in the case of diffusion in an inhomogeneous medium, and we shall therefore call problems of this kind 'of diffusion type'.

In problems of diffusion type another parameter is needed to scale down the fluctuations. The obvious choice is the temperature θ, because the fluctuations are small at low temperatures. That implies of course that one has to know how the coefficients depend on θ, which means that one has to know something about the system under consideration. The kind of system I have in mind is not the general many-body system that Green considered, but a more special type. I take a small system connected with a bath with temperature θ. Examples are the Brownian rotator [11] and a particle diffusing in a three-dimensional inhomogeneous medium.

Suppose then that the system may be described phenomenologically by some Fokker–Planck equation

$$\frac{\partial T(a,t)}{\partial t} = -\sum_i \frac{\partial}{\partial a_i} V_i(a) T + \sum_{i,j} \frac{\partial^2}{\partial a_i \partial a_j} \Xi_{ij}(a) T.$$

I have modified the notation to emphasize that our coefficients V_i, Ξ_{ij} are introduced phenomenologically without claiming any kinship with the underlying microscopic equations of motion. By an algebraic rearrangement of terms this equation can be written

$$\frac{\partial T}{\partial t} = -\frac{\partial}{\partial a_i} F_i T + \frac{\partial}{\partial a_i} P^{\mathrm{eq}} \Xi_{ij} \frac{\partial}{\partial a_j} \frac{T}{P^{\mathrm{eq}}},$$

$$F_i(a) = V_i(a) - \frac{1}{P^{\mathrm{eq}}} \frac{\partial}{\partial a_j} P^{\mathrm{eq}} \Xi_{ij}.$$

Summation is understood. P^{eq} is the equilibrium distribution known from equilibrium statistical mechanics,

$$P^{\mathrm{eq}}(a) = w(a) e^{-U(a)/\theta},$$

where $w(a)$ is the same phase space function that occurred before, $U(a)$ the energy of the small system, and θ the temperature of the bath. The rationale for this rearrangement is that one can now readily apply detailed balance [6] and finds

$$\varepsilon_i F_i(\varepsilon a) = -F_i(a),$$

$$\frac{\partial}{\partial a_i} F_i(a) P^{eq}(a) = 0,$$

$$\varepsilon_i \varepsilon_j \Xi_{ij}(\varepsilon a) = \Xi_{ij}(a),$$

where $\varepsilon_i = \pm 1$ for even and odd variables respectively [12]. Moreover, it turns out by inspection that for many actual systems F_i is independent of θ, while Ξ_{ij} is proportional to it: $\Xi_{ij}(a) = \theta H_{ij}$. Hence F_i is mechanical while Ξ reflects the thermal fluctuations.

After these preliminaries we are now in a position to take the limit [13]. The reason why one cannot simply put $\Xi = 0$ is that for small θ the distribution T becomes very sharp (as exemplified by P^{eq}), so that its derivatives need not be small. Hence one should describe $T(a, t)$ as a peak located at some point $\alpha(t)$ with a width of order $\theta^{1/2}$. Accordingly we transform from a to a new variable ξ by setting

$$a_i = \alpha_i(t) + \theta^{1/2} \xi_i,$$

$$T(a, t) = T(\alpha(t) + \theta^{1/2} \xi, t) \equiv \Pi(\xi, t).$$

The function $\alpha(t)$ is still arbitrary but will be specified presently. Our non-linear Fokker–Planck equation for T transforms into

$$\frac{\partial \Pi}{\partial t} - \theta^{-1/2} \frac{d\alpha_i}{dt} \frac{\partial \Pi}{\partial \xi} = -\theta^{-1/2} \frac{\partial}{\partial \xi_i} F_i(\alpha(t) + \theta^{1/2} \xi) \Pi$$

$$+ \frac{\partial}{\partial \xi_i} P^{eq}(\alpha + \theta^{1/2} \xi) H_{ij}(\alpha + \theta^{1/2} \xi) \frac{\partial}{\partial \xi_j} \frac{\Pi}{P^{eq}(\alpha + \theta^{1/2} \xi)}.$$

In the limit only the terms with $\theta^{-1/2}$ survive; they determine the dependence of α on time:

$$\frac{d\alpha_i}{dt} = F_i(\alpha) - H_{ij}(\alpha) \frac{\partial U}{\partial \alpha_j}.$$

Alternatively this may be written, to the same order in θ,

$$\frac{d\alpha_i}{dt} = F_i(\alpha) + \Xi_{ij}(\alpha) \frac{\partial \log P^{eq}}{\partial \alpha_j}.$$

Thus we arrive at unique, well-defined macroscopic equations, valid when the temperature is low enough for the fluctuations to be negligible.

(In practice usually room temperature is low enough.) The conditions are that the system is connected with a heat bath, that it is of diffusion type and can therefore be described by a Fokker–Planck equation, and that the coefficients depend on temperature in the way stipulated above. It agrees with Green's macroscopic equation in that there is an additional term due to the fluctuation term of the Fokker–Planck equation. It differs, however, in two details: our Ξ_{ij} is symmetric (by construction) in contrast with his ξ_{ij}, and we have no derivatives of Ξ_{ij}, because they do not contribute in this order of θ.

The term $F_i(\alpha)$ in our deterministic equation is reversible in time and represents the mechanical part of the motion. The second term changes sign on reflecting time and is therefore purely irreversible. It represents the damping, as can be seen from

$$\frac{dU}{dt} = -\frac{\partial U}{\partial a_i} H_{ij} \frac{\partial U}{\partial a_j},$$

which is negative because H_{ij} is a positive definite matrix. Moreover, the fact that the coefficients are the same Ξ_{ij} that determine the fluctuations constitutes the generalization of the fluctuation–dissipation theorem to non-linear equations. Finally, the fact that they are symmetric is the non-linear generalization of the Onsager relations.

References

[1] M. S. Green, J. Chem. Phys. **20** (1952) 1281.
[2] M. S. Green, J. Chem. Phys. **22** (1954) 398.
[3] I follow the notation of ref. 1 with minor alterations.
[4] N. N. Bogolyubov, Problems of Dynamical Theory in Statistical Physics, in: Studies in Statistical Mechanics I, eds. G. E. Uhlenbeck and J. de Boer (North-Holland, Amsterdam, 1962);
E. G. D. Cohen, in: Fundamental Problems in Statistical Mechanics II; Proc. of the Sec. NUFFIC Int. Summer Course, ed. E. G. D. Cohen (North-Holland, Amsterdam, 1968).
[5] This reduction of the Chapman–Kolmogorov equation has been discussed in: Advances of Chem. Phys. **34**, eds. I. Prigogine and S. A. Rice (Wiley, New York, 1976);
N. G. van Kampen, Phys. Lett. **62A** (1977) 383.
[6] N. G. van Kampen, Physica **23** (1957) 707, 816.
U. Uhlhorn, Arkiv för Fysik **17** (1960) 361.
R. Graham, in: Ergebn. Exakten Naturw. **66** (Springer, Berlin, 1973).
H. Hasegawa, Prog. Theor. Phys. **55** (1976) 90.
[7] D. K. C. MacDonald, Phil. Mag. **45** (1954) 63, 345.
For a review see N. G. van Kampen, in: Fluctuation Phenomena in Solids, ed. R. E. Burgess (Acad. Press, New York, 1965).

[8] For a review and discussion see R. E. Mortensen, J. Statist. Phys. **1** (1969) 271.
[9] N. G. van Kampen, in: Thermodynamics and Kinetics of Biological Processes, ed. A. I. Zotin (Moscow, to be published).
[10] A Fokker–Planck equation is called 'linear' if the coefficient of the first order derivative is linear in the independent variable and the coefficient of the second order derivative constant. It is of course always linear in the unknown probability distribution.
[11] P. S. Hubbard, Phys. Rev. **A15** (1977) 329;
G. W. Ford, J. T. Lewis and J. McConnell, Proc. Roy. Irish Acad. **76A** (1976) 117.
[12] S. R. de Groot and N. G. van Kampen, Physica **21** (1954) 39.
[13] I am indebted to C. W. Gardiner for an illuminating discussion on this point.
[14] See also Green's recent discussion remark in the Proc. of the Oji Seminar, Suppl. Prog. Theor. Phys. Nr. 64 (1978) p. 434.

CHAPTER 6

H-Theorems for Markoffian Processes

R. KUBO

Department of Physics, University of Tokyo
Japan

© *North-Holland Publishing Company 1981*

Perspectives in Statistical Physics
Ed. H. J. Raveché

Contents

1. Introduction — 103
2. Markoffian processes — 103
3. H-Theorems — 106
4. Comments — 109
References — 110

1. Introduction

The H-theorem was first introduced to statistical mechanics by Boltzmann in 1872 [1]. His original proof of the theorem was based on the assumption of the existence of inverse collisions or the symmetry of the collision kernel with respect to the initial and final states of colliding molecules. As was noticed by Lorentz, this assumption is not generally true for polyatomic molecules or molecules with internal degrees of freedom. In his later treaties of the gas theory [2], Boltzmann himself extended his proof by considering cycles of collisions and thus without the use of the previous assumption. Since that time, probably many people must have noticed that the proof could be simplified. His attention to this problem aroused by Pauli, Stueckelberg [3] noted briefly that the unitarity of the scattering matrix of binary collisions is sufficient to prove the H-theorem. More recently Waldmann [4] discussed this in greater details in a paper given to the centennial celebration of the Boltzmann equation.

In non-equilibrium statistical mechanics, the so-called master equation is often used to formulate the irreversible evolution of a given system, which is assumed to be a Markoffian process. H-theorems for Markoffian processes are essentially much simpler than those for the Boltzmann equation, because the evolution equation for the former is linear whereas that for the latter is non-linear. For pedagogical reasons, however, the proof is usually made with the use of the detailed balance assumption or the symmetry of the transition probability rate. Many years ago Yosida [5] gave a proof of the H-theorem for a Markoffian process without any such assumption. Since Yosida's proof seems to have remained unnoticed by most physicists, I like to present here a simple version of his proof and also some generalizations of it which may be of some relevance to the general aspects of non-equilibrium statistical mechanics.

2. Markoffian processes

We consider a Markoffian process $x(t)$ assuming a continuous time t and a continuous random variable x which may be a multi-component vector. The process is generally non-stationary so that the Chapman–Kolmogorov equation is written as:

$$P(y, t \leftarrow x, s) = \int P(y, t \leftarrow z, \tau) \, dz \, P(z, \tau \leftarrow x, s), \qquad t > \tau > s, \qquad (1)$$

for the probability density of transition from x to y over the time interval (s, t). The transition probability satisfies the conditions,

$$\int dy \, P(y, t \leftarrow x, s) = 1, \qquad P(y, t \leftarrow x, s) \geq 0. \qquad (2)$$

For an infinitesimal time interval Δt, the transition probability is assumed to take the form,

$$P(y, t \leftarrow x, t - \Delta t) = \delta(x - y) + \Delta t \langle y | \Gamma_t | x \rangle + o(\Delta t), \qquad (3)$$

where Dirac's bracket notation is used for convenience to represent the integral kernel of the evolution operator Γ_t. The forward equation is then written as:

$$\frac{\partial \psi(t)}{\partial t} = \Gamma_t \psi(t), \quad \text{or} \quad \frac{\partial}{\partial t} \psi(x, t) = \int \langle x | \Gamma_t | y \rangle \, dy \, \psi(y, t), \qquad (4)$$

and the backward equation as:

$$\frac{\partial \psi^+(s)}{\partial s} = \psi^+(s) \Gamma_s^+, \quad \text{or} \quad \frac{\partial}{\partial s} \psi^+(x, s) = \int \psi^+(y, s) \, dy \, \langle y | \Gamma_s | x \rangle. \qquad (5)$$

The fundamental solution of eq. (4) or eq. (5) with the initial or the final condition,

$$\psi(y, s) = \delta(y - x), \quad \text{or} \quad \psi^+(x, t) = \delta(x - y),$$

is the transition probability $P(y, t \leftarrow x, s)$. The general solution of eq. (4) is given in the form,

$$\psi(y, t) = \int P(y, t \leftarrow x, s) \, dx \, \psi(x) \qquad (6)$$

for a given function $\psi(x)$ at time s. For eq. (5) it is

$$\psi^+(x, s) = \int \psi^+(y) \, dy \, P(y, t \leftarrow x, s), \qquad (7)$$

for a given function $\psi^+(y)$. The expression (7) may be written as

$\langle \psi^+(x(t))\rangle_{x,s}$ which is the expectation of the random variable $\psi^+(x(t))$ at time t when the system has started from x at the initial time s. By the Markoffian property (1), $\psi(y,t)$ and $\psi^+(x,s)$ satisfy the equations,

$$\psi(y,t) = \int P(y, t\leftarrow z, \tau) dz\, \psi(z,t) \quad t > \tau \tag{8}$$

and

$$\psi^+(x,s) = \int \psi^+(z,\tau) dz\, P(z, \tau\leftarrow x, s), \quad \tau > s. \tag{9}$$

If the process is stationary, the transition probability $P(y, t\leftarrow x, s)$ depends only on the time difference $\tau = t - s$. In such a case it is more customary to write the expression (7) as:

$$T_\tau f(x) = f(x,\tau) = \int f(y) dy\, P(y, \tau\leftarrow x, 0), \tag{10}$$

defining a semi-group transformation T_τ. As a function of x, it is equal to the expectation of $f(x(\tau))$ when the initial value $x(0)$ is specified to x. Corresponding to eq. (9) we have:

$$T_t f(x) = \int T_s f(z) dz\, P(z, t-s\leftarrow x, 0). \tag{11}$$

A familiar example of eq. (4) is the diffusion process,

$$\frac{\partial \psi(y,t)}{\partial t} = \frac{\partial}{\partial y} a(y,t)\psi(y,t) + \frac{\partial^2}{\partial y^2} b(y,t)\psi(y,t), \tag{12}$$

for which the backward equation is:

$$-\frac{\partial \psi^+(x,s)}{\partial s} = -a(x,s)\frac{\partial}{\partial x}\psi^+(x,s) + b(x,s)\frac{\partial^2}{\partial x^2}\psi^+(x,s). \tag{13}$$

If the process is stationary, the backward equation can be written as:

$$\frac{\partial}{\partial \tau} f(x,\tau) = -a(x)\frac{\partial}{\partial x} f(x,\tau) + b(x)\frac{\partial^2}{\partial x^2} f(x,\tau). \tag{14}$$

3. *H*-Theorems

Yosida's original *H*-theorem may be stated as follows:

Theorem 1. Assume that the stationary Markoffian process has the invariant measure (equilibrium distribution) $\phi_e(x)$. Then it holds that:

$$\int \phi_e(x) dx\, C(T_s f(x)) \geq \int \phi_e(x) dx\, C(T_t f(x)), \quad s < t, \qquad (15)$$

where $C(\xi)$ is an arbitrary convex function of ξ over the interval which covers the set of all possible values of $T_\tau f(x)$.

For the proof, we first note that the convex property of $C(\xi)$ means the inequality:

$$\sum_i C(\xi_i) w_i \geq C\left(\sum w_i \xi_i\right) \qquad (16)$$

for

$$\sum_i w_i = 1, \quad w_i \geq 0.$$

By taking

$$\xi = T_s f(z) \quad \text{and} \quad w(\xi) = P(z, t - s \leftarrow x, 0) \geq 0,$$

and replacing the weighted sum by a weighted integral, the inequality (16) gives:

$$\int C(T_s f(z)) dz\, P(z, t-s \leftarrow x, 0) \geq C\left(\int T_s f(z) dz\, P(z, t-s \leftarrow x, 0)\right)$$
$$= C(T_t f(x)).$$

Multiplying this with $\phi_e(x)$ and integrating the both sides, we obtain eq. (15) by noticing the equation:

$$\int P(y, \tau \leftarrow x, 0) dx\, \phi_e(x) = \phi_e(y). \qquad (17)$$

This theorem is very general. It holds true for arbitrary functions f and C as long as the required conditions are satisfied. However, it is concerned with the backward equation and is less appealing to physicists than the more familiar types of *H*-theorems concerned with the forward equation. Such a theorem is states as:

Theorem 2. Let $\psi(x,t)$ be a solution of the forward equation of a *stationary* Markoffian process (eq. (4) with a time-independent evolution operator Γ) and $\phi_e(x)$ be its invariant measure satisfying the condition $\phi_e(x)>0$. A generalized *H*-function is defined by:

$$H_C(t)=\int C\left(\frac{\psi(x,t)}{\phi_e(x)}\right)\phi_e(x)\,\mathrm{d}x, \tag{18}$$

using an arbitrary convex function $C(\xi)$. The *H*-function never increases in time, namely:

$$H_C(s)\geq H_C(t), \quad \text{if} \quad s<t. \tag{19}$$

The proof is similar to the previous one. By choosing the weight function,

$$w(x)=P(y,t\leftarrow x,s)\phi_e(x)/\phi_e(y)\geq 0,$$

which satisfies the condition,

$$\int w(x)\,\mathrm{d}x=1,$$

we apply the inequality (16) to:

$$\int C\left(\frac{\psi(x,s)}{\phi_e(x)}\right)w(x)\,\mathrm{d}x \geq C\left(\int \frac{\psi(x,s)}{\phi_e(x)}w(x)\,\mathrm{d}x\right)$$

$$=C(\psi(y,t)/\phi_e(y)),$$

where eqs. (8) and (17) are used. This means that:

$$\int C\left(\frac{\psi(x,s)}{\phi_e(x)}\right)P(y,t\leftarrow x,s)\phi_e(x)\,\mathrm{d}x \geq C\left(\frac{\psi(y,t)}{\phi_e(y)}\right)\phi_e(y).$$

When integrated over y, this yields the required inequality.

The theorem can be generalized to:

Theorem 3. Let $\psi_1(x,t)$ and $\psi_2(x,t)$ be two different solutions of the forward equation (4) of a non-stationary Markoffian process. Assuming that $\psi_2(x,t)>0$, a generalized H-function is defined by:

$$H_{12}(t) = \int dx\, C\left(\frac{\psi_1(x,t)}{\psi_2(x,t)}\right)\psi_2(x,t), \tag{20}$$

with an arbitrary convex function C. Then the H-function never increases in time, namely:

$$H_{12}(s) \geq H_{12}(t) \quad \text{if} \quad s<t.$$

The proof is the same as before. We choose the weight function:

$$w(x) = P(y, t \leftarrow x, s)\psi_2(x,s)/\psi_2(y,t),$$

which satisfies the condition:

$$\int w(x)\,dx = 1.$$

Then the inequality (16) gives:

$$\int C\left(\frac{\psi_1(x,s)}{\psi_2(x,s)}\right) w(x)\,dx \geq C\left(\int \frac{\psi_1(x,s)}{\psi_2(x,s)} w(x)\,dx\right)$$

$$= C\left(\frac{\psi_1(y,t)}{\psi_2(y,t)}\right).$$

Multiplying both sides by $\psi_2(y,t)$ and integrating them over y we get the required inequality. Obviously theorem 2 is a particular case of theorem 3 for the choice of $\psi_2(x,t) = \phi_e(x)$ which is possible for a stationary process having an invariant measure.

The most familiar choice of C is:

$$C(\xi) = \xi \log \xi, \tag{21}$$

which is important and useful because of its extensive property. Then the H-function (18) becomes:

$$H(t) = \int dx\, \psi_1(x,t)\{\log \psi_1(x,t) - \log \phi_e(x)\} \tag{22}$$

for a stationary process. This corresponds to the free energy function generalized to a non-equilibrium system in a stationary environment.

For a more general non-stationary system, eq. (21) gives:

$$H_{12}(t) = \int dx\, \psi_1(x,t) \log \psi_1(x,t)/\psi_2(x,t). \tag{23}$$

If $\psi_1(x,t)$ also satisfies the condition $\psi_1(x,t) > 0$, ψ_1 and ψ_2 can be interchanged so that a symmetrical H-function can be defined as:

$$H(t) = \int dx (\psi_1(x,t) - \psi_2(x,t)) \log \frac{\psi_1(x,t)}{\psi_2(x,t)}, \tag{24}$$

which never increases in time.

4. Comments

Although it is not entirely new, theorem 3 is worth noting. In fact it was first noted more than twenty years ago by Lebowitz and Bergmann [6] who proved the theorem for the H-function of the form (23). As was discussed by these authors, it is an interesting theorem of some importance for statistical mechanics of non-equilibrium systems. It means under some appropriate conditions the asymptotic uniqueness of statistical behavior of a system exposed to a non-stationary environment. This is indeed what we experience in a great many cases of ordinary circumstances. Suppose that a system is subject to external forces or is in contact with external reservoirs which are changing in time. The response of the system to these external conditions is generally dependent on the initial condition in which the system is prepared. As the time goes on, however, the memory of the initial preparation usually is lost. The response becomes asymptotically independent of the initial conditions and is solely determined by the nature of the system and of its interaction with the environments. This asymptotic uniqueness is generally true in near equilibrium situations, where the response is linear in off-equilibrium parameters. Beyond the regime of linear response, the same uniqueness of asymptotic responses is very commonly observed. It should be emphasized, however, that the uniqueness theorem can be violated in some cases which are by no means rare. There are branching phenomena which may delicately depend on the initial states of the system and also on the non-linear interactions within the system as well as those with the environment.

The asymptotic approach to a unique distribution in non-stationary Markoffian processes is a generalization of the existence of an invariant measure and the approach to it in stationary Markoffian processes. Lebowitz and Bergmann proved their theorem much in the same spirit as that of theorem 3. Namely, they noticed that the detailed balance condition or the direct symmetry of transition rate kernel is not necessary but a weaker integral condition is sufficient to prove the asymptotic approach. Also they noted that the decreasing property of the function (23) or (24) is proved under a weak condition.

In order that the theorem be meaningful, the sets of x where the two functions ψ_1 and ψ_2 take non-vanishing values must be identical except a set of zero measure. The set must be indecomposable in the same sense of the word familiar in ergodic theories to guarantee the uniqueness. Furthermore the process must be such that no runaway is possible.

These are only qualitative statements of the condition for the uniqueness of asymptotic distribution which must be formulated in a more rigorous manner. At the same time it will be an important and interesting problem to study the ways how the uniqueness is violated. It may also happen that a solution of the evolution equation ceases to be analytic at a certain time point, in which case a sort of phase change may occur or even an explosion may take place. Such a study will be useful for understanding of various phenomena in non-equilibrium non-stationary physical processes.

References

[1] L. Boltzmann, Wiener Berichte **63** (1972) 275.
[2] L. Boltzmann, Vorlesungen über Gastheorie Bd. II, Kapl VII.
[3] E. C. G. Stueckelberg, Helv. Phys. Actz **25** (1952) 577.
[4] L. Waldmann, in: The Boltzmann Equation, eds. G. D. Cohen and W. Thirring (Springer-Verlag, Wien, 1974) p. 232.
[5] K. Yosida, Proc. Imp. Acad. Tokyo **10** (1940) 43. Functional Analysis (Springer, 1965) p. 379.
[6] J. L. Lebowitz and P. G. Bergmann, Annals Phys. **I** (1957) 1;
P. G. Bergmann and J. L. Lebowitz, Phys. Rev. **99** (1955) 578.

CHAPTER 7

Energy Flow and Thermal Conductivity in One-dimensional, Harmonic, Isotopically Disordered Crystals.

R. J. RUBIN

Polymer Science and Standards Division
National Measurement Laboratory
National Bureau of Standards
Washington, D.C. 2034
U.S.A.

© *North-Holland Publishing Company 1981*

Perspectives in Statistical Physics
Ed. H.J. Raveché

Contents

1. Introduction 113
2. Model and formal solution 114
3. Monte Carlo estimates of thermal conductivity 119
4. Remarks 120
References 122

1. Introduction

The one-dimensional chain of particles coupled by nearest-neighbor harmonic forces is a venerable physical system [1]. In the course of time, only the questions posed for the model have changed. For example, Hamilton [2, 3] studied the propagation of light through the ether or crystals. He obtained a general asymptotic expansion of the time-dependent Green's function solution of the equations of motion of the one-dimensional harmonic crystal, namely Bessel functions of the first kind [2, 3, 4]. In 1914, Schrodinger [5] was interested in modeling the flow of heat in a solid and rediscovered Hamilton's solution. More recently, Hemmer [6] examined the related question of whether Fourier's law of heat conduction held for this model and concluded that it did not. Nevertheless, interest persists in the nature of energy flow in various generalizations of the 1D harmonic crystal [7–21] because explicit solutions (starting with Hamilton's) can be obtained.

In this paper we consider a question raised in some recent investigations of the steady-state energy flow through an isotopically disordered 1D harmonic crystal segment [13, 14]. Two slightly different lattice systems have been considered. In the first system, treated by Rubin and Greer [13] (RG), the isotopically disordered crystal segment containing N defects, is sandwiched between perfect semi-infinite harmonic crystal sections. An initial non-equilibrium ensemble is prepared in which the semi-infinite sections are each initially in equilibrium at a different temperature, T_h and T_c. Subsequently, the combined system is allowed to relax. Explicit formulae for the ensemble-average steady-state energy current through the disordered segment and the local kinetic temperature at either end of the disordered segment are obtained. RG showed that the thermal conductivity, K, defined as the ensemble-average steady-state heat current, $\mathcal{J}_N$, divided by the ensemble average steady-state temperature gradient is abnormal. In particular, in the limit as the length, L, of the disordered section increases at constant overall concentration, $C = N/L$, of mass defects, the conductivity, K, increases at least as fast as $N^{1/2}$.

The second type of disordered crystal model has been treated by Casher and Lebowitz [14] and O'Connor and Lebowitz [17]. In this model, the end particles of a finite isotopically disordered section of harmonic crystal are placed in contact with white-noise heat baths [22] at different temperatures. Particles 0 and $L+1$ are held fixed, and particles 1 and L are each in "contact" with one heat bath. Results for this model are less complete than for the RG model. It has only been established that the steady-state energy current approaches zero as the number of defects in (or length of) the disordered section increases at constant concentration, $C = N/L$. The crucial question of the precise rate of approach to zero remains open. There is some indication from a perturbation calculation [12, 15] and from Monte Carlo simulations [15] that if the boundary particles are free, the energy current is proportional to $N^{-1/2}$ and that if the boundary particles are fixed, the energy current is proportional to $N^{-3/2}$. These asymptotic forms for the energy current imply that the thermal conductivity is proportional to $N^{1/2}$ and $N^{-1/2}$, respectively. The $N^{1/2}$ dependence is consistent with the $N^{1/2}$ dependence in the RG Model [20].

The purpose of this article is twofold. First, we modify the RG model in such a way as to mimic the fixed-end boundary condition of the Casher–Lebowitz model and report on some Monte Carlo results for the two versions of the RG model. Second, we generalize some of the formal results of Rubin [10, 11] and RG [13] to include the case in which there is more than one kind of isotopic defect.

2. Model and formal solution

Consider a one-dimensional harmonic crystal with nearest-neighbor interactions. The particles are labeled consecutively by the index r and all particles have the mass m, except for N isotopic particles at lattice positions $r = A_j, j = 1, \ldots, N$. The mass of the defect particle at A_j is M_j, where $M_j \neq m$ may take one of s values, $M_i, \ldots, M_s$. It is assumed that $A_1 = 0$ and that the subscript j on A_j specifies the order of the defects, i.e., $0 = A_1 < A_2 < \ldots < A_N = L$. Thus the overall concentration of defects of all kinds is $C = N/L$; and the spacings between adjacent pairs of defects, $A_j - A_{j-1}$, are assumed to be independent, identically distributed random variables with a mean value $\overline{A_j - A_{j-1}} = C^{-1}$.

In addition to generalizing the RG model to more than one kind of defect mass, we introduce an additional spring (with force constant f') at lattice point $r = A_{N+1} > A_N$ which connects lattice particle A_{N+1} to a fixed location in space. It will be seen in the following that this "fixed boundary" at only one end of the disordered section of crystal has a profound effect on the dependence of the heat current on the number of defects. The equations of motion of the crystal are:

$$(m_r/m)\ddot{x}(r,\tau) = \tfrac{1}{4}[x(r-1,\tau) - 2x(r,\tau) + x(r+1,\tau)], \quad r \neq A_{N+1},$$

$$\ddot{x}(A_{N+1},\tau) = \tfrac{1}{4}[x(A_{N+1}-1,\tau) - 2x(A_{N+1},\tau)$$
$$+ x(A_{N+1}+1,\tau)] - \tfrac{1}{4}(f'/f)x(A_{N+1},\tau), \quad (1)$$

where $x(r,\tau)$ is the displacement of particle r from its equilibrium position and m_r is the mass of the particle at lattice site r. In eq. (1), each superscript dot denotes differentiation with respect to τ, a dimensionless time such that $\tau = 2(f/m)^{1/2}t$, where f is the nearest-neighbor force constant and $2(f/m)^{1/2}$ is the maximum frequency of the infinite perfect crystal. The force constant associated with a "fixed boundary" at $r = A_{N+1}$ is f'.

Except for the generalization to $s > 1$ values for the mass of the defect particles and the addition of the spring with force constant f', the model is the same as that treated by Rubin [10, 11] and RG [13]. For a particular configuration of the array of defects specified by the sequences $\{A_j\}$ and $\{M_j\}, j = 1,\ldots, N$, we consider the ensemble of initial conditions in which: (i) particle $-R < 0$ is held fixed; (ii) the initial conditions in the perfect portion of the crystal, $r < -R$, are specified by a canonical distribution at temperature T; and (iii) the initial conditions for particles where $-R < r$ are specified by the temperature $T = 0$. Particle $-R$ is then released and the system relaxes to a non-equilibrium steady state. The perfect lattice, initially at temperature T, is an inexhaustible reservoir. In the steady state, there is a local kinetic temperature at each end of the array of defects and there is a steady-state value of the energy current through the array of defects, $\mathcal{J}_N$. For large random arrays of defects, the local kinetic temperatures at the two ends of the array are effectively T and zero [23]. Therefore the thermal conductivity coefficient, defined as the heat current divided by the temperature gradient, is:

$$K_N = -\mathcal{J}_N/(-T/A_N) \qquad (2)$$
$$= NC^{-1}\mathcal{J}_N/T, \quad N \gg 1. \qquad (3)$$

In eq. (3), A_N, which is the sum of $N-1$ independent random spacings of average length C^{-1}, has been replaced by NC^{-1}.

The analysis in RG can be repeated for the present generalized model. The added spring constant, f', only introduces a diagonal perturbation similar to that of an isotope. For that reason, we skip the intervening steps and quote the final expression for the ensemble-average steady-state energy current in the present model,

$$\mathcal{J}_N = \frac{k_B T}{2\pi} \int_0^1 \frac{d\omega}{|\mathcal{D}_{N+1}(i\omega)|^2} = \frac{k_B T}{2\pi} \int_0^1 \mathcal{T}_{N+1}^2(\omega) d\omega \tag{4}$$

where $\mathcal{T}_{N+1}^2(\omega)$ is the square of the transmitted amplitude of a wave of unit amplitude and frequency ω and $\mathcal{D}_{N+1}(i\omega)$ is an $(N+1)$th order determinant which has the structure

$$\mathcal{D}_{N+1}(i\omega) = d(\delta_{jl} + i\Delta_l \exp[-ik|A_j - A_l|]) \tag{5}$$

with δ_{jl} denoting the Kronecker delta and $k = 2\sin^{-1}\omega$. The other parameters in eq. (5) are $\Delta_l = (M_l/m - 1)\omega(1-\omega^2)^{-1/2} = Q_l \omega(1-\omega^2)^{-1/2}$ for $l = 1, \ldots, N$ and $\Delta_{N+1} = -\frac{1}{4}(f'/f\omega)(1-\omega^2)^{-1/2}$. The determinant in eq. (5) can be transformed to tridiagonal form following the steps in Appendix A of ref. [11] (see following scheme for $\mathcal{D}_{N+1}(i\omega)$ which shows the nonzero elements in the first three rows of the upper left hand corner and the nonzero elements in the last two rows of the bottom right hand corner).

$\mathcal{D}_{N+1}(i\omega) =$

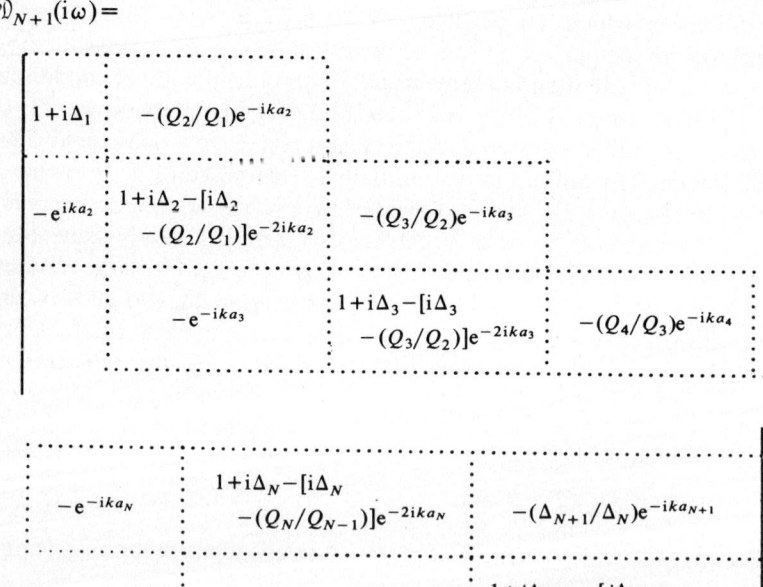

where $a_n = |A_n - A_{n-1}|$. \hfill (6)

In the next section we report the comparison of some Monte Carlo calculations of the thermal conductivity coefficient for disordered arrays in which there is a single type of defect particle ($s=1$) with and without the added fixed boundary spring, f'. Before describing those results, we will obtain a useful relation between the transmitted amplitude $\mathcal{T}_{N+1}(\omega)$ and the transmitted amplitude of the same array of defects without the added spring, $T_N(\omega)$. At the same time we establish a product form which can be used to calculate the value of $T_N(\omega)$ in case there is more than one type of defect mass ($s>1$).

First expand $\mathcal{D}_{N+1}(i\omega)$ in eq. (6) by the last row:

$$\mathcal{D}_{N+1}(i\omega)$$

$$=\left[1+i\Delta_{N+1}+\left(\frac{\Delta_{N+1}}{\Delta_N}-i\Delta_{N+1}\right)e^{-2ika_{N+1}}\right]D_N-\frac{\Delta_{N+1}}{\Delta_N}e^{-2ika_{N+1}}D_{N-1} \tag{7}$$

$$=D_N\left\{1-\frac{(f'/f)i}{4\omega(1-\omega^2)^{1/2}}-\frac{(f'/f)}{4\omega}\left(\frac{1}{Q_N\omega}-\frac{i}{(1-\omega^2)^{1/2}}\right)\right.$$

$$\left.\times e^{-2ika_{N+1}}-\frac{(f'/f)}{4Q_N\omega^2}e^{-2ika_{N+1}}\frac{D_{N-1}}{D_N}\right\}. \tag{8}$$

It follows from eq. (8) that the magnitude of the transmitted amplitude, $\mathcal{T}_{N+1}(\omega)$ of the N-defect array with a "fixed boundary" is proportional to the magnitude of the transmitted amplitude, $T_N(\omega)$, of the N-defect array alone

$$\mathcal{T}_{N+1}(\omega)=T_N(\omega)\left|1-\frac{i(f'/f)}{4\omega(1-\omega^2)^{1/2}}\right.$$

$$\left.-\frac{(f'/f)}{4Q_N\omega^2}(\delta_N-g_N)\exp(-2ika_{N+1}-i\phi_N)\right|^{-1} \tag{9}$$

where

$$1-iQ_N\omega/(1-\omega^2)^{-1/2}=\delta_N e^{-i\phi_N} \quad \text{and} \quad g_N=e^{i\phi_N}D_{N-N}/D_N$$

It can be shown, duplicating the arguments in ref. [11] that:

$$h_N = -(\delta_N - g_N)/|\Delta_N| \tag{10}$$

$$= [1 - T_N^2(\omega)]^{1/2} \exp[i\Omega_N(\omega)] \tag{11}$$

is the amplitude of the reflected wave from the array of N defects alone (without the added spring). Substituting the expression for h_N in (11) for $\mathcal{T}_{N+1}(\omega)$, one can obtain:

$$\mathcal{T}_{N+1}^2(\omega) = \omega^2 T_N^2(\omega) \left| \omega + \frac{(f'/4f)}{(1-\omega^2)^{1/2}} \right.$$

$$\left. \times \left(-i + \frac{|Q_N|}{Q_N} [1 - T_N^2(\omega)]^{1/2} e^{i\Lambda_N} \right) \right|^{-2}, \tag{12}$$

where $\Lambda_N = \Omega_N(\omega) - 2ka_{N+1} + \phi_N$. The factor multplying $T_N^2(\omega)$ in eq. (12), particularly ω^2, results in a dramatic difference between the N-dependence of the energy current $\mathcal{J}_N$ in eq. (4) and the energy current in the original RG model:

$$J_N = \frac{k_B T}{2\pi} \int_0^1 T_N^2(\omega) \, d\omega, \tag{13}$$

without the added spring.

If the expansion of $\mathcal{D}_{N+1}(i\omega)$ initiated in eq. (7) is continued, the general recurrence relation is:

$$D_n = \left[1 + i\Delta_n + \left(\frac{Q_n}{Q_{n-1}} - i\Delta_n \right) e^{-2ika_n} \right] D_{n-1} - \frac{Q_n}{Q_{n-1}} e^{-2ika_n} D_{n-2},$$

$$n = 2, \ldots, N \tag{14}$$

with $D_1 = 1 + i\Delta_1$ and $D_0 = 1$. Using the relations $1 + i\Delta_n = \delta_n e^{i\phi_n}$ and $(Q_n/Q_{n-1}) - i\Delta_n = (Q_n/Q_{n-1})\delta_{n-1} e^{-i\phi_n}$, the function $g_n = e^{i\phi_n} D_{n-1}/D_n$ can be expressed in terms of g_{n-1} from eq. (14) as:

$$g_n = \left\{ \delta_n + \frac{Q_n}{Q_{n-1}} (\delta_{n-1} - g_{n-1}) \exp[-i(2ka_n + \phi_n + \phi_{n-1})] \right\}^{-1}, \tag{15}$$

with $g_1 = \delta_1^{-1}$. Finally the expression for $T_N^2(\omega)$ can be written as:

$$T_N^2(\omega) = \prod_{n=1}^{N} |g_n|^2. \tag{16}$$

3. Monte Carlo estimates of thermal conductivity

The formula which we use for calculating the thermal conductivity of a large disordered array of isotopes with a boundary spring is obtained by combining eqs. (3), (4), (9), (12) and (16). In our numerical calculations, we assume that there is only one kind of isotopic defect with $Q_1 = (M_1/m) - 1 = 1$ and that the inter-defect spacing probability distribution function is $P(a_n) = C(1-C)^{a_n - 1}$ with $C = \frac{1}{2}$. In addition we assume that the added spring constant $f' = f$ and that $a_{N+1} = 1$ (i.e., the added spring has the same strength as the inter-particle springs and is located at lattice particle $A_N + 1$). For these special choices, we have:

$$k_B^{-1} K_N = \pi^{-1} N \int_0^1 F(\omega) \prod_{n=1}^{N} |g_n|^2 d\omega, \tag{17}$$

with

$$F(\omega) = \left| 1 - \frac{i}{4\omega(1-\omega^2)^{1/2}} - \frac{1}{4\omega^2}(\delta - g_n) e^{-i(2k+\phi)} \right|^{-2} \tag{18}$$

$$= \omega^2 \left| \omega + \tfrac{1}{4}(1-\omega^2)^{-1/2} \left\{ -i + [1 - T_N^2(\omega)]^{1/2} e^{i\Lambda_N} \right\} \right|^{-2}, \tag{19}$$

$$g_n = \left\{ \delta + (\delta - g_{n-1}) \exp[-2i(ka_n + \phi)] \right\}^{-1}, \quad n = 2, \ldots, N \tag{20}$$

and

$$g_1 = \delta^{-1}. \tag{21}$$

In the absence of the boundary spring, the factor $F(\omega)$ in eq. (17) is identically equal to unity; and the expression for the thermal conductivity reduces to the one evaluated by Rubin and Greer.

The calculation of the thermal conductivity is straightforward. A random configuration of isotope spacings $\{a_n\}$ is generated and the g_n's are evaluated recursively using eqs. (20) and (21). A typical result for the square of the transmitted amplitude [integrand in eq. (17)] for both versions of the RG model are plotted in fig. 1 for a particular 100-defect array. The integrand in the case of the fixed boundary model is zero at zero frequency whereas the integrand in the original model is unity at zero frequency [see eq. (19)]. It is just this difference which decreases the thermal conductivity of the fixed-boundary model relative to the original model. The results of a series of calculations of the thermal conductivity for N-defect arrays are plotted in fig. 2 for both models for N in the range 25 to 3200.

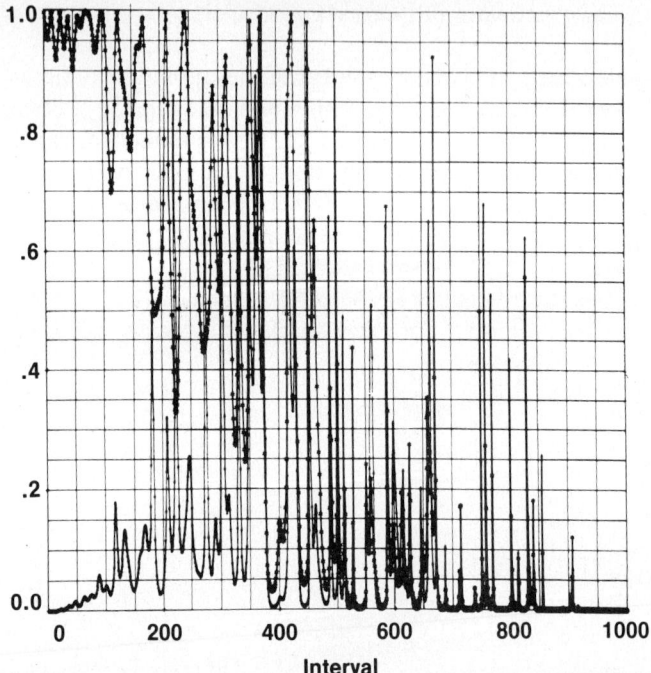

Fig. 1. Computed values of $\mathcal{T}_{101}^2(\omega)$ and $T_{100}^2(\omega)$ are plotted vs. ω and are indicated by dots and asterisks, respectively. The number of intermediate values at which the calculation is made is 1000 and the frequency interval $\Delta\omega = 5.382 \times 10^{-4}$.

4. Remarks

Rubin and Greer [13] established that as the number of defect particles increases, the thermal conductivity increases at least as fast as $N^{1/2}$. Subsequently, Verheggen [20] established that the limiting dependence is exactly $N^{1/2}$. It appears from the Monte Carlo calculations plotted in fig. 2 for the fixed boundary model that the thermal conductivity decreases with increasing N for $N > 800$. However, no well-defined power law dependence on N is evident in the range $800 \leq N \leq 3200$—it is definitely not $N^{-1/2}$.

A number of interesting questions meriting further work are raised by the results reported here: (1) If, in the fixed boundary model, the limiting power-law dependence is $N^{-1/2}$, why is the approach to the limiting law so slow? (2) If a second fixed-boundary spring is added at

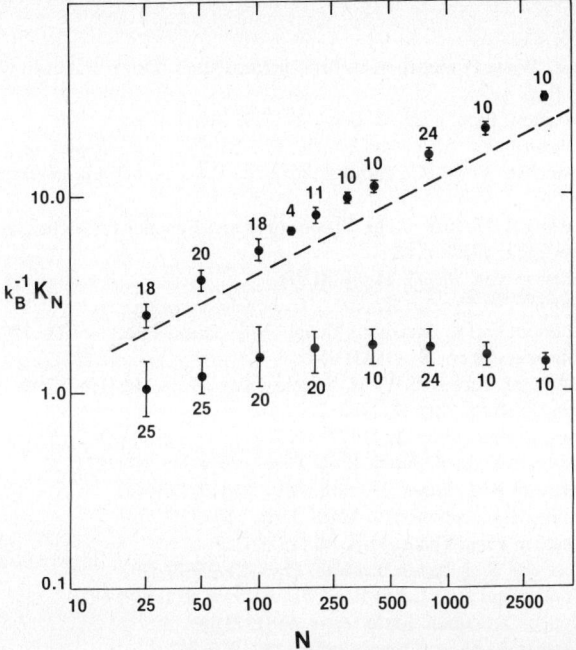

Fig. 2. The averages of the computed values of $k_B^{-1}K_N$ are plotted vs. N. The number of different random isotope arrays on which each average is based is indicated above each plotted point. The range of values entering each average is indicated by the vertical line through each point. For each N, the values of $k_B^{-1}K_N$ for the fixed boundary arrays are necessarily smaller, as explained in the text. The dashed line is a plot of the lower bound derived in RG [13].

the other end of the disordered section of particles, is there a significant change in behavior? (3) What is the limiting power-law dependence in the one and two fixed boundary models?

Acknowledgments

I wish to thank Mrs. M. Menzel who carried out the Monte Carlo calculations at the Los Alamos Scientific Laboratory and Dr. W. M. Visscher for his interest and support. In addition, I acknowledge the hospitality of Dr. Victor M. Blanco, Director of the Cerro Tololo Inter-American Observatory, La Serena, Chile, where this paper was written.

References

[1] L. Brillouin, Wave Propagation in Periodic Structures (Dover Publication Inc., New York, 1953) pp. 1-16.
[2] W. R. Hamilton, Proc. Irish Academy **1** (1839) 267, 341.
[3] A. W. Conway and A. J. McConnell, The Mathematical Papers of Sir William Rowan Hamilton, Vol. **2** (Cambridge University Press, New York, 1940) pp. 451-582, 599.
[4] G. N. Watson, A Treatise on the Theory of Bessel Functions (Cambridge University Press, New York, 1948) p. 12.
[5] E. Schrodinger, Ann. Physik **44** (1914) 916.
[6] P. C. Hemmer, Det Fysiske Seminar i Trondheim **2** (1959).
[7] S. Kashiwamura and E. Teramoto, Suppl. Prog. Theor. Phys. No. **23** (1962) 207.
[8] R. J. Rubin, Phys. Rev. **131** (1963) 964.
[9] D. N. Payton, M. Rich and W. M. Visscher, Phys. Rev. **160** (1967) 706.
[10] R. J. Rubin, J. Math. Phys. **9** (1968) 2252.
[11] R. J. Rubin, J. Math. Phys. **11** (1970) 1857.
[12] H. Matsuda and K. Ishii, Suppl. Prog. Theor. Phys. No. **45** (1970) 56.
[13] R. J. Rubin and R. L. Greer, J. Math. Phys. **12** (1971) 1686.
[14] A. Casher and J. L. Lebowitz, J. Math. Phys. **12** (1971) 1701.
[15] W. M. Visscher, Prog. Theor. Phys. **46** (1971) 729.
[16] W. L. Greer and R. J. Rubin, J. Math. Phys. **13** (1972) 379.
[17] A. J. O'Connor and J. L. Lebowitz, J. Math. Phys. **15** (1974) 692.
[18] A. J. O'Connor, Commun. Math. Phys. **45** (1975) 63.
[19] W. M. Visscher, Methods Comput. Phys. **15** (1976) 371.
[20] T. Verheggen, Commun. Math. Phys. **68** (1979) 69.
[21] M. Toda, Physica Scripta **20** (1979) 424.
[22] (a) P. G. Bergmann and J. L. Lebowitz, Phys. Rev. **99** (1965) 578.
(b) J. L. Lebowitz, Phys. Rev. **114** (1959) 1192.
[23] All steady-state kinetic temperatures in the perfect crystal region near the hot end of the array of defects are equal, as are the steady-state kinetic temperatures near the cold end, see refs. [7] and [13].

CHAPTER 8

Where Do We Go From Here?

R. ZWANZIG

Institute for Physical Science and Technology
University of Maryland
College Park, Maryland 20742
U.S.A.

Contents

1. Introduction — 125
2. Goals — 125
3. Questions — 126
4. Approaches — 127
5. Fokker–Planck and Langevin methods — 128
6. Local equilibrium method — 130
7. Expansion method — 132
8. Summary — 133
References — 134

1. Introduction

It is now twenty-five years since the publication of the second of Melville S. Green's seminal papers on linear transport theory. While the subject has undergone a remarkable evolution, becoming standard graduate course material for many physicists and chemists, it is clear on re-reading his papers that his deep insight and analytical skill led him directly to the heart of the matter.

I do not propose to discuss here the ideas and methods of linear response theory. These seem to be generally accepted. I propose to discuss instead some developments in nonlinear transport theory. It is not so generally recognized that many good ideas and theoretical tools have become available, and that we are now at the threshold of being able to tackle significant nonlinear problems in transport theory. (Thus the title of this article!) I start with a discussion of what I feel are the main goals of a nonlinear transport theory, and follow this with a discussion of some of the important theoretical questions that arise. Then I describe some of the techniques that have become available, and close with a summary of where we stand.

2. Goals

What are the goals of nonlinear transport theory? As is often the case in theoretical physics, different people have different interests; the following is a personal statement about what I would like to see done.

We know that macroscopic transport experiments can often be described by rather simple, though possibly nonlinear, transport equations. One familiar example is the Navier–Stokes equation, including the convective transport term, dependence of the viscosity on local temperature and pressure, and perhaps a nonlinear dependence of the viscosity on the rate of shear. Furthermore, such a description is macroscopically reproducible–a detailed molecular or phase space description does not seem to be needed.

This knowledge may be abstracted as follows. A macroscopic experiment involves measurements of certain dynamical properties of a many-body system, denoted here collectively by $A(\Gamma)$, where Γ is a point in phase space. Suitable candidates for $A(\Gamma)$ may be, for example, the local mass density, energy density and momentum density in a fluid. As time passes, the location of the system point in phase space changes, $\Gamma \to \Gamma_t$, and the numerical values of the dynamical variables $A(\Gamma_t) = A_t$ thereby change with time. The experiment does not follow the evolution of Γ directly; it only follows the numerical values A_t.

The results of such an experiment are described by the transport law

$$\frac{d}{dt} A_t \cong \Phi(\{A_s; 0 \leqslant s \leqslant t\}), \tag{1}$$

where Φ is a functional of the history of the variables A_s. (Often Φ is well approximated by a function of the instantaneous values A_t, rather than by a functional of the history. But we know that "memory effects" can be seen, as in viscoelastic flow, and so the more general functional form is used here.)

We also know from experience that such experiments can be macroscopically reproducible, so that the results of a single experiment are well represented by the average taken over many repetitions of that experiment,

$$A_t \cong \langle A \rangle_t. \tag{2}$$

The goals, then, are first to find the structure of the functional Φ, and to develop methods for computing it explicitly; and second, to see why macroscopic experiments are reproducible.

3. Questions

A number of questions come immediately to mind. One obvious one is: how do we decide what dynamical variables are to be used? or what is $A(\Gamma)$? We surely do not want too many variables, for example the exact location of the system in phase space, and we do not want too few variables, for example leaving out the density or temperature in a hydrodynamic problem. Often this question is answered empirically; we know from experience what variables are macroscopically deterministic. But a satisfactory theory ought to generate this information. One suggestion, due to M. S. Green, is that we must certainly include all "approximate constants of the motion", i.e., the dynamical variables that change very slowly on a molecular time scale. Another suggestion frequently made is that we should include enough variables that the functional Φ is "almost Markoffian", or almost instantaneous in time. (This is done, for example, in explaining viscoelasticity by introducing internal variables such as the population of molecular vibrational states.) At present, there seems to be no wholly satisfactory theoretical method for deciding what to include in $A(\Gamma)$.

Another question is: what is meant by the symbol $\cong$ in eqs. (1) and (2)? These equations are generally not true mathematical identities, but only macroscopic approximations. In a sufficiently careful experiment one always see fluctuations or noise. How is it that one can often find macroscopic transport equations such that the fluctuations are macroscopically negligible? How does it happen that one can successfully replace eq. (1) by an equation for the average behavior,

$$\frac{d}{dt}\langle A\rangle_t \cong \Phi(\{\langle A\rangle_s; 0 \leq s \leq t\}). \tag{3}$$

If the transport equation is linear, the answer is trivial; but for a non-linear transport equation, it is not.

What is meant by the average $\langle \ \rangle$ in eq. (2)? In effect, this is the same as asking: what is meant by "macroscopically reproducible"? To answer this, one must know how experiments are repeated, or perhaps, one must be able to show that details of how experiments are repeated are not so important.

This raises the question of initial conditions. Most modern work in nonequilibrium statistical mechanics is based on following the time evolution of some well defined initial ensemble. But experiments are seldom performed in actuality with the kind of precisely specified repetitions that are implied by most initial ensembles. There seems to be a considerable insensitivity to details of the initial state. How can we see that this is so from purely theoretical arguments?

Most current and recent work in nonlinear transport theory does not address these questions except in a peripheral way. We usually assume that the correct set of variables is known; we don't worry much about the degree of macroscopic reproducibility; and we use initial ensembles that appear to be reasonably simple and well suited to theoretical analysis.

4. Approaches

A number of different approaches to nonlinear transport theory are currently available. Since this is a conference talk and is not intended to be a comprehensive review of the subject, I will not try to give a complete list of approaches or references. I apologize for omitting someone's favorite work.

Quite often, special equations are used to deal with special situations. For example, one can use the nonlinear Boltzmann equation to discuss the nonlinear hydrodynamic behavior of a dilute gas. Because my concern here is with more general procedures, I shall not dwell further on these special equations.

Aside from these, at least three general schemes are available. They involve (1) Fokker–Planck and Langevin equations, (2) a generalization of the Chapman–Enskog local equilibrium approach, and (3) expansions about equilibrium. These schemes have some features in common. They all have a clear and well defined foundation, so that whether or not they are useful, they are manifestly exact. They all require special restrictions on the kinds of initial states that may be studied precisely. The first and second involve projection operator methods, and one way of developing the third scheme is based on projection operators. Each has advantages and disadvantages; I shall comment on these after the main ideas are presented.

5. Fokker–Planck and Langevin methods

The Fokker–Planck method appears first in M. S. Green's work on transport theory. He had the idea of focussing attention on the distribution function $g(a; t)$ in the reduced space of the dynamical variables A, rather than on the complete phase space distribution $f(\Gamma; t)$. The a-space distribution is defined by:

$$g(a; t) = \int d\Gamma \, \delta(A(\Gamma) - a) f(\Gamma; t). \tag{4}$$

By intuitive but essentially correct arguments, Green derived a Fokker–Planck equation for $g(a; t)$, valid for small deviations from equilibrium and for slow processes.

Green's derivation has been generalized several times [1] using projection operator methods, to give an equation for $g(a; t)$ having the general structure;

$$\frac{d}{dt} g(a; t) = \int da' \int_0^t dt' \, K(a, a'; t, t') g(a'; t') + [\ldots] \Delta f(\Gamma; 0), \tag{5}$$

where K is an operator, nonlocal in a-space and in time. The inhomogeneous term, as always in projection operator derivations, comes from the "orthogonal part" of the initial distribution,

$$f(\Gamma;0) = f_{eq}(\Gamma)\delta(A(\Gamma) - a_0) + \Delta f(\Gamma;0). \tag{6}$$

Note that f_{eq} can be chosen either as microcanonical (the original choice) or as canonical; the general structure is not affected. If the initial distribution is chosen so that the orthogonal part vanishes at $t=0$, then the resulting equation is closed in $g(a;t)$. If the initial distribution actually has an orthogonal part, then typically one hopes that its effects will decay to zero very rapidly, so that after a short time it can be ignored. This kind of hope is characteristic of all projection operator derivations, and is seldom substantiated.

When deviations from equilibrium are small, and the dynamical variables vary slowly in time, the operator K reduces to the ordinary Fokker–Planck operator of Green. Otherwise, its structure is much more complicated, but nevertheless well defined.

The generalized Fokker–Planck equation is used to generate equations of motion for the average values:

$$\langle A \rangle_t = \int da\, a g(a;t), \tag{7}$$

and to discuss fluctuations from the average. More will be said about this later.

In conventional Brownian motion theory, there is a one-to-one correspondence between Fokker–Planck and Langevin methods. This is also in formally exact nonequilibrium statistical mechanics. Mori [2] showed first how to construct the Langevin equations corresponding to Green's Fokker–Planck equations. His procedure has been extended to nonlinear processes in two ways [3].

One procedure is to expand the delta function $\delta(A-a)$ in a complete set of polynomials $\varphi_j(a)$, orthonormal with respect to the equilibrium distribution $g_{eq}(a)$,

$$\delta(A-a) \to \sum_j \varphi_j(A)\varphi_j(a). \tag{8}$$

Then Mori's algorithm is used to derive equations of motion (coupled) for all the polynomials. Among these is the set of first powers, or the variables themselves. In this way one has an equation of motion for the variables, with all powers of the variables on the right hand side.

Another procedure is to start with the exact Fokker–Planck equation, taking for the initial state a single point in phase space. Then the state at later times is also a single point, and all averages are obtained trivially because the distribution remains perfectly sharp:

$$f(\Gamma; t) = \delta(\Gamma - \Gamma_t)$$
$$\to \langle A_t \rangle = A(\Gamma_t). \tag{9}$$

Both approaches (which are in fact adjoint representations of the same calculation) lead to the exact nonlinear Langevin equation:

$$\frac{d}{dt} A_t = \Phi_L(\{A_s; 0 \leqslant s \leqslant t\}) + \text{noise}. \tag{10}$$

Note that the functional Φ_L is *not* the one appearing in eq. (1). The noise term, coming from $\Delta f(\Gamma; 0)$ is orthogonal to all functions of the set $A(\Gamma)$.

Both the Fokker–Planck and the Langevin equations pose the same problem in deriving nonlinear transport equations. They both lead to the same equation for the average (over some initial ensemble) $\langle A \rangle_t$,

$$\frac{d}{dt} \langle A \rangle_t = \langle \Phi_L(\{A_s\}) \rangle + \langle \text{noise} \rangle. \tag{11}$$

But when Φ is a nonlinear functional of the variables, one cannot replace the average of the functional by the functional of the average,

$$\langle \Phi_L(\{A_s\}) \rangle \neq \Phi_L(\{\langle A \rangle_s\}). \tag{12}$$

The difference between the right and left-hand sides introduces cumulants of the second and higher order, for example $\langle AA \rangle_t - \langle A \rangle_t \langle A \rangle_t$. These turn out to be of the order of the first moments, and cannot be neglected. In fact, it is just this that leads to the mode-mode coupling theory of renormalized transport coefficients [4]. The initial values of the cumulants also appear, in ways that are probably not relevant to the nonlinear transport equations, but that interfere with clean calculations. The average of the noise term vanishes if the initial distribution has no orthogonal part; otherwise it remains.

6. Local equilibrium method

The local equilibrium method is a far-reaching generalization of the familiar Chapman–Enskog derivation of transport equations in the kinetic theory of gases. It was formulated in an elegant and concise way some years ago by Robertson [5] and rediscovered by Kawasaki and Gunton [6] and by Grabert [7]. Until quite recently, Robertson's work has not received the attention that it deserves.

From the set of dynamical variables $A(\Gamma)$, which may be, e.g., the local density, momentum and energy fields in a fluid, one constructs a time dependent local equilibrium distribution,

$$f_{lt}(\Gamma) = \exp - A(\Gamma) \cdot B_t. \tag{13}$$

The B's have the character of thermodynamic quantities (e.g. local temperature field) that are conjugate to dynamical quantities (e.g. local energy density field). They are determined, as in equilibrium statistical mechanics, by requiring that the local equilibrium averages at time t are identical with the actual averages,

$$\langle A \rangle_t \equiv \langle A \rangle_{lt} = \int d\Gamma A(\Gamma) f_{lt}(\Gamma), \tag{14}$$

so that knowledge of all the averages is equivalent to knowledge of all the B's, and vice versa.

The local equilibrium method is based on the feeling that the actual phase space distribution always stays close to the local equilibrium distribution that gives the same average values. This motivates the construction of a (time dependent) projection operator to separate the exact distribution into its local equilibrium approximation and a remainder,

$$f(\Gamma; t) = f_{lt}(\Gamma) + \Delta f(\Gamma; t). \tag{15}$$

One works out how the remainder term is driven by temporal variations in the B's, and one finally obtains an equation of motion,

$$\frac{d}{dt} \langle A \rangle_t = \Psi(\{B_s; 0 \leqslant s \leqslant t\}) + [\ldots] \Delta f(\Gamma; 0), \tag{16}$$

in which Ψ is a well-defined functional of the conjugate B fields. When the initial distribution has precisely the local equilibrium form, $\Delta f(\Gamma; 0) = 0$ and the last term drops out. Because the B fields are related in a "thermodynamic" way to the averages, this provides closed equations of motion.

Piccirelli [8] has used Robertson's method in a discussion of the structure of hydrodynamics. In particular, he showed that the convective and Eulerian terms in the hydrodynamic equations come out exactly as expected.

The local equilibrium method suffers from the use of time dependent projection operators, which are more clumsy to work with explicitly than the more familiar time independent projection operators. On the other hand, it provides formal results that have a remarkable structural compactness.

Once one has learned the technical details of the manipulations involved in Robertson's method, it is not difficult to investigate the dynamics of fluctuations about the average values. The time dependent fluctuation of the actual variable A_t from its average is

$$\Delta_t A = A_t - \langle A \rangle_t. \tag{17}$$

From these, one constructs the correlation matrix $C(t)$,

$$C(t) = \langle (\Delta_t A)(\Delta_0 A) \rangle, \tag{18}$$

in which the average is taken over the initial distribution, assumed to have the local equilibrium form. Then one finds an equation of motion for the correlation matrix, with the general structure:

$$\frac{d}{dt} C(t) = i\Omega_{lt} \cdot C(t) - \int_0^t ds\, K_l(t,s) \cdot C(s), \tag{19}$$

in which Ω_{lt} and $K_l(t,s)$ are functionals of either $\langle A \rangle_t$ or B_t. These equations are natural generalizations of the ones that arise in the Mori–Zwanzig linear theory of equilibrium time correlation functions. It is interesting that the correlation matrix obeys a linear equation of motion, even for systems very far from equilibrium; but this simplicity is illusory, because Ω and K can be nonlinear functionals of the average values.

7. Expansion-method

The expansion method, due primarily to Oppenheim and co-workers [9], provides a way of expanding transport equations in powers of deviations from equilibrium.

The basic idea is as follows. Suppose that the system is initially in a state of local equilibrium,

$$f(\Gamma; 0) = f_{eq}(\Gamma) \exp[-A(\Gamma) \cdot B_0]. \tag{20}$$

For convenience, the true equilibrium distribution has been separated off, so that all B's vanish at equilibrium. As time passes, $f(\Gamma; 0)$ changes to $f(\Gamma; t)$,

$$f(\Gamma; t) = f_{eq}(\Gamma) \exp(tL) \exp[-A(\Gamma) \cdot B_0]. \tag{21}$$

Now one expands the initial distribution in powers of B_0, and one uses this to calculate the averages $\langle A \rangle_t$ and their rates of change,

$$\begin{aligned} f(\Gamma; t) &= \text{power series in } B_0, \\ \langle A \rangle_t &= \text{power series in } B_0, \\ (d/dt)\langle A \rangle_t &= \langle LA \rangle_t = \text{power series in } B_0. \end{aligned} \tag{22}$$

By eliminating B_0 from the latter two series, an equation of motion is obtained,

$$\frac{d}{dt}\langle A \rangle_t = \Phi(\{\langle A \rangle_s; 0 \leq s \leq t\}), \tag{23}$$

where Φ is a functional of $\langle A \rangle_t$. Also, the local equilibrium correspondence of $\langle A \rangle_t$ and B_t can be used, so that the right hand side of eq. (23) is replaced by another functional, $\Psi(\{B_s\})$.

Recently I have presented a streamlined way of performing the elimination that characterizes the expansion method; this makes use of the Mori algorithm for generating equations of motion. The results are the same [10].

The expansion method can be used quite easily to generate quadratic or even cubic terms in the expansion of the functionals Φ or Ψ. Beyond that, the algebra becomes disagreeable.

There is, however, a problem with the expansion method, that I like to refer to jocularly as Dorfman's theorem [11]. This states: "In non-equilibrium statistical mechanics, all series expansions diverge." Whether or not this is a general phenomena, it is certainly true that divergences often appear when one tries to calculate explicitly coefficients in power series expansions, e.g. the viscosity of a gas as a function of its density, or as a function of the rate of shear. In view of this situation, there may not be much practical use for the expansion method.

8. Summary

Of the three general methods just described, one must conclude that while each one has the potential for providing exact results in nonlinear transport theory, each also has its disadvantages.

The Fokker–Planck–Langevin method places heavy emphasis on the role of fluctuations – an inevitable consequence of breaking averages. This turns out to be very useful in dealing with transport processes near critical points, where fluctuations are of overwhelming importance. But there are serious difficulties in keeping track of higher cumulants when dealing with nonlinear transport theory in a more general context. The potential for obtaining useful results is there, but the road will be rough.

The expansion method is conceptually simple and it is easy to apply – at least in low order. But there is a serious question as to the convergence of the method.

The local equilibrium method has by far the most appealing structure, and may yet become the most effective method for dealing with nonlinear transport processes. But because of its reliance on time dependent projection operators, the method is hard to work with. Perhaps we will learn to cope with these technical difficulties as time passes. After all, we were told by Mel Green twenty-five years ago how to think about linear transport processes, but a lot of time and effort were required before his ideas could be implemented in a practical way.

References

[1] R. Zwanzig, Phys. Rev. **124** (1961) 983.
    L. S. Garcia-Colin and J. L. Del Rio, J. Stat. Phys. **16** (1977) 235.
[2] H. Mori, Prog. Theor. Phys. **33** (1965) 423.
[3] S. Nordholm and R. Zwanzig, J. Stat. Phys. **13** (1975) 347.
[4] For a simple illustration, see R. Zwanzig, K. S. J. Nordholm and W. C. Mitchell, Phys. Rev. **A5** (1972) 2680.
[5] B. Robertson, Phys. Rev. **144** (1966) 151.
    B. Robertson, in: The Maximum Entropy Formalism, eds. R. D. Levine and M. Tribus (The MIT Press, Cambridge, 1978) p. 289.
[6] K. Kawasaki and J. D. Gunton, Phys. Rev. **A8** (1972) 2048.
[7] H. Grabert, Z. Physik **B27** (1977) 95.
[8] R. A. Piccirelli, Phys. Rev. **175** (1968) 77.
[9] J. H. Weare and I. Oppenheim, Physica **72**, 1 (1974) 20.
[10] R. Zwanzig, Prog. Theor. Phys. Supp. No. **64** (1978) 74.
[11] J. R. Dorfman in this volume, ch. 5.

PART II

Phase transitions

CHAPTER 9

*A New Model Hamiltonian
for a Correlated Electron System
within the General Framework
of Critical Phenomena and Phase Transitions*

C. Di CASTRO

*Istituto di Fisica "G. Marconi"-Università di Roma-Italy
Gruppo Nazionale di Struttura della Materia del Consiglio
Nazionale delle Ricerche, Sezione di Roma-Italy.*

© *North-Holland Publishing Company 1981*

*Perspectives in Statistical Physics
Ed. H. J. Raveché*

Contents

1. Introduction 139
2. Phase transitions in some complex systems 142
3. A new model for the metal–insulator transition induced by correlation 144
   3.1. Complications with the standard procedures and physical implications 144
   3.2. The new model 148
References 153

1. Introduction

Some comments on the general scheme of critical phenomena
In writing a chapter in a book to honor Mel Green, I will not enter into his specific contributions to statistical physics and in particular to critical phenomena and phase transitions. I am pleased to remember him as one of those persons of great humanity, who are able to become a center of attraction and establish bridges among different ideas and different schools.

In particular, apart from the series of volumes on "Phase Transitions and Critical Phenomena", I want to mention at least two crucial occasions for the physics of critical phenomena, promoted by him in the last ten years: The Varenna Summer School in 1970 [1] and the Philadelphia Conference on the Renormalization Group Approach in 1973 [2].

At the end of the sixties the phenomenological theory of critical phenomena had found its foundation in the Kadanoff logical sequence of universality, scaling, relevant and irrelevant variables (see Kadanoff lectures in ref. [1]).

The difficult problem of a large number of degrees of freedom strongly correlated within the coherence distance (tending to infinity at the critical point), was thereby reduced to the determination of the critical indices of the few relevant quantities and to find out the mechanism of the approach to their scaling behaviour giving rise to universality.

In this period the field theoretic approach was introduced into this problem simultaneously in the USSR and Italy. Jona-Lasinio and myself [3] noticed in 1969 that the field theoretic renormalization group equations as presented in the Bogolubov and Shirkov book [4], generalize the universality relations in the sense that they relate one model system to another by varying the coupling and suitably rescaling the other variables and the correlation functions. The Kadanoff scaling picture was then obtained under the assumption that the coupling disappears from the equations near to the critical point.

Migdal [5] and Polyakov [6] started from a detailed analysis of the diagramatic structure of the correlation functions rather than from global conditions such as the renormalization group equations and, using Ward identities, tried to build an ad hoc renormalization procedure. This approach, whose connection with the renormalization group was discussed in Varenna, was later developed into the skeleton expansion method for a practical calculation of the critical indices [7].

In spite of the open scepticism of most prominent researchers in the field, Green was immediately open to the renormalization group approach to critical phenomena and gave ample time for its discussion in the Varenna School [1].

At that time the scaling picture had been given a theoretical basis and the renormalization group appeared as the right tool to investigate the scaling behaviour of the correlation functions. Nevertheless the mechanism by which the memory of the original coupling disappears from the equations was not yet clarified and no way to make practical calculations was discovered.

After Wilson [8] had given his great contribution to the understanding of the physics underlying the renormalization procedure by his famous new group transformation, this field developed quite rapidly. By his mathematical realization of the other very physical Kadanoff idea of grouping together degrees of freedom associated with larger and larger cells, he was able not only to clarify the mechanism of disappearance of the original couplings by means of the concept of the fixed point Hamiltonian, but also provided the scheme by which he himself and M. E. Fisher were able to evaluate the critical indices with the ε-expansion [9]

Further developments completed this picture either by clarifying the connection with the thermodynamic quantities [10] or by extending the technique to a real space renormalization group [11].

After Wilson's great advances, it was not difficult to develop the analogous mechanism for the field theoretic renormalization group also, by realizing the crucial role played by the dimension of the coupling constant in a field theory model for a correct description of critical phenomena [12]. For a φ^4 model, this was actually noticed in the classical limit of an interacting Bose gas, studied via the Matsubara technique for the correlation functions. The theory was then completed [13] within the field theoretic renormalization group approach (including the Callan–Symanzik equation) by the introduction of the thermodynamic potential as the generating functional of the relevant correlation function, which permitted the discussion of the thermodynamic scaling.

The Philadelphia meeting was a good occasion for a confrontation of the various methods and a discussion of their wide applications.

I just want to recall a question which arose there (see ref. [2] p. 7–8, 176). It goes back to the very basis of the renormalization group approach. Few theories have been supported by so many successes, nevertheless one may ask to what extent the results rely on what has been introduced since the beginning. A certain knowledge of the fundamental symmetries inherent to the problem, in particular of the order parameter, has to be assumed in order to make the proper choice of the basic variables on which to operate the transformation, (i.e. the partial trace over the degrees of freedom associated with the short range behaviour in the Wilson scheme and the renormalization of the relevant fields in the field theoretic renormalization group).

This question is also related to an open problem of critical phenomena, namely whether the two parameter scaling or hyperscaling is valid as the renormalization group implies or whether some modifications are required as some numerical evidence indicates.

In the field theoretic approach [13] for instance, the differential group equations for the thermodynamic potential fully specifies the theory, since by successive functional derivatives the group equations for all the other physical quantities can be generated from it. The theory is then mainly determined by the "proper choice of the variables" on which the thermodynamic potential may depend and for which variables a renormalization is required. The renormalization parameters in fact determine the coefficients entering the differential group equations. In the simplest case of an ordinary magnetic critical point, the spontaneous magnetization (the order parameter), the deviation from the critical temperature and the coupling constant are considered as the basic variables. Three coefficients are therefore present in the group equation for the thermodynamic potential. One of these is zero at the fixed point. The corresponding scaling invariant theory is fully determined by the remaining two coefficients evaluated at the fixed point. They coincide with two independent critical indices. Hyperscaling then results.

Of course different aspects of critical phenomena manifest themselves when further symmetry constraints derived from phenomenological analysis are taken into account. This supports the fact that our mathematical picture has a physical content.

Preliminarily to any correct application of the renormalization group approach to a given problem one must therefore rephrase it in terms of an effective Hamiltonian expressed in terms of physically significant variables. This effective model already in mean field theory should account for a qualitative thermodynamic behaviour of the system. This is then assumed as the zeroth approximation suitable for the application of the renormalization group transformation.

We shall now discuss how some complex problems of statistical mechanics have recently found model representations which seem to be a good starting point in this direction.

2. Phase transitions in some complex systems

According to the discussion in the previous section, it should now be clear why some problems, where it is difficult to identify the relevant physical quantities and in particular the order parameter, have either only very recently found a satisfactory solution, as in the case of the two-dimensional planar model with topological order [14], or have not yet found a convincing description. For example we mention the following cases:

(a) spin glasses or more generally systems with random quenched impurities [15];
(b) cooperative phenomena in complicated systems as polymer gels [16];
(c) metal–insulator transitions induced by the electron correlation and relative phase diagram [17].

The recent advances in the identification of the essential physical features to be retained in models [18, 19, 21] representative of these systems shall be briefly described for the first two subjects in this section and more extensively for the last one in the next section.

In particular we shall review two papers [20, 21] which have reduced the last problem to a statistical mechanical model more directly related to concepts inherent to the critical phenomena picture discussed so far.

The spin-glass problem, i.e. the problem of magnetic impurities in a normal metal interacting via the oscillatory Ruderman–Kittel interaction is idealized as an Ising system, where the interaction changes in a random fashion along the lattice bonds according to a given probability distribution [22].

Due to the peculiar nature of the interaction, no usual magnetic ordering is present, but lowering the temperature a long range order in time has been proposed:

$$\lim_{t \to \infty} \langle \sigma_i(0) \sigma_i(t) \rangle = Q_i, \qquad (2.1)$$

where σ_i is the spin variable at site i.

It is still not certain that this order parameter represents completely the physical content of the model. To increase the complication, in this case of quenched impurities the average over the distribution of the random interaction has to be taken over the physical quantities and not over the partition function.

To overcome this last difficulty the partition function of n identical replicas of the original system is evaluated and eventually analytically continued to $n=0$. Thus the log of the partition function is reproduced.

By the replica trick the order parameter is reduced to

$$\langle \sigma_i^\alpha \sigma_i^\beta \rangle = Q_i^{\alpha\beta},$$

where α and β refer to different replicas of the same system.

The symmetric choice

$$Q_{\alpha\beta} = \frac{1}{N} \sum_i Q_i^{\alpha\beta} = q$$

appeared to be the most natural one. It leads however to inconsistences in the Sherrington–Kirkpatrick model [23] for which a mean field theory should be exact.

By studying the stability of the thermodynamic potential around the symmetric point, it has been shown that one has to look for non-symmetric order parameter, the replica symmetry being broken [24].

To overcome this difficulty, besides various other proposals for symmetry breaking, Parisi [18] has recently suggested a function in the interval 0–1 as the order parameter.

In spite of these great advances achieved in treating this problem, we are not yet at the point to follow the standard procedure in critical phenomena of writing an effective Hamiltonian in terms of the order parameter, discussing it first in the mean field approximation and then applying the renormalization group approach.

Polymer gels and related complicated systems have been idealized by the relatively simple model of bond percolation, connectivity between monomers being assumed as the relevant feature to be retained [16].

In a given lattice we have a fraction p_b of conducting bonds (chemical bonds between a pair of monomers) and a fraction $(1-p_b)$ of insulating bonds. Above a critical value p_b^c there is a formation of an infinite cluster of conducting bonds giving rise to the gel phase. The weight fraction of monomers belonging to an infinite molecule is the order parameter.

The experiments [25] on the solvent effects in polymer gelation have indicated the existence, in addition to the sol–gel-curve, of a first order phase separation curve with a consolute point.

This has stimulated the presentation of a new model for polymer gelation: the site-bond correlated-percolation problem [19].

The effect of a solvent is idealized by the fact that at each site either a solvent molecule (A site) or a monomer (B site) can be present. Since the experiments indicate a phase separation, A–A, B–B and A–B interactions must be present:

$$H = -W_{AA} \sum_{\langle ij \rangle} \Pi_i^A \Pi_j^A - \varepsilon_{BB} \sum_{\langle ij \rangle} \Pi_i^B \Pi_j^B - W_{AB} \sum_{\langle ij \rangle} \Pi_i^A \Pi_j^B, \qquad (2.2)$$

where $\langle ij \rangle$ are nearest neighbour lattice sites and

$$\Pi_i^A = \begin{cases} 1 & \text{if site A is occupied by a solvent molecule;} \\ 0 & \text{otherwise} \end{cases}$$

and vice versa for Π_i^B, so that

$$\Pi_i^A + \Pi_i^B = 1, \qquad (2.3)$$

ε_{BB} is the usual Van der Waals interaction with weight ρ_u and the chemical bonding energy with weight $1 - \rho_u$.

Due to the interaction between monomers, which introduces a statistical mechanical weighting factor, the bond probability p_b depends on the temperature in addition to ρ_u. Similarly for the volume fraction of monomers.

This percolation problem is solvable on a Bethe lattice and the percolation threshold determines the sol–gel curve. A phase separation curve is also present as a direct consequence of the interaction between the two components. By varying the parameters of the model the sol–gel line can intersect the phase separation curve at the consolute point.

This new model plays the same role in the gelation problem as the model we are going to present in the next section in the metal–insulator transition induced by correlation.
Many physical implications seem to be analogous in the two cases as well as in the model for He^3–He^4 mixtures previously introduced by Blume et al. [27].

3. A new model for the metal–insulator transition induced by correlation

3.1. Complications with the standard procedures and physical implications

Besides the standard mechanism for the metal–insulator transition induced by a change of the band structure in such a way that a gap appears in the density of states in correspondence with the Fermi energy to give an insulating behaviour, two other mechanisms have been proposed for studying this transition. They are related to localization effects induced either by electron correlation [17] or by disorder [26].

In the first case it has been assumed that all the important features are well reproduced by the Hubbard model [28]

$$H_H = \sum_{\langle ij \rangle} \sum_\sigma t_{ij} c_{i\sigma}^+ c_{j\sigma} + U \sum_i n_{i\uparrow} n_{i\downarrow} - \mu \sum_{i,\sigma} n_{i,\sigma}. \quad (3.1)$$

This model essentially represents the competition between the delocalizing hopping term t_{ij} and the localizing on site Coulomb repulsion U. $c_{i\sigma}^+$ and $c_{i\sigma}$ are the fermion creation and annihilation operators of an electron with spin σ in a Wannier state. $n_{i,\sigma}$ are the corresponding occupation number operators. t_{ij} is assumed to be equal to a constant t (proportional to the band width W) for nearest neighbour sites (ij) and zero otherwise. We shall limit ourselves to discuss the case of a half filled band with one electron per site on the average, for which the chemical potential $\mu = U/2$.

The effect of disorder instead is simply reproduced by the non-interacting Anderson model [29]:

$$H = t \sum_{ij} c_i^+ c_j + \sum_i \varepsilon_i c_i^+ c_i, \quad (3.2)$$

where ε_i are independent stochastic variables whose probability distribution was assumed by Anderson to be $\sim 1/V$ for $-V/2 \leq \varepsilon \leq V/2$ and zero otherwise.

The ratio V/t measures the strength of disorder in this model, just as the ratio U/W is a measure of the localizing effect of the Coulomb repulsion. When these ratios are sufficiently high the system becomes an insulator.

We want to stress here that the two mechanisms are in a certain respect complementary to each other and it is difficult to consider them totally independently, especially in the most interesting region $U \sim W$ or $V \sim t$.

This already makes the problem of correlated electron system very complicated. We shall now further explain why in this case it is hard to find the suitable sets of variables, which are physically relevant in the various regions of the parameters U and W [20]. It is in fact a formidable task to derive an effective Hamiltonian in terms of the relevant variables, which already in mean field theory could account for the thermodynamic behaviour of the original Hubbard model in the various regions of the parameter T/W (T being the temperature) and U/W, where very different physical mechanisms take place. As already mentioned this is also the first step towards the correct application of the renormalization group approach to this problem.

It is usually expected [30, 31] that intrinsic to the Hubbard model there is a phase diagram in T/W and U/W plane which shows at least the following features (see fig. 1):

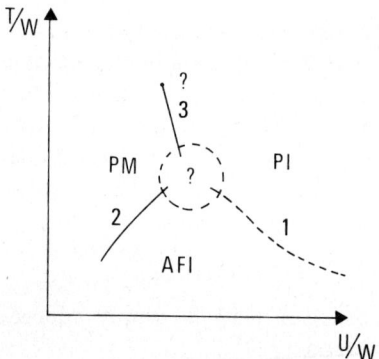

Fig. 1. Schematic phase diagram expected for the Hubbard model (arbitrary units).

(1) A second-order phase transition from a paramagnetic insulator to an antiferromagnetic insulator when the temperature T is lowered at sufficiently high values of U/W.

(2) A first order phase transition line separating the non-magnetic metal from an antiferromagnetic insulator. This separation curve is a continuation of the previous second order line to lower values of U/W. The mechanism by which the second order transition switches into a first order one was still open to discussion.

(3) A direct first order metal–insulator transition line (possibly ending into a critical point) when U is comparable with W.

It is completely uncertain at which point the line-3 intersects the line-1 or the line-2. This may give rise to various polycritical points and to a triple point.

First order phase transitions seem to play an important role in this problem and they have not yet found a satisfactory foundation within the general scheme of the renormalization group approach to critical phenomena.

So far a convincing argument has been given only for the first feature in the large U/W limit. In fact in this limit since it costs an energy U to have a doubly occupied site, the electrons are localized, each site being occupied by one electron with either spin up or down ($|\uparrow\rangle, |\downarrow\rangle$ magnetic sites). The relevant quantities are the local spin operators, and the Hubbard Hamiltonian is equivalent to an Heisenberg Hamiltonian showing the magnetic second order phase transition.

When U/W is not very large, empty or doubly occupied site ($|0\rangle, |\uparrow\downarrow\rangle$ non-magnetic sites) enter into the problem, the local charge operator must become important in establishing the competition between the two sets of sites.

Therefore it turns out to be useful for the future discussion to define the following spin and charge operators:

$$S_i^z = n_{i\uparrow} - n_{i\downarrow}, \quad S_i^+ = c_{i\uparrow}^+ c_{i\downarrow}, \quad S_i^- = c_{i\downarrow}^+ c_{i\uparrow} \tag{3.3}$$

$$\rho_i^z = n_{i\uparrow} + n_{i\downarrow} - 1, \quad \rho_i^+ = c_{i\uparrow}^+ c_{i\downarrow}^+, \quad \rho_i^- = c_{i\downarrow} c_{i\uparrow}. \tag{3.4}$$

The x and y components of S and ρ are defined in the usual way in terms of S^+, S^- and ρ^+, ρ^- respectively.

The spin operators are non-zero only when acting on the magnetic sites ($|\uparrow\rangle, |\downarrow\rangle$); the charge operators are non-zero when acting on the non-magnetic sites ($|0\rangle, |\uparrow\downarrow\rangle$).

Various competing mechanisms in terms of these operatorial variables are responsible for the complex phase diagram of this system. It will be clear at the end that they do not exhaust all the physical content of the original model.

The method commonly used to derive an effective Hamiltonian in terms of physical variables is the functional integral formulation of the Hubbard model, where the classical fields corresponding to spin and charge operators appear [30, 32].

We shall now discuss why in this case the last method is not suitable to accomplish this task. To make the problem amenable one is forced to introduce various approximations and loses some basic properties of the original system [20]. Then we are led to consider a new approach based purely on symmetry considerations on the Hubbard Hamiltonian.

In order to derive a functional integral representation of the Hubbard model, one has first to reduce the interaction term in (3.1) to a sum of squares of local operators. This is usually accomplished by the so-called Schrieffer [33] and Hamann [34] transformations:

$$n_{i\uparrow} n_{i\downarrow} = \tfrac{1}{2} + \tfrac{1}{2}\rho_i^z - \tfrac{1}{2}(S_i^z)^2 \quad \text{Ising type} \tag{3.5}$$

$$= \tfrac{1}{2} + \tfrac{1}{2}\rho_i^z - \tfrac{1}{6}(\mathbf{S}_i \cdot \mathbf{S}_i) \quad \text{Heisenberg type}$$

$$n_{i\uparrow}n_{i\downarrow} = \tfrac{1}{4}(\rho_i^z+1)^2 - \tfrac{1}{4}(S_i^z)^2 \qquad \text{Ising type}$$

$$= \tfrac{3}{8} + \tfrac{1}{2}\rho_i^z + \tfrac{1}{8}(\rho_i^z)^2 - \tfrac{1}{8}(\boldsymbol{S}_i \cdot \boldsymbol{S}_i) \qquad \text{Heisenberg type.} \qquad (3.6)$$

In eq. (3.5) no charge fluctuations are present and spurious terms of equal spins on the same site appear in perturbation theory.

In both cases any polynomial truncation in the spin field of the corresponding classical Hamiltonian leads to a system belonging to different universality classes. They correspond to an order parameter with either one or three components depending on which transformation has been used (Ising type or Heisenberg type). Only the exact theories are equivalent to each other.

More fundamentally since the original model is expressed in terms of the fermion operators c^+ and c, the operators defined in (3.3) and (3.4) must satisfy the following local conditions:

$$(S_i^\nu)^2 + (\rho_i^{\nu'})^2 = 1; \quad S_i^\nu \rho_i^{\nu'} = 0; \quad \nu, \nu' = x, y, z. \qquad (3.7)$$

These local relations have to be satisfied by any effective model if the fermionic character of the original Hubbard model is to be maintained.

This means for instance that the classical spin and charge fields are not completely independent. As already indicated in discussing the competition between magnetic and non-magnetic sites, the charge effects cannot be entirely neglected, though this has usually been done in current literature, where also a polynomial truncation in the spin field is often used together with a static approximation (which is valid only in the two asymptotic limits $V/W \to 0$ or $V/W \to \infty$). This can be made evident [20] by generalizing eqs. (3.5) and (3.6) and studying the classical field theory problem which comes out of the functional integral representations of the Hubbard model.

Furthermore in order to maintain conditions (3.7) i.e. the fermionic character of the Hubbard model, one has to deal with a non-polynomial effective Hamiltonian, which contains all the complications of the original model [20].

We have to look more deeply into the physics in order to derive an amenable effective model Hamiltonian.

3.2. The new model

In presenting our model [21] we start from the physical analogy with the BEG model [27] for the $He^3 - He^4$ mixture.

The experiments [35] show that below a given concentration of He^3 a λ line is present. The system then undergoes a first order phase separation into a super He^4-rich phase and a normal He^3-rich phase. The switching point is a tricritical point. The tendency of superfluid ordering in He^4 is therefore related to this phase separation.

To study this phase diagram the system has been idealized in a spin-1 model. The spin variable S_i at site i assumes the values ± 1 when a He^4-atom is present and 0 when a He^3-atom is present. The BEG Hamiltonian is then

$$H_{BEG} = -J \sum_{\langle ij \rangle} S_i S_j - K \sum_{\langle ij \rangle} S_i^2 S_j^2 + \Delta \sum_i S_i^2. \tag{3.8}$$

The first term leads to the magnetic ordering representative of the superfluid ordering, thus reproducing the λ line. The last two terms represent the He^3-He^3, He^4-He^4 and He^3-He^4 interactions. Δ depends on the difference of the chemical potential of the two species and, together with K at a given temperature determines their relative occurrence. When K is positive two nearest neighbours He^4-atoms are favoured just as in the correlated site-bond percolation problem, when one site is occupied by a monomer a nearest neighbour site is more likely to be occupied by a monomer.

Due to the relation (2.3) between Π_i^A and Π_i^B, the Hamiltonian (2.2) can be rewritten in the form (3.8) without the J term.

By varying the relative values of the parameters J, K, and Δ according to either the mean field approximation [27] or to the renormalization group calculation [36], the model (3.8) undergoes:

(i) A second order phase transition to a magnetically (super) ordered system starting from a normal mixture rich in He^4.
(ii) A first order phase transition with a jump in the magnetization and in the He^3-concentration x.
(iii) A first order transition between two normal phases with a jump in x, ending at a critical point.

The intersection points of the various curves vary in number and nature (critical, tricritical and triple points may be present).

At small values of the ratio K/J (i.e. when the direct interaction among the two components is small with respect to the ordering effect due to J only the first two features are present with a tricritical point. This is the type of phase diagram representative of He^3-He^4 mixtures.

In the opposite limit of $J=0$, the first two features are absent and the phase separation is induced only by the interaction among the two components as in the model (2.2). In this last case the sol–gel transition is related to connectivity properties due to the percolation aspects of the problem.

In the general case when all the previous features of the phase diagram are present, we can state a correspondence with the anticipated phase diagram of the correlated electron system.

The second order transition corresponds to the analogous transition from the magnetically ordered insulator to the paramagnetic insulator.

The first order transition is induced by either the superfluid ordering or by the direct interaction terms. In the first case the first order transition from the magnetic insulator to the paramagnetic metal should be reproduced. The second mechanism could possibly describe the direct metal–insulator transition.

The physical analogy can be stated by the correspondence:

He^4-atom$\leftrightarrow$singly occupied of magnetic sites ($|\uparrow\rangle, |\downarrow\rangle$);
He^3-atom$\leftrightarrow$non-magnetic sites ($|0\rangle, |\uparrow\downarrow\rangle$);
Superfluid ordering$\leftrightarrow$magnetic ordering;
He^3 concentration $x = 1 - \langle S_i^2 \rangle \leftrightarrow$twice the average or doubly occupied sites,

$$x = 1 - \langle (n_{i\uparrow} - n_{i\downarrow})^2 \rangle = 2\langle n_{i\uparrow} n_{i\downarrow} \rangle$$

Due to the rotational symmetry of eq. (3.1) as a first difference from the BEG model the Heisenberg spin, as defined in (3.3), rather than Ising spin variables should be used.

On the basis of the established analogy we can tentatively propose the following trial Hamiltonian as a model for a correlated electron system:

$$H_T = -J \sum_{\langle ij \rangle} S_i \cdot S_j - K \sum_{\langle ij \rangle} S_i^2 S_j^2 + \Delta \sum_i S_i^2. \tag{3.9}$$

Charge effects due to the operators (3.4) are already included in the K and Δ terms via conditions (3.7).

If the analogy works, we should be able to derive the model (3.9) from the Hubbard Hamiltonian H_H (3.1). Additional terms will come out if the analogy does not exhaust the physical content of H_H.

The two subspaces $|\uparrow\rangle, |\downarrow\rangle$ and $|0\rangle, |\uparrow\downarrow\rangle$ have played a crucial role in establishing the physical analogy with the BEG model. We look therefore for the invariance properties of H_H under a rotation in these subspaces:

$$U_S = \prod_i \exp[i\alpha_i K_i \cdot S_i], \quad U_\rho = \prod_i \exp[i\gamma_i q_i \cdot \rho_i]. \qquad (3.10)$$

U_S amounts to a local change of the quantization axis of the local spin. U_ρ corresponds to a generalization of the particle–hole exchange when q_i is in the x–y plane and to a trivial change in the Wannier representation when q_i is in the z-direction. The particle–hole exchange is for $\gamma = \pi/2$ and $q \equiv (0,1,0)$.

Under U_S, H_H is invariant provided the same K_i and α_i are taken for all the lattice sites.

Under U_ρ, H_H is invariant when $q_i \equiv (0,0,1)$ and $\gamma_i = \gamma$ for every lattice site. When q is in the x–y plane, H_H is still invariant provided $\gamma_i = \gamma$ and $\gamma_j = -\gamma$, i and j being the nearest neighbour sites.

Because of the last property, we now decompose the original lattice into two equivalent sublattices, each one of them being formed by the nearest neighbours of the other sublattice.

We now define an effective Hamiltonian by performing partial trace over the variables of one of the two sublattices according to the decimation procedure [37]:

$$\exp(-\beta H_{\text{eff}}) = \text{Tr}_{\text{partial}} \exp(-\beta H_H). \qquad (3.11)$$

The resulting H_{eff} must be globally invariant under the simultaneous action of U_S and U_ρ.

If we limit ourselves to one- and two-site interactions, H_{eff} must include the following linear combination of the invariants which can be built from S_i and ρ_i:

$$\sum_{ij} J_{ij} S_i \cdot S_j - \sum_{ij} K_{ij} S_i^2 S_j^2 + \Delta \sum_i S_i^2 + \sum_{ij} I_{ij} \rho_i \cdot \rho_j. \qquad (3.12)$$

No other invariants are present due to the constraints (3.7) on S_i and ρ_i. This is due to the fermionic character of the original system.

The invariants which cannot be expressed in terms of S_i and ρ_i are still missing. They must be expressed in terms of the original c^+ and c operators.

The most general combination invariant under U_S and U_ρ is

$$\sum_{ij} \sum_\sigma D_{ij} (c_{i\sigma}^+ c_{j\sigma} + c_{j\sigma}^+ c_{i\sigma})(1 - n_{i,-\sigma} - n_{j,-\sigma}). \qquad (3.13)$$

The full effective Hamiltonian is then:

$$H_{\text{eff}} = -J \sum_{\langle ij \rangle} \mathbf{S}_i \cdot \mathbf{S}_j - K \sum_{\langle ij \rangle} S_i^2 S_j^2 + \Delta \sum_i S_i^2$$
$$+ I \sum_{\langle ij \rangle} \boldsymbol{\rho}_i \cdot \boldsymbol{\rho}_j + D \sum_{\langle ij \rangle} \sum_\sigma (c_{i\sigma}^+ c_{j\sigma} + c_{j\sigma}^+ c_{i\sigma})(1 - n_{i,-\sigma} - n_{j,-\sigma})$$

(3.14)

where we have assumed only constant nearest neighbour interactions.

In order to obtain the new coupling constants J, K, Δ, I, D in terms of the original parameter T/W and U/W, the decimation has to be carried out explicitly.

If we neglect the additional terms proportional to I and D in eq. (3.14), the BEG Hamiltonian is recovered and all the previous results can be tentatively transferred to the correlated electron system as idealized by the Hubbard model.

The term associated with the coupling I is indeed of no importance in describing the critical properties of our system. In fact in the present model there is an internal structure of the non-magnetic sites which distinguishes between empty sites and doubly occupied sites. This feature was absent in the BEG model. The I term acts within this internal structure in the same way as the J term acts on the singly occupied sites giving rise to the magnetic order. No charge ordering can occur in the physically relevant region ($U \geq 0$) and this term can be neglected. The relevant charge effects have already been included in the K and Δ terms giving rise to a competition between sites of different type.

Particular care has to be given to the term proportional to D. It is a hopping term, which interchanges magnetic sites with non-magnetic sites allowing for their relative motion.

It must be important therefore to describe the proper nature of the metallic phase. In fact the correspondence between the direct phase separation into a He^3-rich phase and a He^4-rich phase and the direct metal–insulator transition is not straightforward due to the fact that the metal phase should at most be a mixture of magnetic and non-magnetic sites with equal weight. The D term can introduce particular aspects related to the conductivity of the system, which are more directly connected with percolation model (in analogy with the sol–gel transition) and with time order parameters of the type (2.1).

Even if all the competing mechanisms allowing for the various transitions appearing in the anticipated phase diagram of a correlated electron system have been clarified (at least for the magnetic transitions in mean field approximation when the D term does not contribute) by the analogy with the BEG model, a complete understanding of the system can only come out from the application of the renormalization group approach to the full effective Hamiltonian (3.14).

Acknowledgements

I would like to express my deep gratitude to C. Castellani and J. Ranninger. Most arguments presented here have been discussed together.

References

[1] Proceedings of the International School of Physics "E. Fermi", Course LI, ed. M. S. Green (Academic Press, 1971).
[2] Renormalization group in critical phenomena and quantum field theory: proceedings of a conference, eds. J. D. Gunton and M. S. Green (Temple University, Philadelphia, 1973).
[3] C. Di Castro and G. Jona-Lasinio, Phys. Lett. **29A** (1969) 322.
[4] N. N. Bogolubov and P. V. Shirkov, Introduction to the theory of quantized fields (Intersciences Publishers, New York, 1959).
[5] A. A. Migdal, Zurn Eksp. Theor. Fiz. **28** (1969) 1036.
[6] A. M. Polyakov, Zurn. Eksp. Theor. Fiz. **28** (1969) 533; **30** (1970) 151.
[7] T. Tsuneto and E. Abrahams, Phys. Rev. Lett. **30** (1973) 217.
F. De Pasquale, P. Tombesi, Nuovo Cimento **12B** (1972) 43.
[8] K. G. Wilson, Phys. Rev. **B4** (1971) 3174, 3184.
[9] K. G. Wilson and M. E. Fisher, Phys. Rev. Lett. **28** (1972) 240;
K. G. Wilson, Phys. Rev. Lett. **28** (1972) 548.
[10] F. Wagner, Phys. Rev. **B5** (1972) 4529.
[11] Th. Niemijer and J. M. J. van Leuwen, Phys. Rev. Lett. **31** (1973) 1411.
[12] C. Di Castro, Lett. Nuovo Cimento **5** (1972) 69.
G. Mack in "Lecture Notes in Physics" Vol. 17 (Springer-Verlag, Berlin, 1972).
[13] E. Brezin, J. C. Le Guillou and J. Zinn-Justin Phys. Rev. D8, **434** (1973) 2418.
C. Di Castro, G. Jona-Lasinio and L. Peliti, Ann. Phys. **87** (1974) 327.
[14] J. M. Kosterlitz and D. J. Thouless, J. Phys. C. **6** (1973) 1181.
J. M. Kosterlitz, J. Phys. C. **47** (1974) 1046.
[15] C. De Dominicis, Invited paper at the Geneva Conf. on "Critical Dynamics, April 2nd, 1979".
[16] H. E. Stanley et al., in: International Symposium on Synergetics, ed. H. Haken (Springer Verlag, 1980).
[17] N. F. Mott, Metal–Insulator Transitions (Taylor and Francis, 1974);
S. Doniach, Adv. Phys. **18** (1969) 1819.
[18] G. Parisi, Phys. Rev. Lett. **43** (1979) 1754; J. Phys. A (in print) (1980).
[19] A. Coniglio, H. E. Stanley and W. Klein, Phys. Rev. Lett. **42** (1979) 518.

[20] C. Castellani and C. Di Castro, Phys. Lett. **70A** (1979) 37.
[21] C. Castellani, C. Di Castro, D. Feinberg and J. Ranninger, Phys. Rev. Lett. **43** (1979) 1957.
[22] S. F. Edwards and P. W. Anderson, J. Phys. F **5** (1975) 965.
[23] D. Sherrington and S. Kirkpatrick, Phys. Rev. Lett. **35** (1975) 1972.
[24] J. R. L. de Almeida and D. J. Thouless, J. Phys. A**11** (1978) 983.
[25] T. Tanaka, G. Swislow and A. Ohmine, Phys. Rev. Lett. **42** (1979) 1566.
[26] D. J. Thouless, Phys. Reports **13** (1974) 94.

D. J. Thouless, Les Honches 1978, eds. R. Balian, R. Maynard and G. Toulouse (North-Holland, Amsterdam, 1979).
[27] M. Blume, V. J. Emery and R. Griffiths, Phys. Rev. A**4** (1971) 1071.
[28] J. Hubbard, Proc. Roy. Soc. A **276** (1963) 238.
[29] P. W. Anderson, Phys. Rev. **109** (1958) 1492.

For a review of the results for this model see ref. [26].
[30] J. M. Cyrot, J. Physique **33** (1972) 125.
[31] E. N. Economou and C. T. White, Phys. Rev. Lett. **38** (1977) 289;

R. De Marco, E. N. Economou and D. C. Licciardello, Solid State Commun. **21** (1977) 687.
[32] J. R. Schrieffer, C. A. P. Summer School Notes Banff (1969).
[33] S. Q. Wang, W. E. Evanson and J. R. Schrieffer, Phys. Rev. Lett. **23** (1969) 92.
[34] D. R. Hamann, Phys. Rev. Lett. **23** (1969) 95.
[35] E. H. Graf, D. M. Lee and J. D. Reppy, Phys. Rev. Lett. **19** (1967) 417.
[36] A. N. Berker and M. Wortis, Phys. Rev. B **14** (1976) 4946.
[37] L. P. Kadanoff, Ann. Phys. **100** (1976) 359.

CHAPTER 10

Membrane Flux:
Conditions for Limit Cycle Oscillations

A. G. De ROCCO

Trinity College
Hartford, Connecticut 06106
U.S.A.

G. L. CLARK

School of Medicine
University of Maryland
Baltimore, Maryland 21201
U.S.A.

© *North-Holland Publishing Company 1981*   *Perspectives in Statistical Physics*
Ed. H. J. Raveché

Contents

1. Introduction — 157
2. The mathematical structure of the model — 160
3. A physiological interpretation of J_{AT}: metastability and hysteresis — 168
References — 171

1. Introduction

In this paper we shall examine the conditions under which a reasonably general model for membrane flux will exhibit limit cycle oscillations in the direction of net transport. The motivation for this study was the suggestion of Sweeney [1] concerning a binary switching model for the flux of an essential ingredient in the photosynthetic clock of the marine alga *Acetabularia*. We begin by listing those features of Sweeney's model which we regard as crucial for an understanding of its phase plane behavior:

(1) The fluxing molecule 'X' undergoes both active transport and passive diffusion;
(2) There is an essentially instantaneous switching on and off of the active transport process;
(3) The switching on and off of the active transport process depends upon the attainment of certain critical concentrations of 'X';
(4) The switching on and off of the active transport process is hypothesized to be a result of changes in membrane conformation.

The idea that critical concentrations induce changes in membrane conformation (or, alternatively, in membrane-bound enzymes) suggests that a system of two stable states exists, and that critical concentrations somehow trigger transitions between these two states.

The physiological model, one notices, can best be described as a relaxation oscillator, and a central requirement of a mathematical model is that it leads to such oscillations: in particular, the variable characterizing the active transport must be discontinuous in time. Furthermore, one would like the model to possess a solution that is a limit cycle, i.e., independent of initial conditions and stable to the small fluctuations that are likely to occur in a complex biochemical environment. Feature (2) might be generalized slightly to say that instead of two states of active transport, corresponding to "on" and "off", the states correspond to a high transport rate and a slow transport rate. (Obviously the slow transport must be slow enough to permit outward passive diffusion to predominate.) To facilitate the analysis, it is desirable to amend feature (3) to read: "certain critical *internal* concentrations of 'X'..." instead of the original specifications set down by Sweeney, which were overly specific. Comment on feature (4) will be reserved for a later, more appropriate, place. We assume that only one molecule (the simplest possible assumption consistent with the data), 'X', participates crucially in the cyclic process; thus the same substance 'X' is being actively transported as well as passively diffusing across the membrane.

The preceding considerations lead us to a model which closely follows that of Sweeney except for the mentioned modifications. To begin development of the model we first obtain an equation for passive diffusion across the membrane. Define the diffusional flow vector $\boldsymbol{J}$ whose magnitude is

$$|\boldsymbol{J}| = \frac{1}{A}\frac{dn}{dt}, \tag{1}$$

that is, the number of moles n of a substance crossing a unit area in unit time. Conservation of mass implies

$$\frac{\partial s}{\partial t} = -\nabla \cdot \boldsymbol{J}, \tag{2}$$

where s is the concentration of the substance 'X' fluxing across the membrane. We further assume a phenomenological relation in which the flow vector $\boldsymbol{J}$ is proportional to the spatial gradient of the concentration of s, in particular

$$\boldsymbol{J} = -D'\nabla s, \tag{3}$$

where the diffusion constant D' is related to the mobility of 'X' in the conventional manner.

We model diffusion across the membrane as a steady-state process, assuming that the membrane connects two large resevoirs of 'X', whose concentrations are sufficiently slowly varying on the time scales of interest that they can be considered constant. Assuming a steady-state distribution of 'X' across the membrane, we have everywhere in the membrane

$$\frac{\partial s}{\partial t} = 0 \quad \text{and} \quad -\nabla \cdot \boldsymbol{J}_\mathrm{D} = 0, \tag{4}$$

implying that the steady-state flow of 'X' is constant. Assuming neutrality, fixed volumes and spatial homogeneity of the membrane, eq. (3) reduces to

$$J_\mathrm{D} = -D'\frac{ds}{dx}. \tag{5}$$

Assigning the membrane a thickness h, we may easily integrate (5) across the membrane to obtain:

$$J_D = \frac{D'}{h}(s-s_0)\hat{i}. \tag{6}$$

Clearly the rate of change of the internal concentration of 'X', s, resulting from passive diffusion will be proportional to the flow, J_D, i.e., $\dot{s}_D \propto J_D$. Thus we can write:

$$\dot{s}_D = D(s-s_0), \tag{7}$$

where D is an 'effective' transmembrane diffusion constant, while s and s_0 are the internal and external concentrations of 'X' respectively. This equation is the version of Fick's Law appropriate to describe the transmembrane passive diffusion of the protein.

To model the active transport we consider a process with a very special dependence of flow, J_{AT}, on the internal concentration s. The net rate of flow across the membrane will be:

$$J_{net} = J_{AT} - J_D. \tag{8}$$

We are led to fig. 1A which details the idealized form of J_{AT} and J_D.

The diagram supplies us with the following qualitative picture of the process. When J_{AT} and J_D intersect, then $J_{net}=0$ and we are in a steady-state situation. Suppose that the intersection occurs on either the upper or lower horizontal branch of J_{AT}. In particular, consider the point B in fig. 1A. Consider a fluctuation tending to increase s from point B. Clearly, as s is increased, $J_D > J_{AT}$ and the reactions proceed so as to return the concentration s back to its value at B. A similar argument applies for fluctuations tending to decrease s from its steady-state value at point B. Hence one concludes that steady-state points on the upper or lower branches of J_{AT} are stable to fluctuations in concentration s or velocity v. However, suppose that the steady-state concentration is a point other than on the stable branches already discussed (as for example point B, fig. 1B), then for fluctuations towards greater s away from the point B, we have $J_{AT} > J_D$ and s will continue to increase. It follows that the branch EC of J_{AT} is unstable to fluctuations in s or v. Furthermore, the concentration s as well as J_{AT} will continue to increase along BC. At point C, where $J_{AT} > J_D$ still, further increase in s is impossible if one is still to follow the active transport rate curve. Hence the point jumps to D, and the active transport suddenly drops to a new level, with the concentration decreasing. A study of the diagram reveals that relaxation oscillations along the path CDEA are established.

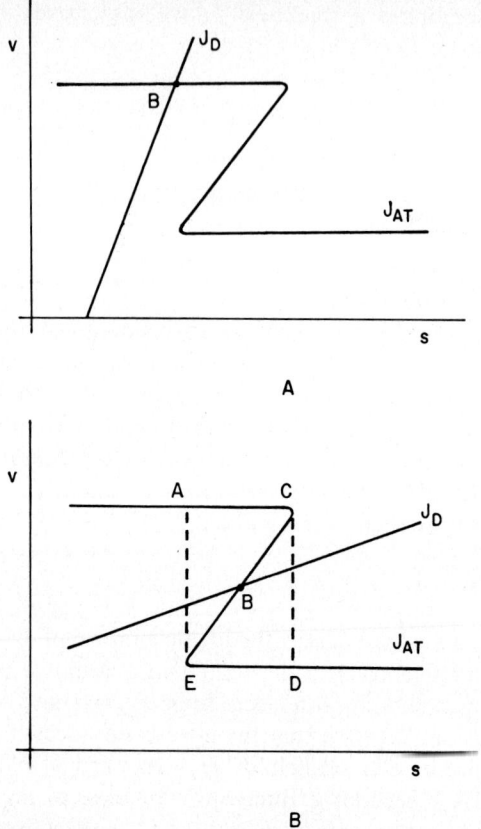

Fig. 1. Schematic illustrations of the form of J_{AT} and J_D needed to realize the oscillation described by Sweeney.

It is clear that this model satisfies all the criteria listed, at least qualitatively. The switching on and off (or more generally sudden changes in rate) of the active transport arises in a relatively natural way from the form of the kinetics chosen for the transport process.

2. The mathematical structure of the model

We first derive the phenomenological equation which will govern the dynamics of the model. For the moment we denote the explicit form of J_{AT} simply by $F(v)$, where $v = \dot{s}$ is the velocity of the active transport process. The rate equation describing the model is then:

$$\dot{s}_{\text{total}} = F(v) - D(s - s_0). \tag{9}$$

The steady state, a state of dynamic equilibrium between the two opposing diffusion processes, is defined by:

$$\dot{s}_{\text{total}} = F(v) - D(s - s_0) = 0.$$

Since at steady state all processes are stationary in time, eq. (9) becomes an algebraic equation relating the steady-state velocities and concentrations. Indeed, for arbitrary s, eq. (9) defines a curve that is the locus of steady-state points in the v–s plane. This curve will have two stable branches and one unstable branch, as noted in the introduction.

Suppose now that we have eq. (9) satisfied for a point on the unstable branch of J_{AT}. As long as no perturbations occur to cause fluctuations in the concentration of s away from its steady-state value, the system will remain stationary in time. However, if an arbitrary fluctuation in either concentration or velocity occurs, eq. (9) will no longer be satisfied, and the system will begin to undergo steady oscillations in time as previously described. Thus, if eq. (9) describes the steady state, then adding a small term $\mu \dot{v} = \mu \ddot{s}$, corresponding to a small acceleration in reaction velocity, will be sufficient to ensure that from any steady state which lies on the unstable branch we will be able to generate the oscillatory motion we seek.

We call attention to the fact that we have represented J_{AT} by a function whose argument is a 'velocity', $\dot{s}$. This is merely an anticipation of the fact that a simple finite power series expression for J_{AT} in terms of s will not lead to the necessary multiple valued form. On the other hand it is easy to see that a power series in $\dot{s}$ will easily provide a function that is multivalued in s. As we are primarily concerned with exhibiting the mathematics of the system, we choose the latter course, even though the mechanism behind such a form for J_{AT} is not immediately clear. Such a choice for J_{AT} suggests that we formally choose s to be the dependent variable, and $\dot{s}$ to be the independent variable. Carefully following the implications of such a choice, we are led to the following equation of motion:

$$\mu \ddot{s} - F(\dot{s}) + k\dot{s} + s - s_0 = 0 \tag{10}$$

where $k = D^{-1}$.

Here we can assume that μ is such that $\mu \ddot{s}$ is small compared to the rest of the terms in the equation.

Let us now turn our attention towards obtaining an explicit functional form for $F(v)$. We have seen that a curve of the sort sketched in fig. 1B for $F(v)$ will qualitatively lead to the anticipated results. A simple algebraic expression for a curve of precisely this form is not trivial to improvise, and in fact, as mentioned earlier, it can be shown that no power series in s, whether fractional or integral, will lead to it. However, it is clear that if $F(v)$ were chosen cubic in v, a reasonable approximation to the curve could be obtained, and this is precisely what we shall do. It should be noted that a major task in completing the model will be to devise a reasonable active transport model which leads to the kinetics postulated in the explicit form of $F(v)$. But first, we must investigate whether the mathematical structure of the model provides meaningful results, namely, limit cycle oscillations of the desired sort.

We label the coordinates of such a point as B in fig. 1B by (v_1, s_1) and the slope of the linear segment which connects CBE by m, then:

$$F(v) = -a(v-v_1)^3 + 2s_1 + m^{-1}(v-v_1). \tag{11}$$

Upon simplification and collection of terms, one finds that $F(v)$ can be written as:

$$-F(v) = a_3 v^3 + a_2 v^2 + a_1' v + a_0$$

with

$$a_3 = a > 0$$
$$-a_2 = 3v_1 a > 0$$
$$-a_1' = m^{-1} - 3v_1^2 a$$
$$-a_0 = v_1^3 a - (m^{-1})v + 2s_1.$$

Setting $\mu = 1$, and substituting for $F(v)$ into eq. (10), we obtain as the final equation to analyze

$$\ddot{s} + a_3 \dot{s}^3 + a_2 \dot{s}^2 + (a_1' + k)\dot{s} + s - s_0 = 0. \tag{12}$$

Defining $s' = s - s_0$ and $a_1 = a_1' + k$, we obtain eq. (12) in the form

$$\ddot{s}' + a_3 \dot{s}'^3 + a_2 \dot{s}'^2 + a_1 \dot{s}' + s' = 0. \tag{13}$$

From here on, we drop the primes, and remember that we have redefined the value of s. For the moment, we also do not restrict ourselves to μ small, but investigate the more general case.

Let us now introduce some nomenclature. Equation (13) reduces to Rayleigh's equation when $a_2 = 0$, hence it could be appropriately termed as the generalized Rayleigh equation. It can be shown that eq. (13) is equivalent to the Lienard equation, of which the van der Pol equation is a special case. Theorems concerning the existence of limit cycles are well established for the Lienard equation, hence, it is helpful to transform eq. (13) to the form of the Lienard equation so that we might investigate whether it possesses a solution which is a limit cycle.

To this end, consider Lienard's equation in the form

$$\ddot{x} + (1 + x + x^2)\dot{x} + x = 0. \tag{14}$$

Setting $\dot{y} = x$, one obtains from eq. (13)

$$\dddot{y} + (1 + \dot{y} + \dot{y}^2)\ddot{y} + \dot{y} = 0$$

or

$$\frac{d}{dt}\ddot{y} + \frac{d}{dt}\dot{y} + \frac{1}{2}\frac{d}{dt}\dot{y}^2 + \frac{1}{3}\frac{d}{dt}\dot{y}^3 + \frac{d}{dt}y = 0. \tag{15}$$

Integrating, and setting constants to zero,

$$\ddot{y} + \dot{y} + \tfrac{1}{2}\dot{y}^2 + \tfrac{1}{3}\dot{y}^3 + y = 0. \tag{16}$$

One sees that eq. (16) is equivalent to eq. (13). It follows that since eq. (13) and the Lienard equation are related by a change of dependent variable, then if the Lienard form of eq. (13) can be shown to possess a limit cycle as a solution, eq. (13) will possess a limit cycle as well. Setting $x = \dot{s}$ in eq. (13) one obtains:

$$\ddot{x} + \left[3a_3 x^2 + 2a_2 x + a_1\right]\dot{x} + x = 0, \tag{17}$$

which is of the Lienard form, that is of the form

$$\ddot{x} + f(x)\dot{x} + g(x) = 0. \tag{18}$$

We now set about proving that eq. (17) has a solution that is a limit cycle. According to Dragilev [2], the solution to eq. (18) has at least one limit cycle (although it may not be unique) under the following conditions: define

$$F(x) = \int_0^x f(x)\,dx \quad G(x) = \int_0^x g(x)\,dx.$$

then, if:

(1) $g(x)$ satisfies the Lipschitz conditions $xg(x)>0$, $x=0$; $G(\infty)=\infty$;
(2) $F(x)$ is defined uniquely in the interval $-\infty<x<\infty$, and in any finite interval it satisfies the Lipschitz conditions; moreover, for sufficiently small x, $F(x)<0$ if $x>0$ and $F(x)>0$ if $x<0$;
(3) There exists a number M and numbers k and k', $k'<k$ such that:

$$F(x) \geqslant k \quad \text{when} \quad x>M$$
$$F(x) \leqslant k' \quad \text{when} \quad x<-M,$$

then eq. (18) will possess at least one limit cycle, although it is not necessarily unique. Referring to eq. (17) we note:

$$f(x)=3a_3x^2+2a_2x+a_1 \quad F(x)=a_3x^3+a_2x^2+a_1x$$
$$g(x)=x \qquad\qquad\qquad G(x)=x^2/2.$$

Clearly condition (1) is satisfied, as can be seen by inspection. To investigate conditions (2) and (3), we must further consider $F(x)$. The roots of $F(x)=0$ are:

$$x=\left(a_2 \pm \sqrt{a_2^2-4a_3a_1}\right)/2a_3,$$
$$x=0.$$

For the roots to be non-imaginary we require (recall $a_3>0$)

$$a_2^2-4a_3a_1>0. \tag{19}$$

Furthermore, the condition that one root be negative is

$$4a_3a_1<0,$$

which leads to

$$a_1' \leqslant -k,$$

which is consistent with eq. (19).

From these facts, coupled with the behavior of $F(x)$ for small x, it becomes clear that conditions (2) and (3) are satisfied. $F(x)$ is a cubic, one root of which is zero and the two extreme branches of which are unbounded in the first and third quadrants. Hence we conclude that the solution to eq. (13) under the condition (19) is a limit cycle. Accordingly, our model exhibits a limit cycle.

We next proceed to examine the behavior of eq. (13) in the phase plane. At this point we restrict ourselves to the form:

$$\mu\ddot{s} + a_3\dot{s}^3 + a_2\dot{s}^2 + a_1\dot{s} + s = 0 \tag{20}$$

with $\mu \ll 1$.

Set $\zeta = \dot{s}$, one can then see that

$$\mu\dot{\zeta}/\zeta = -\left(a_3\zeta^3 + a_2\zeta^2 + a_1\zeta + s\right)/\zeta \tag{21}$$

so that

$$\mu\,d\dot{s}/ds = -\left(a_3\dot{s}^3 + a_2\dot{s}^2 + a_1\dot{s} + s\right)/\dot{s}. \tag{22}$$

We have thus derived an expression for the slope of the tangents to the trajectories in the $\dot{s}$–s phase plane. Equation (22) tells us that the slope of the tangents in the phase plane is very large for μ small. In fact the tangents must be everywhere very nearly vertical except along the curve

$$s = -\left(a_3\dot{s}^3 + a_2\dot{s}^2 + a_1\dot{s}\right) \tag{23}$$

where the slope of the tangents to the trajectories vanishes identically. The curve defined by eq. (23) divides the phase plane into two distinct regions. In region I, as can be seen by inspection of eq. (20), $\ddot{s} < 0$. In region II, we have that $\ddot{s} > 0$. Thus, in region I, the time development of the motion is so as to decrease $\dot{s}$. Likewise, in region II, the time development of the motion must be so as to increase $\dot{s}$. Furthermore, in the region $\dot{s} > 0$ the motion must be to the right; in the region $\dot{s} < 0$ the motion must be to the left. Combining all of these facts together, we can draw the phase portrait shown in fig. 2.

As noted earlier, the tangents are nearly vertical everywhere, except near the curve defined by eq. (23), where the tangent must gradually slope to zero, hence the small bends in the vertical lines. We can follow the motion of a point starting at A by examining the phase portrait. Such a point would drop nearly instantaneously to point B, where it would intersect the curve, and move horizontally across it to the left in accordance with its negative velocity. Once below the curve, it would find itself in a region where the direction field guides it towards the left and up, as indicated by the direction of the arrows. Thus it would follow a trajectory slightly below the curve and upwards towards point D. At

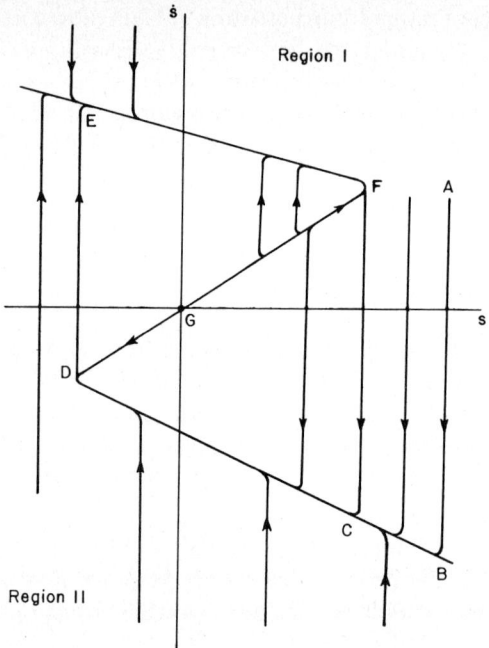

Fig. 2. The phase plane trajectory for the relaxation oscillator described in the text.

point D the guiding field is vertical again, hence, it moves nearly instantaneously to point E, where it again crosses the curve to the right, since now it has a positive velocity. Once above the curve, it would continue to be guided to the right and downwards, in accordance with the direction field, until it reached point F, where it would drop to point C and continue the cycle. A similar path is projected for a point starting within the interior of the loop CDEF.

Thus we see that the phase portrait of eq. (20) is such that the preferred trajectory is slightly exterior to the loop CDEF. It has been shown (by Pontryagin) that the true limit cycle for the equation approaches the loop CDEF arbitrarily closely under appropriate conditions. From this discussion then, we find a mathematically sound picture which is equivalent to the qualitative picture sketched earlier [3].

In the particular case considered here, an analytic proof of a limit cycle could be provided. Unfortunately, equations for which such analytic proofs of limit cycle behavior may be provided are the exception rather than the rule. In cases where an analytic proof is not possible, analysis of phase plane trajectories is indispensable to obtain insight concerning the behavior of the solution. Because the model we ultimately present is too complicated mathematically for an analytic proof of limit cycle behavior, we shall summarize briefly the phase plane conditions that we expect to lead to a limit-cycle relaxation oscillator [4].

(1) A necessary condition for a singly periodic solution is that the trajectory be closed. A necessary condition for a closed trajectory is that it must contain at least one singular point that is a node, a focus, or a center. *Comments*: a singular or equilibrium point is defined by $\dot{s}=\ddot{s}=0$. Reference to eq. (13) demonstrates that such a place is the origin of the phase plane, which is point G in fig. 2. Linearization of eq. (13) about the singular point indicates that the point corresponds to an unstable focus. Here, infinitesimal perturbations will drive the system from point G to the limit cycle trajectory.

(2) For an equation of the form $\ddot{x}+\phi(\dot{x})+x=0$, $\phi(\dot{x})$ must not be single valued everywhere. This is an obvious topological requirement needed to guarantee that the solution corresponds to a closed trajectory enclosing a finite area.

(3) Consider a set of first order equations of the sort $\mu\dot{x}=P(x, y)$ $\dot{y}=Q(x, y)$ ($\mu \ll 1$). It is possible to divide the phase plane into two regions, a slow region and a fast region, such that: (i) for the slow region $P(x\,y)\leqslant\theta(\mu)$; (ii) for the fast region $P(x, y)\geqslant\theta(\mu^a)$, $0 < a < 1$. *Comments*: the slow region corresponds to a narrow band contiguous to the curve $P(x, y)$. In this region $\dot{x}=(1/\mu)P(x, y)=(1/\mu)\theta(\mu) \sim \theta(1)$ and $\dot{y} \sim \theta(1)$. This is the region where the trajectory slowly parallels the curve $P(x, y)=0$, or, as in fig. 2, the branches EF and CD of the loop EFCD. The fast region, which fills the part of the phase plane not occupied by the slow region, is characterized by $\dot{x} \sim \theta(\mu^{a-1}) \to \infty$ as $\mu \to 0$, and $\dot{y} \sim \theta(1)$ (bounded by definition). From these characterizations, it can be seen that the time of traversal of the fast region tends to zero as $\mu \to 0$, and furthermore $\Delta y \to 0$ as $\mu \to 0$. This region clearly corresponds to the vertical lines in fig. 2. If these slow and fast regions alternate, and are of the proper sense, then a closed oscillation will result.

Conditions (1) and (2) are topologically necessary for a periodic solution, while condition (3) ensures that departures from the trajectory, which is uniquely defined by $P(x, y)=0$, will be quickly damped. Thus, what we have are the topological requirements for a limit-cycle relaxation oscillator. Any equation demonstrating these characteristics in the phase plane will, we can be reasonably sure, exhibit a limit cycle solution.

We conclude that the type of oscillation Sweeney postulated can be provided a satisfactory mathematical description, and since we have shown that it can be described as a non-linear relaxation oscillator, we have immediately at our disposal the tools for providing quantitative foundations for many of the physiological proposals made by Sweeney. Next we must provide a biochemically acceptable mechanism which will produce J_{AT} of the necessary sort. The remainder of this essay will be devoted to an exposition of that problem.

3. A physiological interpretation of J_{AT}: metastability and hysteresis

The relaxation oscillator introduced in the preceding section is strongly reminiscent of a hysteresis loop. With this idea in mind, we undertake a discussion of molecular hysteresis, and metastable states, following to a degree the exposition by Neumann [4].

The most familiar example of metastable states and their accompanying hysteresis loop is that of ferromagnetic materials. However, there are many other physical examples, the most notable being melting–crystallization cycles of ammonium salts and in adsorption–desorption processes in porous materials. It was not until fairly recently that so-called 'biocolloids' were found frequently to have the remarkable capability of developing long-lived metastable states. There are now many known examples of such a phenomenon. In nerves, membrane conductance as a function of the mole fraction of external univalent cation demonstrates a clear hysteresis. Acid-base titration of membrane fragments of Dead Sea halobacteria yields a hysteretic proton binding curve. At constant pH, cyclic variation of NaCl will cause a hysteresis loop in the relative activity of a membrane-bound enzyme [4]. In this latter case, the hysteresis likely arises from long-lived metastable states in the association–dissociation reaction between the dimer subunits of the enzyme complex [5]. Metastable intramolecular conformational states have been observed in *E. coli* ribosomal RNA to last for days [6]. This finding leads to the speculation that such metastabilities may have a biological significance. Finally, the helix–coil transition in polynucleotides also demonstrates a pH hysteresis loop [7]. We shall study this last example in some detail to present a thorough description of the metastabilities involved. We first discuss the nature of metastable states.

Long lived hysteresis usually indicates thermodynamically metastable states and cooperative non-equilibrium transitions. Cooperative structural transitions in biopolymers and membranes are generally classified as 'diffuse' phase transitions, and can be treated as first order phase transitions. Such metastable states are frequently associated with energy barriers that impede the 'phase change' by inhibiting, for example, nucleation or propagation.

Thermodynamically, the transition behavior of a cooperative system with instabilities may be described with a state function $G(x)$ of the van der Waals type [8]. Between the two stable branches there is a region that is thermodynamically not realizable. The equilibrium path between stable branches avoids the metastable region. Supposing, however, that sufficiently high energy barriers exist, the change of state can then occur only via metastable states. In such a case, the phase changes would be unidirectional and irreversible, and would take place at different values of x, leading to the hysteresis. The irreversibility also leads to a fixed cycling direction. It is clear that the hysteresis phenomenon is characterized by a state function G that is not everywhere a single valued function of its independent variables. The metastable state also clearly is a non-equilibrium situation.

Suppose that a time $t=0$, the system is at a point x_1, and that the variable x is suddenly changed from $x=G^{-1}(x_1)$ to $x=G^{-1}(x_2)$. If the value of G remains constant for observation times greater than a characteristic time $\bar{t}, t \gg \bar{t}$, we consider the metastable state to be long-lived [4]. In the absence of this, we have a short-lived metastable state. It is as a consequence of the relative time independence of G that metastable states can be described by thermodynamic state functions. From these arguments, one recognizes that the size of the hysteresis loop will generally depend on the cycling rate if the underlying metastabilities are short-lived.

Let us now focus attention on a particular biological example to gain some physical insight into the origin of the barrier and the hysteresis process. The example we choose is that of Weisbuch and Neumann [7] who studied the helix–coil transition of the triple-stranded helix polyriboadenylate (poly (A))–2 polyribouridylate (poly (U)), which demonstrates hysteresis as a function of pH. In the helix form of poly A–2 poly U, the nucleotide bases can be in either of two forms: *closed* or *open*. In a closed state an adenine base is hydrogen bonded to two uracils. The open state finds the hydrogen bonds interrupted, and stacking interactions reduced. The free energy difference between a closed and open state comes from two terms: (1) a stabilizing term arising from hydrogen bonding and base stacking and (2) a destabilizing term resulting from electrostatic repulsion between the charged phosphate groups of opposing strands. These two terms do not vary in the same way as the helix unwinds. When the strands are open, the stabilization energy is completely lost. The electrostatic repulsion term, on the other hand, decreases significantly only when the chains can move far apart.

If we suppose experimental conditions such that only a few base residues within the structure can open, then the dissociated chain residues can not separate significantly because their closed neighbors prevent them from moving far apart. But because of electrostatic repulsion a small separation is energetically unfavorable and dissociation encounters a barrier which can create metastability of the closed state relative to the open state. The barrier, however, can be removed by protonation of the negatively charged phosphate groups (and consequent conversion to equilibrium) which is a physical explanation for the pH dependence of the loop.

To treat the phenomenon theoretically we may employ a molecular field theory which takes into account only long range interactions. The enthalpy of the closed state is taken as $-A$, while the enthalpy of the open state is taken as Bf, where f is the fraction of closed nucleotides. The Gibbs free energy of the system is then:

$$G = -Af + Bf(1-f) + kT[f\ln f + (1-f)\ln(1-f)]. \tag{24}$$

We find the extrema of G with respect to f which leads to the following implicit equation for f:

$$f = \frac{1}{1+\exp(-\beta f + \alpha)} = \psi(f) \tag{25}$$

where

$$\alpha = (-A+B)/kT, \quad \beta = 2B/kT. \tag{26}$$

The solutions to eq. (25) can be found by graphical means. In particular, if $\phi(f)$ is plotted for increasing values of α, it can easily be shown that as α increases there are successively one, two or three extrema. The profiles of $G(f)$ thus correspond to both a single minimum and to dual minima separated by a continuously varying barrier; in short, to metastable states.

From this discussion we see that metastability can be well understood even for an apparently complex biological process. Furthermore the discussion strongly suggests to us that the type of active transport curve we seek could have its roots in some sort of long-lived metastable states of the active transport protein involved. In particular, our example suggests that we look at a cooperative process. Such cooperative processes involving the binding of ligands to congenial substrates are well known, and the allostearic transitions which result have been implicated in a variety of regulatory processes in molecular and cellular biology. What is not immediately obvious, however, is that such phase transitions possess metastable states, and in their absence the net flux, which combines linear Fick's Law with the allostearically regulated active transport, will everywhere exhibit stable steady state solutions, i.e., departures from the focus will everywhere have a guiding field which restores the steady state (fig. 1A). If, on the other hand, solutions to the cooperative transition exist which exhibit hysteresis, then an intersection of the passive flow and the metastable region of the active transport will lead in a natural way to oscillations in the net flux (fig. 1B). We have investigated this problem and shown that such solutions exist. Thus the limit cycle implications adumbrated above have found expression in a cogent biophysical model, upon which we shall report in a subsequent publication.

Acknowledgements

We are pleased to acknowledge valuable discussions with Professors Beatrice Sweeney, Philip Sokolove and Arthur Winfree. In its earliest version the work reported here was completed when the authors were resident at the University of Maryland, College Park.

References

[1] B. M. Sweeney, Int. J. Chronobiol. **2** (1974) 25.
[2] N. Minorsky, Nonlinear Oscillations (D. Van Nostrand, New York, 1962) p. 103.
[3] H. Lashinsky, Nonlinear Periodic Phenomena (North-Holland, Amsterdam, in press).
[4] E. Neumann, Angen. Chem. Internat. Edit. **12** (1973) 356.
[5] F. Snell, Principles of Biological Membranes (Gordon Breach, New York, 1970) p. 24.
[6] A. Revzin, E. Neumann and A. Katchalsky, J. Mol. Biol. **79** (1973) 95.
[7] G. Weisbuch and E. Neumann, Biopolymers **12** (1973) 1470.
[8] T. L. Hill, J. Chem. Phys. **17** (1949) 520.

CHAPTER 11

*Critical Phenomena –
A Model Illustration
of Scientific Method.*

C. DOMB

*Wheatstone Physics Laboratory
University of London King's College
Strand, London, W.C. 2
U. K.*

© *North-Holland Publishing Company 1981* *Perspectives in Statistical Physics
Ed. H. J. Raveché*

Contents

1. Mel Green's contributions to critical phenomena — 175
2. Task of physics — 176
3. The classical period — 179
   3.1. Liquid–gas critical point — 179
   3.2. Curie point of a ferromagnet — 180
   3.3. Microscopic critical behaviour of fluids — 182
   3.4. Critical behaviour of binary alloys — 184
   3.5. Universality of second order transitions (Landau) — 185
4. The Onsager revolution — 187
5. Reconciliation — 189
6. Renormalization group: respectability — 192
7. Conclusion — 196
References — 198

1. Mel Green's contributions to critical phenomena

During the past 20 years the field of critical phenomena has been transformed from a relatively obscure and specialist area into one of central interest in condensed matter physics. A major part in this development has been played by cross-fertilization, between theory and experiment, between exact and approximate methods, between statistical mechanics and field theory, etc. The great progress achieved in this many body problem was in no small measure due to "many mind" interactions, and I can think of no single person who did more to facilitate these interactions than Mel Green.

During the past 15 years there have been important "get togethers" for workers in the field at crucial stages in the development of new ideas, and for three of these Mel Green was in the forefront of the initiative. The conference at NBS in 1965 can reasonably be termed the founding conference of Critical Phenomena, since for the first time all the different strands of the subject were woven into a coherent fabric. Mel was host to the conference, and edited the conference proceedings with Jan Sengers. He did a superb job and I still recommend them to my graduate students who wish to learn the fundamentals of the subject. At two dollars a bound copy, NBS Miscellaneous Publication 273 must surely be the cheapest conference proceedings in recent scientific history. It was shortly after this conference that the idea of scaling emerged, providing a unified description of the new theoretical and experimental results.

A few years later, in 1970, Mel was host again at the Summer School on Critical Phenomena in Varenna, Italy. It was his task to organize the programme of courses and lectures, and the result was a resounding success. Among the new ideas highlighted at the School were Leo Kadanoff's principle of "universality", Michael Fisher's theory of finite size scaling, and the first hints from the Italian field theorists that the Renormalization Group might have applications to critical phenomena. The volume which records the Summer School Proceedings has become part of the library of every serious worker in the field.

Only a year or two elapsed before Ken Wilson's ideas on the use of the Renormalization Group catalysed the most rapid progress which the subject had seen. Again Mel Green (who had now moved to Temple University) took the initiative in hosting a conference in 1973 at which the many new paths which had been explored were made available for public use. The Conference Proceedings, edited by Mel Green and Jim Gunton, appeared quickly and were undoubtedly a powerful factor in the very fruitful developments which have followed.

With characteristic modesty and self-effacement, Mel Green did not give a talk himself at the first two meetings, and presented only a small paper at the third. This does not mean that he had nothing worth saying, on the contrary, his wide grasp of all that was going on was clearly demonstrated by the excellent programmes which he helped to arrange. But he preferred to stay in the background, devoting his energies to ensuring that the programme ran smoothly so that participants derived maximum benefit, and to keeping himself up to date with current developments.

Mel's personal contributions to critical phenomena, particularly in fluids, are of the greatest importance and have taken their place in the literature. But it is more appropriate in this memorial volume to draw attention to his public activities, for which all who have been engaged in research in critical phenomena must be deeply grateful.

2. Task of physics

One of the "get togethers" on critical phenomena referred to in the previous section was the Battelle Colloquium held in September 1970 in Geneva and Gstaad, Switzerland. The after-dinner address was given by Van Hove [1], a leading CERN scientist, on "The changing face of physics". Van Hove laid special emphasis on what he regarded as the basic task of physics – "we should constantly recall and reassert that elucidation of the fundamental laws remains the most essential task of physics". Hence, despite the staggering increase in costs associated with the building of new and more powerful machines, we must not be diverted, since the mainstream of physics is in this direction. He took particular exception to remarks made a few months previously by Dyson [2] who speaking at the dedication of a new physics laboratory in Princeton had recommended cosmic rays, biophysics and astrophysics as fruitful directions for research.

In relation to "critical phenomena" and topics of a similar kind Van Hove said the following: "It seems to me that physics now looks more like chemistry in the sense that, in percentage, a much larger fraction of the total research activity deals with complex systems, structure and processes, as against a smaller fraction concerned with the fundamental laws of motion and interaction. This colloquium is a good example. Surely, we all believe that the fundamental laws of classical mechanics, of the electromagnetic interaction and of statistical mechanics dominate the multivarious transition and critical phenomena you discuss this week; and I presume that none of you expects his work on such problems to lead to modifications of these laws. You know the basic equations better than the phenomena. You are after the missing link between them, i.e. the intermediate concepts...which should allow a quantitative understanding and prediction of the phenomena...".

A few years later Ziman [3] took the very different view that confining physics to the search for fundamental laws is grossly restrictive; if physics is to have a secure future it must concern itself with natural phenomena in a much broader sense.

"In the education of a physicist, we recount the bold voyages of the great explorers – Newton and Einstein, Faraday and Bohr – in search of new laws of nature. They found and charted the continents on which we have built our cities of the mind and of art. Does anyone really suppose that similar vast and fertile territories are still waiting to be discovered and colonized?... The unaccustomed rules that govern black holes and quasars in the cosmic deeps affect our lives no more than the icy crags of the Himalayas or the conjunctions of the planets...".

"Think of physics simply as the 'fundamental' science and it is oversubscribed almost to bankruptcy. But define it as the science whose aim is to describe natural phenomena in the most mathematical or numerical language, and you will understand its past and have confidence in its future.... The task of the modern physicist is to determine the mathematically comprehensible characteristics of the natural world and of human artifacts...".

Our own attitude to physics follows the direction of Ziman rather than that of Van Hove. But we would go further than Ziman in asserting that even the great explorers of the past did not concentrate their efforts on the direct search for fundamental laws. They tried to investigate and understand all natural phenomena which interested them, and the fundamental laws emerged in a quite unpredictable manner. Clerk Maxwell (whose name should surely be included in any short list of great explorers of physics) tried to construct a mechanical model to illustrate Faraday's law of induction. He drew heavily on the theory of vortices in a perfect fluid and classical elasticity, and was thereby led to introduce his famous displacement current. It was only later that he removed the "scaffolding" to reveal the beautifully symmetric pattern of Maxwell's equations.

The passage of ideas from one branch of physics to another is one of the most powerful stimulants to progress, and like the analogous cross-fertilization between mathematics and physics, the traffic is two-way. One must always be prepared to find a concept which has evolved in one area of investigation being surprisingly fruitful in a completely different area.

It is our aim in the present article to illustrate these thoughts with a brief survey of the history of critical phenomena. Although the properties under investigation are not "fundamental" they succeeded in engaging the attention of some of the greatest physicists of the past 100 years, and can serve as a model illustrating the application of scientific method to the exploration of a new class of phenomena.

We shall divide this history into 4 periods:

1869–1944, The classical period;
1944–1965, The Onsager revolution;
1965–1971, Reconciliation of Onsager with van der Waals;
1971–   Renormalization group. Respectability.

In the first period the term classical is used in the sense of well established [4] (and not with any relationship to quantum theory). The subject starts with the introduction of the term "critical point" by Thomas Andrews in his Bakerian Lecture of 1869. This period is characterized by the interplay of experiment and theory, and the introduction of new concepts which seemed adequate to provide a qualitative and quantitative description of the phenomena under consideration.

The second period was initiated by a mathematical *tour de force*, the exact solution by Onsager of a realistic two dimensional model, giving results which were in basic disagreement with the classical calculations. This period was characterized by further exact calculations of the Onsager kind and also by down to earth numerical calculations which made extensive use of computers; in addition new experimental techniques were exploited to furnish more accurate data than had been available previously. As a result the classical theories were completely discredited in relation to quantitative predictions of critical behaviour. However, no coherent new pattern was formulated to put in its place.

This led Uhlenbeck [4] to suggest in 1965 that the most important task facing workers in the field was to reconcile Onsager with van der Waals. And indeed shortly after this a wider framework emerged which the new theoretical and experimental results seemed to fit. During the period 1965–1971 this framework was explored in more detail, and the pattern of behaviour was summarized most succinctly in the hypothesis of universality [5]. But the pattern was obtained empirically – there was no theoretical explanation.

The latest phase supplied the theoretical understanding which has made the field "respectable". It was initiated in 1971 by a brilliant application [6] by Kenneth Wilson of the renormalization group, a concept taken from field theory, to behaviour near a critical point. The burst of research activity stimulated by Wilson's ideas has opened up many new avenues of exploration; but there are also indications that substantial progress in field theory may result from the "feedback" from critical phenomena.

3. The classical period

3.1. Liquid-gas critical point

The relation between gases and liquids was the subject of much attention during the first half of the 19th century. It was realized quite early on that the liquid phase ceased to exist above a certain temperature [7], but the general assumption was that it disappeared into the gaseous phase. Thus Faraday talked of the "disliquefying point", while Mendeleev used the term "absolute boiling temperature" for the point at which the latent heat of evaporation becomes zero.

In addition to careful and accurate experiments on carbon dioxide described in the Bakerian Lecture of 1869, Thomas Andrews introduced a new concept of symmetry between the liquid and gaseous phases; the two phases merged at the critical point into one fluid phase, "but if any one should ask whether it is now in the gaseous or liquid state, the question does not, I believe, admit of a positive reply". He emphasized this feature in the title of his lecture "On the Continuity of the Gaseous and Liquid States of Matter", and pointed out how it was possible, by suitable choice of path, to pass from the liquid to the gaseous phase without any discontinuity.

Only four years elapsed before van der Waals used the newly developing ideas on the kinetic theory of gases to give a plausible theoretical explanation of Andrew's experimental data. Van der Waals assumed that a gas is made up of molecules with a hard core and a long range mutual attraction. The attractive forces give rise to a negative "internal pressure" which he calculated from the virial of Clausius as being equal to a/v^2; the hard core leads to a reduction in available volume from v

to $(v-b)$. Hence the equation of state he put forward was

$$p_{\text{total}} = p + p_{\text{internal}} = p + a/v^2 = RT/(v-b). \tag{1}$$

Van der Waals was a doctoral student, and the above ideas were put forward in his thesis in Dutch. Their importance was quickly recognized by Maxwell who reviewed the thesis in *Nature* in 1874, and in a lecture to the Chemical Society in 1875 where he wrote as follows:

> "The molecular theory of the continuity of the liquid and gaseous states forms the subject of an exceedingly ingenious thesis by Mr. Johannes Diderick van der Waals, a graduate of Leyden. There are certain points in which I think he has fallen into mathematical errors, and his final result is certainly not a complete expression for the interaction of real molecules, but his attack on this difficult question is so able and so brave, that it cannot fail to give a notable impulse to molecular science. It has certainly directed the attention of more than one inquirer to the study of the Low-Dutch language in which it is written."

It was in this lecture that Maxwell put forward his famous "equal-area" construction which completes the van der Waals treatment of liquid–gas equilibrium. The new concept of "internal pressure" which van der Waals introduced was to bear fruit 30 years later in a completely different area of physics.

3.2. Curie point of a ferromagnet

The fact that a magnet loses its magnetic power at high temperatures was noted by Gilbert [8] in his famous treatise "De Magnete" in 1600. A more detailed quantitative investigation started in the 19th century involving Faraday in the 1830's, Barrett in 1874, Bauer in 1880 and Hopkinson [9] in 1889 who was the first to introduce the term "critical temperature" for "the temperature at which the magnetism disappears". But the definitive paper for magnets comparable to that of Andrews for fluids was written by Curie [10] in 1895 before he started his more famous investigations of radioactivity.

One of the most interesting new ideas put forward in this paper is the analogy between magnets and fluids. Taking pressure p as the analogue of magnetic field and density ρ as the analogue of magnetization, Curie points out the close similarity between the p–ρ and M–H isothermals. The paramagnetic state at high temperatures corresponds to the gaseous phase, and ferromagnetic state at low temperatures to the liquid phase.

Curie says that this analogy could be used to suggest new and useful experiments, and he poses the question whether there exists a precisely defined critical point with associated critical constants for a ferromagnet analogous to a fluid.

It was this analogy which led Pierre Weiss in 1907 to postulate his *Molecular Field* hypothesis [11], in which he assumed that the mutual interactions between the molecules could be replaced by a uniform field nM proportional to the magnetization in the same direction. He states, "One may give to nM the name *Internal Field* to mark the analogy with the internal pressure of van der Waals".

For the magnetic susceptibility of a gas Curie had discovered experimentally the inverse temperature dependence

$$\chi = C/T, \tag{2}$$

and Langevin had applied the techniques of statistical mechanics to explain this result theoretically. For an ideal gas of molecules each having magnetic moment m, Langevin derived the magnetic equation of state (analogous to $pV = RT$ for an ideal gas)

$$M = L(mH/kT), \tag{3}$$

where $L(x)$ is the Langevin function coth $x - 1/x$.

For a ferromagnet Weiss put forward the simple modification

$$M = L[m(H + H_{\text{int}})/kT] \quad (H_{\text{int}} = nM), \tag{4}$$

and this led to far reaching conclusions. There is indeed a sharply defined critical point analogous to that for a fluid; Weiss [12] later termed this the Curie point in memory of Curie who had been killed in an accident in 1906. Below the Curie temperature there is a non-zero spontaneous magnetization; in the paramagnetic state above the Curie temperature T_c, Curie's relation (2) for the susceptibility is modified to

$$\chi = C/(T - T_c), \tag{5}$$

and this is usually called the Curie–Weiss Law.

The magnet–fluid analogy which proved so fruitful seems to have been forgotten for more than 30 years until it was rediscovered in the lattice–gas model by Cernuschi and Eyring [13] in 1939. It was fruitful again in subsequent developments after 1944.

3.3. Microscopic critical behaviour of fluids

The experiments of Andrews were concerned with the thermodynamic properties of bulk fluids. A new experimental phenomenon which excited interest at the beginning of the 20th century was critical opalescence, in which a colourless transparent fluid suddenly becomes opaque in a narrow region of temperatures near T_c; the onset of opacity is accompanied by dramatic changes of colour. It was clear that a considerable increase in light scattering takes place near the critical point, and this was explained by Smoluchowski and Einstein as arising from large fluctuations in density as follows.

In 1910 Einstein showed how Boltzmann's formula relating the probability of any state of a system to its entropy could be used to calculate fluctuations of thermodynamic quantities. For the density he derived the formula

$$\langle \Delta \rho^2 \rangle = \frac{\rho^2 kT}{V} K_T, \tag{6}$$

where K_T is the isothermal compressibility,

$$K_T = \frac{1}{\rho}\left(\frac{\partial \rho}{\partial p}\right)_T = -\frac{1}{V}\left(\frac{\partial V}{\partial p}\right). \tag{7}$$

Assuming that a change in density is accompanied by a change of refractive index according to the Lorentz-Lorenz law

$$\frac{\varepsilon - 1}{\varepsilon + 2} = A\rho \tag{8}$$

and that scattering takes place randomly, Einstein came to the conclusion that the scattering of light of wavelength λ should be proportional to K_T/λ^4. Since van der Waals' equation leads to an infinite value of K_T at T_c, this seemed to provide a satisfactory explanation of critical opalescence.

It was remarkably perceptive of Ornstein and Zernike [14] to note an inconsistency in the treatment arising from the assumption that the fluctuations in all elements of volume are independent of one another. In fact they pointed out that there must be a correlation between different elements which increases indefinitely in range as the critical point is approached.

To treat this correlation Ornstein and Zernike introduced a basic new function afterwards called the "pair distribution function" which has played a central role in the theory of liquids ever since. Let $\nu(d\mathbf{r})$ be a random variable representing the number of molecules in a volume $d\mathbf{r}$ centered at $\mathbf{r}$. Since $d\mathbf{r}$ is very small, the probability of occupation by more than one particle being of order $d\mathbf{r}^2$ can be neglected. Hence we can write:

$$\langle \nu(d\mathbf{r}) \rangle = n_1(\mathbf{r}) \, d\mathbf{r}, \tag{9}$$

where $n_1(\mathbf{r})$ is the density which they took to be a constant ρ. Similarly for the correlation between particles at points $\mathbf{r}_1, \mathbf{r}_2$,

$$\langle \nu(d\mathbf{r}_1) \nu(d\mathbf{r}_2) \rangle = n_2(\mathbf{r}_1, \mathbf{r}_2) \, d\mathbf{r}_1 \, d\mathbf{r}_2, \tag{10}$$

and for a homogeneous isotropic fluid this is of the form $n_2(r)$ ($r = |\mathbf{r}_1 - \mathbf{r}_2|$).

Ornstein and Zernike introduced the function

$$g(r) = n_2(r)/\rho^2, \tag{11}$$

which approaches 1 at large distances where the correlation becomes negligible. They used the fluctuation relation (6) to derive a fundamental identity between K_T and $g(r)$. But they also obtained an integral equation for $g(r)$ which they were able to solve explicitly.

The basic physical idea behind Ornstein and Zernike's treatment was to differentiate between the direct influence of molecular interactions which should be short ranged and was represented by a function $f(r)$, and the correlation between densities, represented by $g(r)$ above, which should become long ranged as the critical temperature is approached. The integral equation relates $g(r)$ and $f(r)$.

The original paper of Ornstein and Zernike makes difficult reading, and certain aspects of their treatment are obscure. (The whole subject has been beautifully clarified in a classic review paper by Fisher [15].) But if I were asked to nominate the contribution in the classical period which gave greatest insight into the nature of critical behaviour I would choose that of Ornstein and Zernike.

Their calculations suggested that the correlations fall off asymptotically as

$$\frac{1}{r} \exp - \kappa r, \tag{12}$$

where the value of κ can be determined from van der Waals' equation ($\kappa \sim (T - T_c)^{1/2}$). Their detailed conclusions about light scattering differed from those of Einstein; near the critical temperature the λ^{-4} Rayleigh dependence on wavelength ceases to be valid, and there is a "whitening" of the scattered light. By temperature T_c the wavelength dependence has become of the form λ^{-2}.

3.4. Critical behaviour of binary alloys

The early years of the 20th century saw the development of X-ray diffraction as a powerful tool in the investigation of crystal structures, and compounds such as NaCl were found to have a regular ordered structure. However the ionic bond between Na and Cl is so strong that no significant disordering occurs when the temperature is raised; the crystal melts before it disorders.

*In 1919 Tamman [16] suggested that similar ordering might occur in metallic alloys, and this was demonstrated experimentally a few years later [17] by the existence of superlattice lines in the X-ray diffraction pattern of copper–gold alloys. But for such systems there were soon indications that significant disordering takes place with increase in temperature, and that this is accompanied by an anomalous specific heat.

The mathematical description of the disordering process is usually associated with the names of Bragg and Williams. In their classic paper [18] in 1934 they introduced a parameter S to characterize the *degree of order*, and used Boltzmann's principle to calculate its behaviour as a function of temperature. They found a pattern closely analogous to the Weiss theory of ferromagnetism, with S falling rapidly to zero at a critical temperature T_c, and remaining zero for $T > T_c$. In fact for a binary alloy with equal concentrations of constituents the equation derived for S was

$$S = \tanh(ST_c/T), \tag{13}$$

which is of the same form as eq. (4) with $H = 0$ and $\tanh x$ replacing the Langevin function $L(x)$.

In 1935 Bragg and Williams wrote a second paper [19] elaborating their ideas. Because of the analogy with ferromagnetism they called T_c the Curie temperature of the alloy; S was now termed long-distance order to differentiate it from another parameter which had been introduced by Bethe to characterize the short-range order which persists

---

*I am indebted to Dr. L. Muldawer of Temple University for a detailed account of the historical background.

above T_c. But they also apologized in this paper for having ignored the work of other investigators, notably Gorsky [20], Borelius [21] and Dehlinger [22], who had been thinking along similar lines.

Some years ago shortly before Sir Lawrence Bragg's death, I had an opportunity to speak to him about the background to the above two papers. Sir Lawrence told me that in 1933 he had given a seminar in Manchester describing qualitatively how he thought ordering takes place in binary alloys. E. J. Williams was in the audience, and at the end of the seminar he presented Sir Lawrence with pencilled notes on a sheet of paper, which, he claimed, gave numerical substance to the qualitative ideas. Sir Lawrence was naturally impressed, but suggested that if the mathematics was really so simple, someone must surely have done it before. But no one in the audience knew of any such calculations in the literature, and Bragg and Williams proceeded to write a paper which was presented to the Royal Society.

A few days after the corrected proofs of the paper had been returned to the Royal Society, Sir Lawrence on clearing up his desk was shocked to find a preprint by Borelius developing ideas very similar to those which he had just sent off. He told me that he had been very embarrassed in subsequent years to find that all the credit for the development had been given to him and Williams; he would be pleased if a more correct perspective could be introduced.

Sir Lawrence added that the reason for their obtaining the credit was probably because the Bragg–Williams papers described the ideas more clearly than those of any of the other investigators; anyone looking at the literature today will readily confirm that this is so. The important new concept of long-range order is clearly described in the second paper [19]; and a significant distinction is drawn between long-range order and long-range forces. It is stated clearly that short-range forces can give rise to long-range order, a conclusion exactly parallel to that of Ornstein and Zernicke for fluids described above of which Bragg and Williams were apparently unaware.

3.5. Universality of second order transitions (Landau)

It was Landau [25] who first attempted to provide a unified description for all second order transitions. In addition to the phenomena described above, experimental evidence was accumulating about specific heat anomalies in liquid helium, ammonium chloride and a number of other substances which were termed λ-point transitions, and the superconducting transition was also in this category. Ehrenfest had introduced a thermodynamic classification of higher order transitions [23], but there were difficulties associated with his treatment [24].

Landau generalized the ideas introduced in the theory of alloys to all systems manifesting λ-point transitions. He suggested that for every such system one must identify an order parameter analogous to the long-range order in an alloy, which would be zero on the high-temperature side of the transition and non-zero on the low temperature side. He emphasized the important role which symmetry plays in phase transitions, and suggested that the important features of behaviour in the vicinity of a λ-point could be determined by expanding the free energy in a power series as a function of the order parameter η. Landau argued that for reasons of symmetry the form of the expansion of the Gibbs function would be:

$$\Phi(p,T,\eta) = \Phi_0(p,T) + A(p,T)\eta^2 + B(p,T)\eta^4 + \dots \qquad (14)$$

and the Curie temperature would correspond to:

$$A(p,T) = 0, \quad B(p,T) > 0. \qquad (15)$$

Above the Curie temperature $A > 0$ and the solution corresponding to a minimum of Φ is $\eta = 0$; this is the phase of higher symmetry. Below the Curie temperature the minimum of Φ corresponds to a non-zero value of η given by

$$\eta^2 = -A/2B, \qquad (16)$$

and this is the phase of lower symmetry.

Landau's theory enabled one to understand why all the systems discussed previously had the same essential pattern of critical behaviour; even though equations of state like eqs. (1) and (4) look very different, they both conform to eq. (15). Critical exponents are the same for all λ-point transitions, and in later terminology the transitions would be described as universal.

The behaviour of typical thermodynamic quantities for a ferromagnet near T_c were as follows ($t = T/T_c - 1$):

Spontaneous magnetization	$M_0 \sim (-t)^{1/2}$	$(t<0)$;
Initial susceptibility	$\chi_0 \sim t^{-1}$	$(t>0)$;
Critical isotherm	$H \sim M^3$	$(t=0)$;
Derivative of susceptibility	$d\chi_0/dH \sim t^{-4}$	$(t>0)$;
Specific Heat	$C_H = \begin{cases} C_- \\ C_+ \end{cases}$	$(t \to 0_-)$ $(t \to 0_+)$.

The transcription to appropriate thermodynamic variables can readily be made for other systems.

The correlation behaviour in classical systems could be dealt with by arguments similar to those of Ornstein and Zernike, and in the critical region followed the universal pattern (12).

4. The Onsager revolution

An extremely simple atomic model of ferromagnetism [26] was suggested by Lenz to his student Ising in 1925. Each spin in a lattice can orient parallel or antiparallel to an external field. Does the resultant lattice show a non-zero spontaneous magnetization when the field is reduced to zero? Ising was able to solve the problem only in one dimension, where there is no spontaneous magnetization at any non-zero temperature. In 1928 Heisenberg introduced his vector-coupled model which seemed to have a sounder quantum mechanical basis, and this model commanded the attention of most physicists in the field. But the Ising model continued to attract the interest of theoreticians.

In 1936 Peierls showed [27] that the two-dimensional Ising model did indeed have a spontaneous magnetization. In 1944 Onsager published [28] an exact solution of the partition function of this model for the simple quadratic lattice in zero-field. The result was a shattering blow to classical theory. The specific heat was not discontinuous but logarithmically infinite (this had been conjectured previously by Kramers and Wannier [29]). But more important, the partition function was non-analytic at T_c so that an expansion of the type used by Landau was completely invalid.

In a subsequent paper with Kauffman [30], Onsager calculated the ordering, and this did not fit in with the results of the Ornstein–Zernike treatment; finally he was able to calculate the spontaneous magnetization (later calculated independently by Yang [30a]) and this was of the form $(-t)^{1/8}$ in the critical region, very different from $(-t)^{1/2}$ of the classical Weiss theory.

Experimental evidence on critical behaviour also began to accumulate in conflict with classical predictions. Annake Sengers has pointed out [31] that such evidence for fluids was clearly recognized by Verschaffelt [32] around 1900, but his work did not attract the attention it deserved. No direct comparison of Onsager's results with experiment was possible since his treatment applied only to a two-dimensional system. It was most important to obtain theoretical results for three-dimensional systems. But the exact techniques of Onsager were specific to two dimensions.

During the two decades which followed the publication of Onsager's solution, the critical properties of model systems had to be established on an individual basis – each model and each property entailed a separate calculation. The most useful tool turned out to be the generation of lengthy perturbation series expansions at high and low temperatures, and the present writer and his research group at King's College London, were involved in this development.

There were a number of novel features in the series approach [33]. The actual generation of terms required a close familiarity with graph theory and the programming of sophisticated graphical enumeration problems on computers. Then there were problems of interpretation, and here the Onsager solution served as a valuable guide; it was usual to assume branch point singularities, and when the terms were consistent in sign, the asymptotic behaviour of the coefficients could be related directly to critical points, exponents and amplitudes. But in many important cases, particularly for three-dimensional models, the coefficients were not consistent in sign, and spurious unphysical singularities masked the true critical behaviour. Baker's application of the Padé Approximant, a piece of mathematics which had lain dormant since the end of the 19th century, led to remarkable progress, and sparked off similar applications in many other fields concerned with perturbation expansions [34].

Gradually a body of reliable information was assembled on the critical properties of different theoretical models. At the same time the emergence of new experimental techniques, and the development of new magnetic materials provided experimental data on critical behaviour of much greater accuracy than had been available previously. It was gratifying to find good agreement between calculation and experiment, even though it was well understood that the theoretical models were simple and crude, and did not adequately map the true physical interactions. It seemed as if critical behaviour was insensitive to the details of the interaction mechanism.

As the available data increased certain regularities were obtained empirically; critical behaviour depended very significantly on dimension, but rather little on lattice structure in a given dimension; critical exponents varied from the Ising to the Heisenberg model but seemed to be independent of spin value for a given model.

Further exact calculations were also useful in filling in the background. Extensions of Onsager's work to other two-dimensional lattices gave identical results for critical exponents, but small differences in critical amplitudes. A new method of calculation for forces of very long range introduced by Kac, Uhlenbeck and their collaborators [35] gave precisely the results of the classical theory, a van der Waals type of equation for fluids, the Weiss equation of state (4) for ferromagnets, and the Bragg–Williams relation (13) for long-range order in alloys. Classical theory was thus partially reinstated—it is a valid theory for long-range forces, but does not accord with experiment because intermolecular forces in nature are usually short range.

One additional piece of exact information was available, but did not seem to fit anywhere in the picture. In 1952 Berlin and Kac [36] considered an Ising model in which the spins could have any value subject to the condition that the sum of their squares remained equal to N. They were able to provide an exact solution for this, the "spherical model". But the interaction seemed so remote from physical reality that I did not include it in a comprehensive review on critical phenomena [37] which I published in 1960. Nevertheless, as we shall find shortly even theoretical models with no possibility of physical realization have played their part in achieving a proper understanding of critical behaviour.

5. Reconciliation

We have referred in our opening remarks to the important role of the conference at N.B.S. in April 1965. The keynote address was given by Uhlenbeck [4], and it is particularly fascinating now to quote the actual words of one of his concluding paragraphs:

"If there is such an universal, but not-classical behaviour, then there must be an universal explanation which means that it should be largely independent of the nature of the forces. The only corner where this can come from is I think the fact that the forces are not long range. The Onsager solution gives I think a strong hint. It may well be so that away from the critical points the classical theories give a good enough description but that they fail close to the critical point where the substance remembers so to say Onsager. I think that to show something like this is the central theoretical problem. One can call it the reconciliation of Onsager with van der Waals."

The first requirement in effecting the reconciliation was a coherent description of the "non-classical universality" analogous to the van der Waals description of classical behaviour. Such a description in the form of a non-classical equation of state emerged within a few months of the above conference. It was suggested independently by three different groups; Widom [38] in the USA who searched for a generalization of van der Waals' equation which could accommodate non-classical exponents; Domb and Hunter [39] in the UK as a result of an analysis of the behaviour of series expansions of higher derivatives with respect to magnetic field at the critical point; and Patashinskii and Pokrovskii [40] in the USSR who considered the behaviour of multiple-correlations near the critical point. The results were tied together neatly by Griffiths [41] in the form:

$$H = M^\delta h(tM^{-1/\beta}) \qquad (18)$$

as the equation of state of a ferromagnet. (For a fluid, H was replaced by $p-p_c$ or $\mu-\mu_c$ and M by $\rho-\rho_c$.) Here β and δ are two parameters which determine all the exponents, and $h(x)$ is an analytic function. Classical theory corresponds to $\delta=3$, $\beta=\frac{1}{2}$, $h(x)$ linear. Non-classical results could be accommodated with different values of δ, β and a different function $h(x)$.

The two characteristic features of eq. (18) are firstly that critical exponents for any system are determined by two parameters; and secondly that critical data satisfy a "scaling" relation, i.e. if $HM^{-\delta}$ is plotted against $tM^{-1/\beta}$, the two-dimensional data will all fall on a single curve $h(x)$. These two predictions were subjected to experimental tests for a large number of systems both fluids and magnets, and were found to be well satisfied by experimental data.

But eq. (18) says nothing about possible values of β, δ and $h(x)$. We have already mentioned empirical information accumulating about regularities in the pattern of critical exponents. These were collected together in the hypothesis of universality put forward by Kadanoff [42] in 1970 (an analogous smoothness postulate was advanced independently by Griffiths [43]). This non-classical universality is more sophisticated than had been envisaged by Uhlenbeck; different classes are defined by the space-dimension d, and the spin dimension* n, and within a given class critical behaviour is universal. If a third parameter σ is introduced to take account of the range of the intermolecular forces, both classical and non-classical behaviour can be taken into account in a wider pattern of universality. Once d, n and σ have been specified the exponents δ, β and the function $h(x)$ are determined, and the behaviour is smooth and universal within the class. Changes in critical exponents with accompanying discontinuities occur in the cross-over between one universality class and another (e.g. in the transition from a two-

---

*The letter D was introduced originally to denote spin dimension by Stanley [44]. But Wilson subsequently used n, and this has now become widely accepted.

dimensional to a three-dimensional system by the introduction of a new interaction).

Two highly theoretical ideas were now pursued which, although remote from physical reality, were subsequently to make an important contribution. Stanley [44] investigated the behaviour of Ising models as a function of spin dimension n. An increase of n gives greater freedom to the spin and corresponds to a decrease of "co-operative strength". He found that as $n \to \infty$ the spherical model solution [36] is retrieved. Hence this model has a place in the general framework.

Joyce [45] had already examined the behaviour of the spherical model with varying dimensions and ranges of force. He found that classical behaviour results with long-range forces in low space-dimensions d, but also with short-range forces if $d \geqslant 4$ (with possible logarithmic correction terms). Hence one could reasonably conclude that the same result will also hold for finite n since the co-operation is then stronger. This conclusion was demonstrated more convincingly by diagrammatic methods [46], which give precise results when $d \geqslant 4$.

We have noted that the behaviour of critical correlations as determined from model calculations differs from the classical results of Ornstein and Zernicke [46a]. But the non-classical correlation exponents and functions fit in with the universality class picture above. Are the correlation exponents determined by δ and β of eq. (18)? In regard to this question the treatment of Patashinskii and Pokrovskii [40] went further than the others mentioned above and suggested the relation:

$$d\nu = \beta(\delta + 1), \tag{19}$$

where ν is the non-classical exponent which characterizes the range of correlation ($K \sim t^\nu$) in (12). But the argument was put more cogently by Kadanoff [47] in a key paper in which he tried to find a theoretical basis for the scaling properties which had been discovered in critical behaviour.

Kadanoff noted that near T_c the coherence length ξ becomes very large, and hence it is possible to find a length L large compared with the lattice spacing but small compared with ξ. He then considered replacing the interaction of individual spins by the interaction of blocks of L^d spins. One might perhaps expect that in each block the spins would be nearly all up or nearly all down, and the original Ising model of spin σ and interaction J could then be replaced by a new model with block spin $\tilde{\sigma}$ and interaction $\tilde{J}$. If the new block spin model was effectively the same as the original model, the free energies would be related by

$$f(t, h) = L^d f(\tilde{t}, \tilde{h}) \qquad (h = \beta H). \tag{20}$$

But how would $\tilde{h}$ and $\tilde{t}$ be related to h and t? Kadanoff suggested that

$$\tilde{h}=L^x h, \quad \tilde{t}=L^y t;$$

it is then easy to show that all the exponents (including ν) can be expressed in terms of x and y, and that both eqs. (18) and (19) result.

Relation (19) is satisfied by the Ising model in two dimensions, yet numerical calculations for the three-dimensional model show a small but persistent discrepancy. The discrepancy has still not been resolved satisfactorily to the present day. Also there are several features of Kadanoff's argument which do not stand up to critical examination. But his ideas stimulated the next major theoretical advance which completed the reconciliation sought by Uhlenbeck.

6. Renormalization group: respectability

In a talk [48] on "The Curie Point" given in honour of Uhlenbeck's 70th birthday in October 1969, I summarized the current situation as follows: "From the point of view of practical calculations of the behaviour near the Curie point we are well on the way to satisfying the needs of the experimentalist for most models of interest. Unifying features have been discovered which suggest that the critical behaviour of a large variety of theoretical models can be described by a simple type of equation of state. But the rigorous mathematical theory needed to make the above developments respectable is still lacking."

The application of the renormalization group which Wilson introduced so effectively a year or two later was not rigorous by mathematical standards, and the search for a proper mathematical description of his ideas still continues. But he was able to account convincingly for the striking empirical discoveries of the previous section, and the "respectability" which his work achieved is adequate for most theoreticians in the field.

A suggestion that the RG could be relevant to critical phenomena was made [49] at the summer school which Mel Green organized in 1970 in Varenna. However, no precise indication was forthcoming as to how it should be used.

The RG had been developed nearly 20 years earlier [50] in connection with field theory, but it had no apparent practical consequences, and had not been taken very seriously. Wilson saw that the theory was capable of providing an understanding of universality and a framework for the detailed calculation of critical behaviour. He acknowledged his debt to Kadanoff for setting his thoughts in the right direction, but he converted Kadanoff's vague block spin mappings into a precise tool for calculation.

We can give only a brief indication of the flood of new ideas and projects which stemmed from Wilson's work. For more details we refer to reviews and texts [51]; we particularly recommend anyone interested in the development of scientific concepts to read Wilson's first two papers [52] and his first definitive review [53], where he describes the evolution of his own ideas, explains the philosophy of the RG approach and shows with simple illustrated examples how it can be applied exactly to simple artificial models and approximately to real systems.

The RG is a transformation of the original Hamiltonian with N degrees of freedom into a new Hamiltonian H',

$$H' = \mathbf{R}[H], \tag{21}$$

with a reduced number of degrees of freedom N'. The transformation is chosen so that the partition function is preserved,

$$Z_{N'}(H') = Z_N(H). \tag{22}$$

A wide choice of operators $\mathbf{R}$ is possible satisfying these conditions. The block spin device of Kadanoff can be used by eliminating the internal interactions in the block in some suitable manner; or half the spins can be eliminated by partial summation in a decimation procedure. Choices of this type are usually referred to as "real-space renormalization". Alternatively, the transformation can be carried out in momentum space, and high momentum variables corresponding to short range fluctuations can be integrated out.

The most important feature of the transformation is that it can be iterated,

$$H' = \mathbf{R}[H], \quad H'' = \mathbf{R}[H'], \ldots \tag{23}$$

and the universality properties follow from the limiting behaviour of such iterative processes. Elementary iterative processes of the form

$$x_{n+1} = f(x_n) \tag{24}$$

are well known in numerical analysis [54] as a useful method of approximating to a root of the equation

$$x = f(x). \tag{25}$$

They have the advantage that the final solution is within wide limits independent of the starting point. A root of eq. (25) is called a fixed point, x^*, of the transformation (24). If eq. (25) has a number of different roots, each of these fixed points will have a characteristic "range of attraction", i.e. range of starting points within which convergence will occur to the particular root.

Equation (23) involves the generalization of the above process (24) to a multi-dimensional space

$$\mathbf{K}_{n+1} = f(\mathbf{K}_n), \tag{26}$$

corresponding to a number of parameters in the Hamiltonian, but many of the general characteristics remain the same, including the possibility of convergence to fixed points $\mathbf{K}^*$.

We can now see qualitatively how the above properties provide a basis for the explanation of universality classes. Each fixed point $\mathbf{K}$ corresponds to one universality class, and Hamiltonians with a wide variety of parameters all converge to the same fixed point.

Critical behaviour is determined by the behaviour of eq. (26) near $\mathbf{K}^*$, and if a linear expansion is undertaken near this point, a linear operator can be derived whose eigenvalues are related to critical exponents. A scaling equation of state equivalent to eq. (18) can readily be deduced. The non-analytic critical behaviour arises from analytic functions $f(\mathbf{K})$ as a result of the limiting procedure as $n \to \infty$.

Although Kadanoff's idea led naturally to real-space renormalization, the major progress in actual calculations was achieved by applying perturbation methods which had been developed in quantum field theory to momentum-space renormalization. To make this application possible Wilson went back to Landau's ideas, not in the original macroscopic form of eq. (14), but in the later microscopic Ginzburg–Landau formulation [55]. This corresponds to a continuous spin with local Hamiltonian

$$H = (\nabla s)^2 + Rs^2 + Us^4 - Hs \tag{27}$$

where R and U are analytic functions of T. Few of the older workers in the field (remembering the failure of the macroscopic Landau theory) would have been prepared to believe that an expansion of the form (27) was sufficient to lead to non-classical results of Ising model type. But Wilson was able to indicate that any higher order terms in s^6, s^8 etc. make no significant contribution to critical behaviour (the coefficients are irrelevant parameters).

The parameter chosen [56] for the expansion of critical exponents and scaling functions was

$$\varepsilon = 4 - d, \tag{28}$$

where d is the space dimension, following earlier ideas by Fisher and Gaunt [57] which had seemed remote from reality when they were published. $\varepsilon = 0$ corresponds to classical theory.

Only a few terms of the ε expansion were available, it turns out to be an asymptotic expansion, and the value $\varepsilon = 1$ corresponding to 3-dimensional systems is not small. But the results were in satisfactory agreement with those of series expansions, and later work has derived more terms and devised improved methods of summation. As mentioned above, there are some small discrepancies with the results of series expansions, but hopefully these should be ironed out in the course of time.

Real-space renormalization methods worked well in two dimensions and were able to reproduce the Onsager solution to a high degree of accuracy. More recently methods have been devised by which they can be applied in three dimensions, and they are now contributing accurate numerical information.

Wilson [53] has described his own work as a second stage of the Landau theory, and has shown precisely where the original Landau theory breaks down [58]. The microscopic Hamiltonian (27) is correct. But the macroscopic form of the free energy obtained by averaging over a region

$$F = \int d^3x \{ [\nabla M(x)]^2 + RM^2(x) + UM^4(x) - B(x)M(x) \} \tag{29}$$

is incorrect since it ignores the variation of R and U with L, the size of the region, which is non-analytic (as had already been suggested previously by Kadanoff [47]).

The RG theory did not stop at explaining the empirical results of the previous section. Like all good scientific theories it suggested new avenues of exploration, and was able to deal with problems for which previous methods had not been successful. Perhaps the most notable are tricritical and multicritical points [59], systems with dipolar forces [60] and corrections to the scaling equation of state, eq. (18) [61,62].

7. Conclusion

We have endeavoured in this article to provide a summary of the main stream of development of critical phenomena as a model illustration of scientific method in practice. Particular emphasis has been laid on *concepts* which led to substantial progress. We have not attempted to provide an account of all contributions which have led to significant progress, and there are many independent developments which would need to be described in detail to provide a proper history. We shall briefly mention a few of them.

Exact solutions have played a role of special importance in critical phenomena. We have seen that the modern era was initiated by Onsager's exact solution, which revealed the defects of the classical theories and served as a test for new techniques of calculation. For about 20 years after Onsager's work there were no new solutions for realistic interactions which differed in a substantial way from that of Onsager. Then in 1967 Lieb produced a variety of new solutions for two-dimensional ferroelectric models [63]. The critical behaviour followed a completely different pattern from the Ising model, and it seemed that new ground was being broken.

Lieb was followed by Baxter whose solution of another ferroelectric model, the eight vertex model [64], gave rise to exponents which vary continuously with the strength of the interaction. This was a major challenge to the idea of universality, since the model is equivalent to an Ising model with a two and four spin interaction. The issue was resolved by an analysis by Kadanoff and Wegner [65], who indicated that the model possesses a special symmetry which gives rise to continuously varying exponents; this particular feature would not be expected to arise in normal physical systems.

Baxter subsequently derived a number of other new solutions [66], and as a by-product extended very significantly exact series expansions for two-dimensional models which he was unable to solve exactly [67].

It should be emphasized that although the RG transformation (21) is exact, the process of solution always involves an approximation, e.g. stopping eq. (26) with a finite number of parameters, or expanding as a perturbation series. It is therefore important for a test of the RG to have as many exact solutions as possible.

An interesting point which has led to fruitful ideas is the behaviour of two-dimensional models when $n=2$. Peculiarities in this model were noted some years ago [67a], and their investigation by Kosterlitz and Thouless [68] has suggested a new type of critical phenomenon related to topological order. Experimental work has already confirmed some of the theoretical predictions.

Other problems in condensed matter physics which have been profoundly influenced by progress in critical phenomena are the shape and size of polymer chains in solution and percolation processes. The model of a self-avoiding walk on a lattice [69] had for many years been regarded as a useful representation of a polymer chain with a realistic hard core "excluded volume". It had been noted empirically that the properties of such SAW's paralleled those of the Ising model. The precise connection was revealed by de Gennes [70] who noted that the SAW configurations correspond to a ferromagnetic model with spin dimension n when n is allowed to take the value zero.

A theoretical model which burst into prominence after being ignored for nearly two decades is the Potts model [71]. This was first suggested in 1952 as a generalization of the Ising model having q orientations but only 2 different energies of interaction; the critical point could be located exactly. At the Kyoto Conference on Statistical Mechanics in 1968, Kasteleyn and Fortuin [72] showed that percolation processes associated with random mixtures which have applications to dilute magnets, amorphous semi-conductors and a number of other problems in condensed matter physics correspond to the Potts model with $q=1$! A number of physical systems have been identified for which the Potts model is a reasonable representation, and the difference in symmetry between the n-vector ferromagnetic model and q-component Potts model give rise to important differences in critical behaviour. The literature on percolation and the Potts model has escalated sharply in the past few years.

We should like now to return to the theme of sect. 2, the relation between research in different branches of physics. We have seen how the RG, an idea introduced in the area which investigates the basic laws of nature, has been instrumental in re-vitalizing a branch of condensed matter physics which tries to understand critical effects which arise from the co-operation of a large number of molecules. This application to critical phenomena has given increased insight into the RG itself, and has already generated feed-back into the fundamental area from which it was drawn [73].

We conclude with a quotation from Medawar [74] on the variety of types who engage in the pursuit of science. "Scientists are people of very dissimilar temperaments doing different things in very different ways. Among scientists are collectors, classifiers and compulsive tidiers-up; many are detectives by temperament and many are explorers; some are artists and others artisans. There are poet-scientists and philosopher-scientists and even a few mystics." In the illustrative model with which we have been concerned there are examples of the different types listed by Medawar. Their common quality has been the ability to assimilate and build on the work of others, a quality possessed to such a marked degree by Mel Green.

References

[1] L. Van Hove, in: Critical Phenomena in Alloys, Magnets and Superconductors, eds. R. E. Mills, E. Ascher and R. I. Jaffee (McGraw-Hill, 1971).
[2] F. J. Dyson, The Future of Physics, Physics Today (1970) 23.
[3] J. M. Ziman, Physics is Dead: Long Live Physics, Physics Bulletin **25** (1974) 280.
[4] G. E. Uhlenbeck, The Classical Theories of Critical Phenomena, in: N.B.S. Misc. Publ. **273**, eds. M. S. Green and J. V. Sengers (N.B.S., Washington, D.C., 1965) p. 3.
[5] L. P. Kadanoff, in: Proceedings of the Enrico Fermi Summer School on Critical Phenomena, ed. M. S. Green (Academic Press, New York, 1971) p. 100.
[6] K. G. Wilson, Phys. Rev. B **4** (1971) 3174, 3184.
[7] J. S. Rowlinson, Thomas Andrews and the Critical Point, Nature **224** (1969) 541.
[8] W. Gilbert, De Magnete [translated by P. F. Mottelay, New York (1893) and for the Gilbert Club London, 1900] 1600 p. 66.
[9] J. Hopkinson, Phil. Trans. Roy. Soc. (1889) p. 443.
[10] P. Curie, Ann. Chim. Phys. **5** (1895) 289; J. de Phys. **4** (1895) 263.
[11] P. Weiss, J. de Phys. **6** (1907) 661.
[12] P. Weiss and H. Kamerlingh Onnes, J. de Phys. **9** (1910) 555.
[13] F. Cernuschi and H. Eyring, J. Chem. Phys. **7** (1939) 547.
[14] L. S. Ornstein and F. Zernike, Proc. Akad. Sci. Amsterdam **17** (1914) 793; F. Zernike, Proc. Akad. Sci. Amsterdam **18** (1916) 1520.
[15] M. E. Fisher, J. Math. Phys. **5** (1964) 944.
[16] G. Tammann, Z. Anorg. Chem. **107** (1919) 1.
[17] E. C. Bain, Trans. AIME **68** (1923) 625; C. H. Johanson and J. O. Linde, Ann. de Phys. **78** (1925) 439.
[18] W. L. Bragg and E. J. Williams, Proc. Roy. Soc. A **145** (1934) 699.
[19] W. L. Bragg and E. J. Williams, Proc. Roy. Soc. A **151** (1935) 540.
[20] W. Gorsky, Z. Phys. **50** (1928) 84.
[21] G. Borelius, Ann. der Phys. **20** (1934) 57.
[22] U. Dehlinger, Z. Phys. **64** (1930) 359; **74** (1932) 267; **83** (1933) 832.
[23] P. Ehrenfest, Proc. Kon. Akad. Wetenschap, Amsterdam **36** (1933) 147.
[24] A. B. Pippard, Classical Thermodynamics, Cambridge (1957) ch. 9.
[25] L. D. Landau, Phys. Z. Sovietunion **11** (1937) 26, 545; L. D. Landau and E. M. Lifshitz, Statistical Physics (Pergamon, London, 1959).
[26] E. Ising, Z. Phys. **31** (1925) 253.
[27] R. E. Peierls, Proc. Camb. Phil. Soc. **32** (1936) 477.
[28] L. Onsager, Phys. Rev. **65** (1944) 117.
[29] H. A. Kramers and G. H. Wannier, Phys. Rev. **60** (1941) 252, 263.
[30] B. Kauffman and L. Onsager, Phys. Rev. **76** (1949) 1244.

[30a] C. N. Yang, Phys. Rev. **85** (1952) 808.
[31] J. M. H. Levelt Sengers, Proc. van der Waals Centennial Conference on Statistical Mechanics (North Holland, Amsterdam, 1974) p. 73.
[32] J. Verschaffelt, Proc. Kon. Akad. Sci. Amsterdam **2** (1900) 588.
[33] Phase Transitions and Critical Phenomena, Vol. 3, eds. C. Domb and M. S. Green (Academic Press, London, New York, 1974).
[34] The Padé Approximant in Theoretical Physics, eds. G. A. Baker and J. L. Gammel (Academic Press, London, New York, 1970).
[35] See e.g. P. C. Hemmer and J. L. Lebowitz, in: Phase Transitions and Critical Phenomena, Vol. 5b, eds. C. Domb and M. S. Green (Academic Press, London, New York, 1976).
[36] T. H. Berlin and M. Kac, Phys. Rev. **86** (1952) 821.
[37] C. Domb, Advances in Physics **9** (1960) 149, 245.
[38] B. Widom, J. Chem. Phys. **43** (1965) 3892.
[39] C. Domb and D. L. Hunter, Proc. Phys. Soc. **86** (1965) 1147.
[40] A. Z. Patashinskii and V. L. Pokrovskii, Zh. eksp. teor. Fiz. **50** (1966) 439; Sov. Phys. JETP **23**, 292.
[41] R. B. Griffiths, Phys. Rev. **158** (1967) 176.
[42] L. P. Kadanoff, in: Proceedings of the Enrico Fermi Summer School on Critical Phenomena, ed. M. S. Green (Academic Press, New York, London, 1971) p. 103.
[43] R. B. Griffiths, Phys. Rev. Lett. **24** (1970) 1479.
[44] H. E. Stanley, Phys. Rev. **176** (1968) 718.
[45] G. S. Joyce, Phys. Rev. **146** (1966) 349.
[46] A. I. Larkin and D. E. Khmelnitskii, Sov. Phys. JETP **29** (1969) 1123.
[46a] See also M. E. Fischer and R. V. Burford, Phys. Rev. **156** (1967) 583.
[47] L. P. Kadanoff, Physics **2** (1966) 263.
[48] C. Domb, in: Statistical Mechanics at the Turn of the Decade, ed. E. G. D. Cohen (Marcel Dekker, New York, 1971) p. 81.
[49] F. de Pasquale, C. di Castro and G. Jona-Lasinio in ref. [5] p. 123.
[50] E. C. G. Stueckelberg and A. Petermann, Halv. Phys. Acta **26** (1953) 499; M. Gell-Mann and F. E. Low, Phys. Rev. **95** (1954) 1300.
[51] M. E. Fisher, Rev. Mod. Phys. **46** (1974) 597;
K. G. Wilson, Rev. Mod. Phys. **47** (1975) 773;
Phase Transitions and Critical Phenomena, Vol. 6, eds. C. Domb and M. S. Green (Academic Press, London, New York, 1976);
S. K. Ma, Modern Theory of Critical Phenomena (W. A. Benjamin, Reading, Mass. 1976);
P. Pfeuty and G. Toulouse, Introduction to the Renormalization Group and to Critical Phenomena (Wiley, London, New York, 1977).
[52] K. G. Wilson, Phys. Rev. **B4** (1971) 3174, 3184.
[53] K. G. Wilson and J. Kogut, Phys. Rep. **12C** (1974) 77.
[54] D. R. Hartree, Numerical Analysis (Oxford, 1958) p. 211.
[55] V. L. Ginzburg and L. D. Landau, JETP **20** (1950) 1064.
[56] K. G. Wilson and M. E. Fisher, Phys. Rev. Letts. **28** (1972) 240.
[57] M. E. Fisher and D. S. Gaunt, Phys. Rev. **133A** (1964) 224.
[58] K. G. Wilson, Physica **73** (1974) 119.
[59] M. E. Fisher, Bakerian Lecture (1979), Proc. Roy. Soc. A. (in press).
[60] E.g. A. Aharony, in: Phase Transitions and Critical Phenomena, Vol. 6, ref. [51].
[61] J. C. Le Guillou and J. Zinn Justin, Phys. Rev. Lett. **39**, (1977) 95.

[62] E.g. E. Brezin, J. C. Le Guillou and J. Zinn-Justin, in: Phase Transitions and Critical Phenomena, Vol. **6**, ref. [51].
[63] E. Lieb and F. Y. Wu, in: Phase Transitions and Critical Phenomena, Vol. 1.
[64] R. J. Baxter, Phys. Rev. Lett. **26** (1972) 832; Ann. Phys. **70** (1972) 193.
[65] L. P. Kadanoff and F. J. Wegner, Phys. Rev. B **4** (1971) 3989.
[66] R. J. Baxter, Phil. Trans. Roy. Soc. **289** (1978) 315; see also Hard Hexagons, Exact solution, J. Phys. A. (in press).
[67] R. J. Baxter and I. G. Enting, J. Stat. Phys. **21** (1979) 103.
[67a] H. E. Stanley and T. A. Kaplan, Phys. Rev. Letts. **17** (1966) 913.
[68] J. M. Kosterlitz and D. J. Thouless, J. Phys. C **6** (1973) 1181.
[69] C. Domb, Adv. Chem. Phys. **15** (1969) 229.
[70] P. G. de Gennes, Phys. Lett. **38A** (1972) 339.
[71] R. B. Potts, Proc. Camb. Phil. Soc. **48** (1952) 106.
[72] P. W. Kasteleyn and C. M. Fortuin, J. Phys. Soc. Japan Suppl. **26** (1969) 11.
[73] See e.g. L. P. Kadanoff, Rev. Mod. Phys. **49** (1977) 267.
[74] P. B. Medawar, The Art of the Soluble (Methuen, London, 1957) p. 132.

CHAPTER 12

Coarse-grained Helmholtz Free Energy Functional

K. KAWASAKI and T. IMAEDA

Department of Physics, Kyushu University
Fukuoka 812, Japan

J. D. GUNTON

Department of Physics, Temple University
Philadelphia, Pa. 19122, U.S.A.

© *North-Holland Publishing Company 1981*

Perspectives in Statistical Physics
Ed. H. J. Raveché

Contents

1. Introduction — 203
2. Renormalization group equations and the Helmholtz functional — 206
3. An approximate analytical solution for the coarse grained free energy — 213
4. Discussion and concluding remarks — 218
Appendix — 220
References — 223

1. Introduction

There are many important problems in physics related to two phase coexistence in which the so-called "coarse-grained" Helmholtz free energy functional plays a fundamental role [1]. These include the kinetics of first order phase transitions such as the dynamical behavior of metastable and unstable states, which involve nucleation [2–6] and spinodal decomposition [7–11] processes, respectively. Examples of such phenomena include alloys, fluids, binary fluids and glasses. In the case of spinodal decomposition the functional derivative of this Helmholtz functional is assumed, for example, to be the driving force which produces the ultimate two phase equilibrium from an initial quenched, unstable one phase state. Other problems in which this Helmholtz functional appears include the surface tension and interface profile of gas–liquid, liquid–liquid interfaces [12–15] and wetting [16–22]. In many applications in these problems one minimizes suitably chosen free energy functionals. In this case the proper free energy functionals to be used should be the so-called fully renormalized functionals such as the Γ functional used by Brezin et al. [23], where all the fluctuation effects are already taken into account. These should be distinguished from the partially renormalized free energy functionals that enter the dynamical problems mentioned earlier, a point often stressed by Langer [5]. It must be emphasized, however, that although fully renormalized free energy functionals have been calculated for liquids [24, 25], the partially renormalized one is in general never explicitly calculated. Instead, quite often some phenomenological form is assumed for this coarse grained free energy such as the Ginzburg–Landau "Hamiltonian". A second point to be noted is that in the two phase region the role of the cut-off (or inverse coarse-graining cell size) is in a sense both more important and more subtle than in the one phase region above the critical point, where the functional has been extensively studied in critical phenomena [23]. In order to make this point more precise we limit ourselves to an initial free energy functional which is of the Ginzburg–Landau form $H(\{\sigma\}, \Lambda_0)$ for a local scalar order parameter with Fourier components $\{\sigma_k; k \leq \Lambda_0\}$ where Λ_0 is the initial cut-off. We choose Λ_0 to satisfy

$a_0^{-1} \gg \Lambda_0 \gg \xi^{-1}$ where a_0 is some microscopic length and ξ is the correlation length. The condition restricts us to the close vicinity of a critical point. We assume that $H(\{\sigma\}, \Lambda_0)$ has been obtained from some microscopic model by an averaging procedure involving very short wavelength fluctuations. The averaging here is, roughly speaking, over a "finite", small cell size and involves relatively small numbers of degrees of freedom.

Our main concern in this work is to study the effects of further coarse graining involving fluctuations with wavelength reaching ξ and beyond. The averaging involves a great number of degrees of freedom and in general is a far more difficult task. For the purpose of carrying it out we employ a renormalization group method in which one further coarse grains (reduces the cut-off) by integrating out some of the short-wavelength components [26–29]. To indicate the crucial dependence of this free energy functional on the cut-off, consider the standard form for $H(\{\sigma\}, \Lambda_0)$ which consists of a spatial integral of an integrand consisting of two terms, $|\nabla \sigma|^2$ and a free energy density $\tilde{A}(\sigma)$ for a completely uniform state. Then this latter quantity will in general be a nonconvex function of the order parameter (in contrast to the equilibrium free energy) which will have two minima and a maxima for a suitable coarse graining size Λ_0. If one asks what such a size should be for nucleation or spinodal decomposition, say, then on one hand Λ_0^{-1} must be sufficiently large that the continuum approximation makes sense, but on the other hand it cannot be much larger than the correlation length ξ, since otherwise phase separation can occur within a cell. In this case, one would have lost the interesting physics of metastable or unstable states which one wishes to describe, by choosing too large a coarse graining size. One therefore usually chooses Λ_0^{-1} to be proportional to ξ and makes some ansatz as to what the actual Helmholtz functional is for this choice [7]. It is therefore of some interest to carry out the coarse-graining procedure explicitly in order to actually evaluate the functional and in particular to examine the ansatzes which have been made in previous approximation schemes when $\Lambda_0^{-1} \simeq \xi$. The renormalization group obviously provides a natural method of studying the effects of changing the cut-off Λ_0. In order to obtain an explicit realization of the renormalization group approach, we employ the ϵ-expansion (where $\epsilon = 4 - d$ and d is the dimensionality). One word of warning is appropriate here. It is well known that finite order ϵ-expansions cannot correctly explain all aspects of first order phase transitions. For example, it does not, due to its inability to correctly describe finite amplitude fluctuations, yield the essential singularity in the equilibrium free energy which is believed to exist on the coexistence curve. Thus the equation of state which one obtains for the Ising model by this method has no singularity and can be smoothly continued into the "metastable" region, eventually terminating at a "spinodal" point at which the susceptibility diverges. This latter feature, i.e. the existence of a sharp spinodal curve, is also thought to be incorrect for real systems and is an artifact of the approximation scheme. On the other hand, for the problems described above and in particular for $\Lambda_0^{-1} \ll \xi$ one would expect the ϵ-expansion to yield correct results for the Helmholtz functional. One must clearly be careful, however, in the interpretation of the results for coarse graining sizes much larger than ξ, as we indicate later.

The outline of the paper is as follows. In sect. 2 we discuss our basic method, which involves the differential renormalization group equations of Wegner and Houghton [30] as reformulated by Nicoll, Chang and Stanley [26–28] for the Helmholtz free energy functional. We summarize the exact solution to order ϵ of the renormalization group equations for the quadratic and quartic coupling constants which occur in the differential equation for the coarse grained Helmholtz functional. This free energy functional is then solved numerically in the two phase region for various values of the rescaled cut-off, $\Lambda_l = \Lambda_0 e^{-l}$, where e^{-l} is the length rescaling parameter. We find that in general this free energy functional is real and retains its double well structure for $l < l_c(M, T)$ for fixed M and T, where l_c is the value of l for which the renormalized propagator diverges. For $l > l_c(M, T)$ however, the Helmholtz free energy functional becomes complex. We interpret this behavior as due to the intrinsically unstable, small amplitude fluctuations which are responsible for spinodal decomposition. Further discussion of this subtle point is given at the end of sect. 2. In general, the value $l = l^*$ for which $\Lambda_{l^*}^{-1} = \xi$ is less than l_c, so that we conclude that it is reasonable to coarse-grain on a scale of the correlation length, as done for example by Langer, Bar-on and Miller (LBM) [7] in their theory of spinodal decomposition. We are also able within the framework of the ϵ-expansion to determine the functional form of the coarse-grained free energy for $l = l^*$ and calculate terms which are ignored in the LBM theory. In sect. 3 we use some results of Rudnick and Nelson [29] to obtain an approximate analytical solution for the Helmholtz free energy functional valid for $l \leq l^*$. This allows us to calculate to order ϵ the scaling function approximated by LBM by a polynomial in M^2 of order two. Finally, in sect. 4 we summarize our results and in particular make explicit comparisons with the phenomenological forms for the Helmholtz free energy functional assumed by LBM and others.

2. Renormalization group equations and the Helmholtz functional

Our starting point is the differential renormalization group equation of Wegner and Houghton [30] as reformulated by Nicoll, Chang and Stanley [26-28]. Here one deals with the coarse-grained free energy functional $H_l\{\sigma^<, \sigma^>\}$ constructed as follows: Let $H_0\{\sigma^<, \sigma^>\}$ be the initial Hamiltonian with the upper wave number cut-off equal to 1 where $\sigma^< \equiv \{\sigma_k\}$ with $0 \leq k \leq e^{-l}$ and $\sigma^> \equiv \{\sigma_k\}$ with $e^{-l} < k < 1$, σ_k being the Fourier component of the local order parameter $\sigma(r)$. The thermodynamic potential $W_l\{\sigma, h\}$ in the presence of an external field $h(r)$ is defined by:

$$W_l\{\sigma^<, h\} = \ln \int d\{\sigma^>\} \exp\left[-H_0(\sigma^<, \sigma^>) + \int h(r)\sigma^>(r)dr\right]. \tag{2.1}$$

H_l is defined as the Legendre transform [23] of W_l:

$$H_l\{\sigma^<, \sigma^>\} = \int dr\, h(r)\sigma^>(r) - W_l\{\sigma^<, h\}, \tag{2.2}$$

where $\sigma^>$ is given by:

$$\sigma^>(r) = \delta W_l(\sigma^<, h)/\delta h(r). \tag{2.3}$$

The fact that eq. (2.3) contains only short wavelength components follows if we note that eq. (2.1) contains only short wavelength components of $h(r)$. H_l is the coarse-grained free energy where fluctuations with $k > e^{-l}$ are averaged out. In particular, H_∞ is just the Γ function of Brezin et al. [23].

An exact differential RG equation for H_l is obtained in refs. [27] and [28] as:

$$\partial H_l/\partial l = \tfrac{1}{2} Tr\ln\left\{H_l^\ll - H_l^{<>}(H_l^\gg)^{-1} H_l^{><}\right\}. \tag{2.4}$$

Here $H_l^\ll$ etc. are the matrices with elements $\delta^2 H/\delta\sigma_k \delta\sigma_{k'}$. For $H_l^\ll$, $|k| = |k'| = e^{-l}$; for $H_l^\gg$, $|k|, |k'| > e^{-l}$; for $H_l^{<>}$, $|k| = e^{-l}$, $|k'| > e^{-l}$; for $H^{><}$, $|k| > e^{-l}$, $|k'| = e^{-l}$. We now make an ansatz:

$$H_l\{\sigma\} = \int dr\left[\tfrac{1}{2}|\nabla\sigma|^2 + \hat{A}_l(\sigma(r))\right], \tag{2.5}$$

where σ contains both short and long wavelength components. Equation (2.5) is considered to be generally valid in the first order in $\epsilon = 4-d$

where d is the dimensionality of space*. Then eq. (2.4) can be shown to reduce to:

$$\frac{\partial \hat{A}_l(\sigma)}{\partial l} = \frac{K_d}{2} e^{-dl} \ln\left[e^{-2l} + \frac{\partial^2 \hat{A}_l(\sigma)}{\partial \sigma^2} \right], \qquad (2.6)$$

where $K_d = S_d/(2\pi)^d$ with S_d the surface area of a d-dimensional hypersphere. If we introduce the function $A(M)$ by:

$$A(M) = (2/K_d)\hat{A}(\sigma), \quad M = (2/K_d)^{1/2}\sigma, \qquad (2.7)$$

eq. (2.6) reduces to:

$$\frac{\partial A_l(M)}{\partial l} = e^{-dl} \ln\left[e^{-2l} + \frac{\partial^2 A_l(M)}{\partial M^2} \right]. \qquad (2.8)$$

The solution correct up to one-loop order is given by:

$$A_l(M) = \tfrac{1}{2}\tilde{r}_l M^2 + \tfrac{1}{4}\tilde{u}_l M^4 + \int_0^l \mathrm{d}p\, e^{-dp}$$
$$\times \left\{ 2 - \ln g_p(M) - 3\tilde{u}_p g_p(0) M^2 + \tfrac{9}{2}\tilde{u}_p^2 g_p(0) M^4 \right\}, \qquad (2.9)$$

where

$$g_l(M)^{-1} \equiv e^{-2l} + \tilde{r}_l + 3\tilde{u}_l M^2. \qquad (2.10)$$

Here $\tilde{r}_l$ and $\tilde{u}_l$ are related to the usual renormalized parameters of the Wilson Hamiltonian through:

$$\tilde{r}_l = e^{-2l} r_l, \qquad (2.11)$$
$$\tilde{u}_l = e^{-\epsilon l} u_l;$$

r_l and u_l satisfy the RG equations which are, up to $O(\epsilon)$,

$$\dot{r}_l = 2r_l + 6u_l/(1+r_l), \qquad (2.12\mathrm{a})$$
$$\dot{u}_l = \epsilon u_l - 18(u_l/(1+r_l))^2. \qquad (2.12\mathrm{b})$$

---

*One must add a word of warning about this assumption. Some recent studies of the interface order parameter profile (ref. [14]) revealed the inadequacy of eq. (2.5) even in $O(\epsilon)$. This seems to be the case when one is dealing with a spatially inhomogeneous finite order parameter. We hope to investigate this aspect of the problem on another occasion.

The full solution of this equation was obtained by Nicoll, Chang and Stanley [26], [31] and will be described in the appendix. Note that in the power series expansion in M^2, the integral in M^2 in eq. (2.9) contributes M^6 and higher powers. $\tilde{r}_\infty$ and $\tilde{u}_\infty$ remain finite since $A_\infty(M)$ is just the true thermodynamic potential.

In this work, we are particularly interested in the coarse-grained free energy when the cut-off wavelength coincides with ξ:

$$e^{+l^*} = \xi, \quad \Lambda_0 = 1. \tag{2.13}$$

Denoting this free energy by $A^*(M)$ we can now write:

$$A^*(M) = \frac{2}{K_d} \frac{f_0}{a^d} \phi(M/M_s), \tag{2.14}$$

where a gives a measure of the size of a coarse graining cell and is chosen to be:

$$a = \frac{2\pi}{V_d^{1/d}} \xi, \tag{2.15}$$

where V_d is the volume of a d-dimensional unit sphere, f_0 is a dimensionless number given by:

$$f_0 = B^2 r_-^* \xi_0^{2\beta/\nu} V_d^{-1} (2\pi)^d, \tag{2.16}$$

where B, r_-^* and ξ_0 are the amplitudes defined through the following (M_s being the value of the order parameter on the coexistence curve):

$$M_s = (K_d/2)^{1/2} B |t|^{\beta/\nu},$$

$$\xi = \xi_0 |t|^{-\nu},$$

$$\tilde{r}_{l^*} = r_-^* \xi^{-\gamma/\nu}, \quad (T < T_c). \tag{2.17}$$

Finally, ϕ is the scaling function which takes the following form for $T < T_c$:

$$\phi(x) = -\frac{x^2}{2} + \frac{\tilde{u}_{l^*} M^2}{4|\tilde{r}_{l^*}|} x^4 + \frac{1}{|\tilde{r}_{l^*}| M_s^2} \int_0^{l^*} dp\, e^{-pd}$$

$$\times \left\{ \ln\left[1 + e^{2p}\tilde{r}_p + 3e^{2p}\tilde{u}_p M_s^2 x^2\right] - 3X_p(x) + \tfrac{9}{2} X_p^2(x) \right\}, \tag{2.18}$$

where

$$X_p(x) \equiv \frac{\tilde{u}_p M_s^2 e^{2p}}{1+e^{2p}\tilde{r}_p} x^2. \qquad (2.19)$$

That $\phi(x)$ in fact is a scaling function can be seen as follows. First, use of eq. (2.17) and

$$\tilde{u}_{l*} = u^* \xi^{(2\beta-\gamma)/\nu}, \qquad (2.20)$$

with constant u^* gives for the coefficient of x^4 in eq. (2.18) the following:

$$\tilde{u}_{l*} M_s^2/4|\tilde{r}_{l*}| = \frac{2}{K_d} B^2 \frac{u^*}{r_-^*} \xi_0^{2\beta/\nu}. \qquad (2.21)$$

Next, noting that the integral term of eq. (2.18) is higher order in ϵ than the first two terms, we can use the mean field results for this term such that:

$$|\tilde{r}_{l*}| M_s^2 \simeq 2B^2 r_-^* \xi_0^2/K_d,$$
$$\tilde{u}_p \simeq \tilde{u}^*, \qquad (2.22)$$
$$M_s^2 \simeq \frac{2}{K_d} B^2 (\xi_0/\xi)^2.$$

Furthermore, we change the integration variable p to $\zeta = e^{-p}\xi$, which permits us to write:

$$\tilde{r}_p \simeq r_-^* \xi^{-2} \zeta^2 R(\zeta), \qquad (2.23)$$

where R is some scaling function. Thus the third term of $\phi(x)$ now becomes:

$$(\xi^{-d} K_d/2B^2 r_-^* \xi_0^2) \int_1^\xi d\zeta\, \zeta^d \left\{ \ln\left[1+r_-^* R(\zeta) + \frac{6}{K_d} B^2 \tilde{u}^* \xi_0^2 \zeta^{-2} x^2\right] \right.$$
$$\left. - 3\overline{X}_\zeta + \frac{9}{2} \overline{X}_\zeta^2 \right\}, \qquad (2.24)$$

where

$$\overline{X}_\zeta(x) \equiv 2B^2 \tilde{u}^* \xi_0^2 x^2/K_d \zeta^2 [1+r_-^* R(\zeta)]. \qquad (2.25)$$

As far as the part of eq. (2.24) which depends on x^2 is concerned, the integral converges at large ζ and the upper bound ξ of the integral can be extended to infinity. Thus ϕ is indeed a scaling function if we disregard a term which contains ξ but not x.
If we note that:

$$A^*(M) = |\tilde{r}_{l^*}| M_s^2 \phi^*(M/M_s), \tag{2.26}$$

and introduce the coarse-grained susceptibility χ^* given by:

$$\chi^{*-1} = (\partial^2 A^*/\partial M^2)_{M=M_s} = |\tilde{r}_{l^*}| \phi^{*\prime\prime}(1), \tag{2.27}$$

eq. (2.16) can be reduced to:

$$f_0 = \frac{(2\pi)^d}{V_d} \frac{B^2}{C^* \phi''(1)} \xi_0^d, \tag{2.28}$$

where C^* is the critical amplitude of χ^*:

$$\chi^* = C^*|t|^{-\gamma}. \tag{2.29}$$

This is basically equivalent to eq. (4.9) of LBM except that χ^* is not equal to the thermodynamic susceptibility χ and $x=1$ is not in general equal to the minimum of $\phi(x)$. Thus if we denote the corresponding quantities of LBM with suffices LBM, we find:

$$f_0/(f_0)_{\text{LBM}} = \phi''_{\text{LBM}} \chi/\phi^{*\prime\prime}(1)\chi^*. \tag{2.30}$$

Next we present some numerical results of eq. (2.9) with the approximations described in the appendix. First we note that the argument of the logarithms in $A_l(M)$ must be positive in order to obtain a real coarse grained free energy. The boundary for this region is given by the equation:

$$e^{-2l_c} + \tilde{r}_{l_c} + 3\tilde{u}_{l_c} M^2 = 0. \tag{2.31}$$

For fixed $M, t_0, \exp(l_c)$ gives the maximum coarse graining length, which should characterize the thermodynamic instability of the system.

In fig. 1, we show the lines in the M, t_0 plane corresponding to fixed values of l_c in three spatial dimensions. As M approaches zero, we should have the following behavior:

$$t_0 \propto -M^{1/\beta}, \quad l_c \to \infty;$$
$$t_0 \propto -(M^2+c), \quad l_c < \infty, \tag{2.32}$$

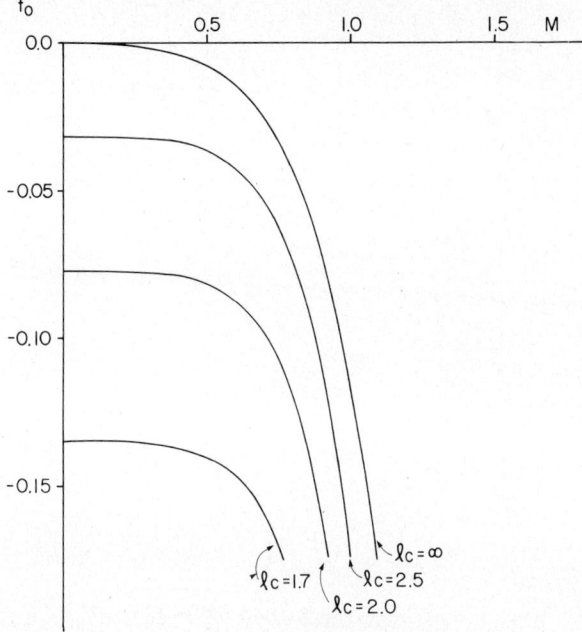

Fig. 1. The boundary value of l for $\epsilon = 1$. The points on the curve give the same value of l_c, 1.7, 2.0, 2.5 and ∞, respectively.

where c is some positive number. Numerical analysis of eq. (2.31) become cumbersome for very small values of M. Thus we have extrapolated the lines in fig. 1 toward $M = 0$ whenever slopes get negligibly small.

It is interesting to note that for the case of $\epsilon = 0$ this curve reduces to the van der Waals spinodal line. On the other hand, the line for $l_c \to \infty$ gives the effective spinodal line. Outside of this curve the system is stable for fluctuations of infinitesimal amplitude. This is an artifact of the ϵ-expansion. In fig. 2, we present $A_l(M)$ for some values of l at three spatial dimensions. It is important that $l^*(t_0) = \ln \xi(t_0)$ is less than $l_c(M = 0, t_0)$. As a consequence it is safe to coarse grain for all values of $l \leq l^*$. The effect of taking fluctuations into account is that $A_l(M)$ becomes flatter, as one would expect. Here $\bar{l}$ is given by $\bar{l} = \ln \xi_+(-t_0)$ where ξ_+ is the correlation length above T_C. In fig. 3, we display $\phi^*(x)$ defined by eq. (2.26) together with LBM's choice $\phi_0(x)$ given by eq. (3.29).

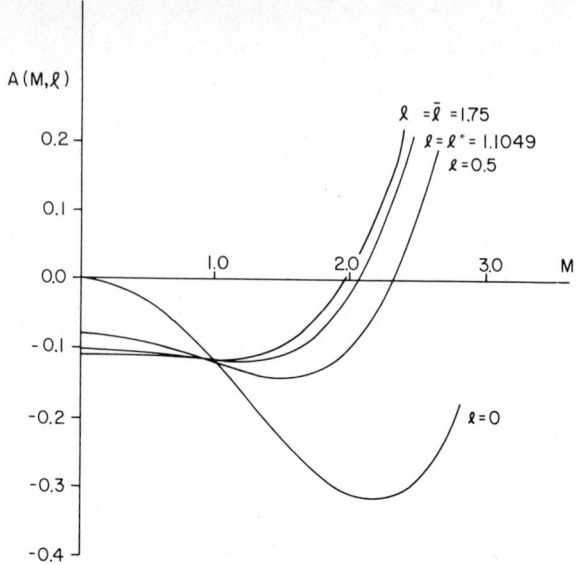

Fig. 2. The coarse grained free energy for $l=0$, 0.5, l^*, and $\bar{l}$ at $t_0 = -0.0537$ and $\epsilon = 1$.

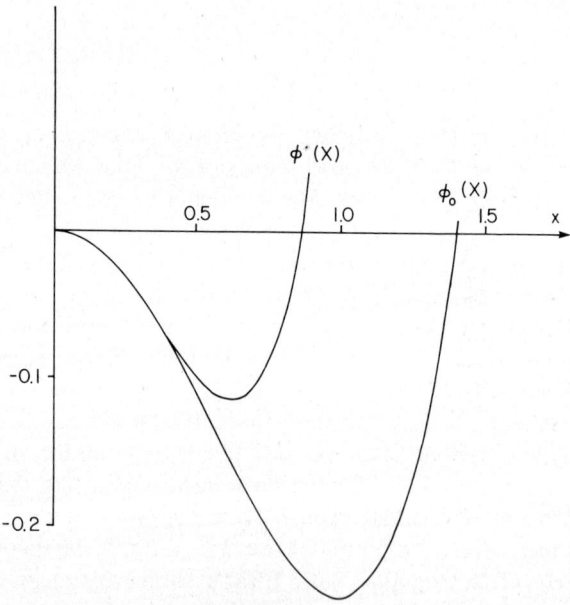

Fig. 3. Graphs of $\phi^*(x)$ (obtained numerically) and $\phi_0(x)$ for $\epsilon = 1$ and $l^* = 1.1049$.

Before leaving this section we wish to further discuss the result that the coarse grained free energy becomes complex in some domain of M and t_0. It is true that for whatever the values of M and t_0, the coarse grained free energy would remain real if we were able to carry out the coarse graining exactly. In particular, for those values of M and t_0 which are inside the coexistence curve, the free energy will eventually approach the well known convex envelope obtained in exact calculations, as l increases indefinitely. Thus, the complex free energy we are finding is clearly an artifact of our approximate way of coarse-graining using the finite order ϵ-expansion. Indeed, a complex part of similar nature already appears if one tries to continue the ϵ-expanded equation of state into the unstable region of the variables. We believe that our results are meaningful as long as we stay in the region where the coarse-grained free energy is real.

3. An approximate analytical solution for the coarse grained free energy

As we have seen in sect. 2 a numerical solution for $A_l(M)$ can be obtained using the Nicoll–Chang equation (2.9) together with the solution of eqs. (2.12a) and (2.12b). However, it is useful for the purpose of comparison with earlier phenomenological expressions for $A_{l*}(M)$ to have an analytic expression for $A_l(M)$ which is valid away from the stability boundary of fig. 1. In this section we provide an approximate analytic solution of the original differential eq. (2.8) which is based on a method used by Nicoll, Chang and Stanley [27] for computing the equation of state together with the approximate solution of the renormalization group equation due to Rudnick and Nelson [29]. The method is, however, limited to those values of l^* for which $1 + r_{l*}$ is positive and not near zero.

Our starting point is eq. (2.9) and we approximately evaluate the integral there up to $l = l^*$ defined by eq. (2.13). As long as we are near criticality where ξ and hence l^* are much greater than unity, the integral can be approximately evaluated by noting that the main contributions come from p near l^* and hence replacing $\tilde{r}_l$ and $\tilde{u}_l$ by $\tilde{r}^* = \tilde{r}_{l*}$ and $\tilde{u}^* = \tilde{u}_{l*}$ respectively. In fact it is easier to perform this integration first for $\partial A_l/\partial M$ as was done for the equation of state [27]. The result becomes, apart from a term independent of M,

$$A^*(M) = \tfrac{1}{2}\tilde{r}^* M^2 + \tfrac{1}{4}\tilde{u}^* M^4 - \tfrac{1}{2}|\tilde{r}^*|^{1-\epsilon/2}$$

$$\times \left\{ \frac{(\tilde{r}^*+y)^2 - e^{-4l^*}}{2r^*} \ln(\tilde{r}^* + \tilde{y} + e^{-2l^*}) - \tilde{y}\ln(\tilde{r}^* + e^{-2l^*}) \right.$$

$$- \tfrac{1}{2}(1 - e^{-2l^*}/\tilde{r}^*)\tilde{y}$$

$$\left. + \frac{1}{4\tilde{r}^*} \left[\frac{3\tilde{r}^* + e^{-2l^*}}{\tilde{r}^* + e^{-2l^*}} - 2\ln(\tilde{r}^* + e^{-2l^*}) \right] \tilde{y}^2 \right\}, \qquad (3.1)$$

where $\tilde{y} \equiv 3\tilde{u}^* M^2$. We now make use of the approximate solution of eq. (2.12) obtained by Rudnick and Nelson [29].

$$r_l = \bar{t}_l - 3u_l + 3u_l \bar{t}_l \ln(1 + \bar{t}_l), \qquad (3.2a)$$

$$u_l = u e^{\epsilon l}/Q_l, \qquad (3.2b)$$

where

$$Q_l = 1 + u(e^{\epsilon l} - 1)/u^*, \qquad (3.3)$$

$$\bar{t}_l = \bar{t}_0 e^{2l}/Q_l^{1/3}, \qquad (3.4)$$

$$\bar{t}_0 = r + 3u, \qquad (3.5)$$

$$u^* = \epsilon/8; \qquad (3.6)$$

$\bar{t}_l$ reduces to t_l of the appendix for small values of t_l, and the critical line is given by $\bar{t}_0 = 0$. u^* is the fixed point value of u_l. For $l = l^*$ we will have $Q_{l^*} \simeq y_0 e^{\epsilon l^*}$ with $y_0 = u_0/u^*$; y in fact reduces to $y_{l=0}$ of (A.2) for $|r_{l=0}| \ll 1$. Therefore we obtain:

$$\bar{t}_{l^*} \simeq -(\bar{\xi}_0)^{1/\nu} y^{-1/3}, \qquad (3.7)$$

where we have put

$$\xi = \bar{\xi}_0 |\bar{t}_0|^{-\nu}. \qquad (3.8a)$$

We can then express eq. (3.1) in the scaling form if we further note the following:

$$\xi^d M_s^2 |\tilde{r}^*| = (\bar{B})^2 |r^*| (\bar{\xi}_0)^{2/\nu} \equiv a_1. \tag{3.8b}$$

$$\frac{\tilde{u}^* M_s^2}{|\tilde{r}^*|} = \frac{u^*(\bar{B})^2(\bar{\xi}_0)^{2\beta/\nu}}{|r^*|} \equiv a_2, \tag{3.8c}$$

$$\frac{|\tilde{r}^*|^{-\epsilon/2} e^{-2l^*}}{M_s^2} = \frac{|r^*|^{-\epsilon/2}}{(\bar{B})^2(\bar{\xi}_0)^{2\beta/\nu}} \equiv a_3, \tag{3.8d}$$

$$e^{-2l^*} \tilde{u}^* M_s^2 = u^*(\bar{B})^2 |\bar{\xi}_0|^{2\beta/\nu} \equiv a_4, \tag{3.8e}$$

where we have put:

$$M_s = \bar{B}(\bar{t})^\beta, \quad r^* = r_{l^*}, \tag{3.9}$$

and have used the relation $(2-\epsilon)\nu = 2\beta$ valid up to $O(\epsilon)$. Notice that all the quantities on the right hand side of eq. (3.8b) are dimensionless and finite at T_C. Equation (3.1) thus becomes:

$$A^*(M) = \xi^{-d} a_1 \phi^*(M/M_s), \tag{3.10}$$

where

$$\phi^*(x) = -\frac{x^2}{2} + a_2 \frac{x^4}{4} - \frac{a_3}{2} \left\{ \frac{(r^* + 3a_4 x)^2 - 1}{2r^*} \ln(r^* + 3a_4 x + 1) \right.$$

$$- 3a_4 x \ln(r^* + 1) - \frac{3a_4 x}{2}(1 - r^{*-1})$$

$$\left. - \frac{(r^{*2} - 1)}{2r^*} \ln(r^* + 1) + \frac{(3a_4 x)^2}{4r^*} \left[\frac{3r^* + 1}{r^* + 1} - 2\ln(r^* + 1) \right] \right\}. \tag{3.11}$$

From the scaled equation of state [27] we find:

$$\bar{B}^2 u^* = y_0^{(2\gamma - 3)/3} [1 - \epsilon(2\ln 2 - 3)/6]. \tag{3.12}$$

The equation of state also yields for the magnetic susceptibility above T_C

$$\chi_+ = y_0^{\gamma/3} \bar{t}_0^{-\gamma}. \tag{3.13}$$

We choose the length scale such that above T_C the correlation range of fluctuation ξ_+ is related to χ_+ by $\chi_+ = \xi_+^2$ or $\bar{\xi}_{0+} = y_0^{\gamma/3}$. Then, we have for below T_C, [23]

$$\bar{\xi}_0 = y_0^{\gamma/6} 2^{-\nu}(1 - a_5 \epsilon), \qquad (3.14)$$

where $a_5 = \frac{5}{24}$ or $(\pi/\sqrt{3} - 1)/4$ depending on the definition of ξ through the second moment of the pair correlation function in real space (to be denoted as ξ_1) or through the zero point of the inverse of the wave number dependent susceptibility (denoted as ξ_2) respectively. We then obtain for a_4:

$$a_4 = 2^{-1}\left[1 + \left(\tfrac{1}{2} - 2a_5\right)\epsilon\right]. \qquad (3.15)$$

Other a's are obtained in terms of a_4 as:

$$a_1 = a_4/u^*, \quad a_2 = a_4/|r^*|, \quad a_3 = u^*|r^*|^{-\epsilon/2}/a_4. \qquad (3.16)$$

We further obtain using eq. (3.14):

$$\bar{t}^* = 2^{-1} + a_5 \epsilon. \qquad (3.17)$$

Hence by eq. (3.2a) we have with $u_{l^*} \simeq u^* = \epsilon/18$;

$$r^* = -2^{-1} + \left[a_5 - \tfrac{1}{6} + \tfrac{1}{12}\ln 2\right]\epsilon. \qquad (3.18)$$

The scaling function ϕ^* thus becomes:

$$\begin{aligned}\psi^*(x) = &-\frac{x^2}{2} + \frac{1}{4}\left[1 + \frac{(1+\ln 2)}{6}\epsilon\right]x^4 \\ &+ \frac{\epsilon}{18}\left\{\frac{3}{4}(x^2-1)(3x^2+1)\ln\left(\frac{3x^2+1}{2}\right)\right. \\ &\left. + \frac{3}{2}\left(\frac{3}{2} - \ln 2\right)x^2 + \frac{9}{8}(2\ln 2 - 1)x^4 - \frac{3}{4}\ln 2\right\}. \end{aligned} \qquad (3.19)$$

It is interesting to note that up to $O(\epsilon)$ this scaling function does not depend on a_5, i.e. on the choice of cut-off l^* corresponding to the two different definitions of ξ. For small x eq. (3.19) gives the following expansion:

$$\phi^*(x) = -\frac{x^2}{2} + \frac{1}{4}\left[1 + \frac{(1+\ln 2)}{6}\epsilon\right]x^4 + \frac{3}{8}\epsilon x^6 + \ldots \qquad (3.20)$$

For comparison we here obtain the free energy corresponding to the fully renormalized equation of state [23], [27]. This can be obtained by integrating the equation of state and using:

$$\tilde{u}_\infty M_s^2/|\tilde{r}_\infty| = 1 - \tfrac{1}{6}\epsilon(2\ln 2 - 3), \qquad (3.21a)$$

$$|\tilde{r}_\infty|^{1-\epsilon/2}/M_s^2 \simeq u^*, \qquad (3.21b)$$

$$a_1' = \xi_0^d y^{-\gamma/3}(\bar{B})^2 \simeq (1-\epsilon/3)a_1, \qquad (3.21c)$$

as

$$A_\infty(M) = \xi^{-d} a_1' \phi_\infty(M/M_s), \qquad (3.22)$$

where

$$\phi_\infty(x) = -\frac{x^2}{2} + \frac{1}{4}\left[1 - \frac{\epsilon}{6}(2\ln 2 - 3)\right]x^4$$

$$+ \frac{\epsilon}{72}\left\{(3x-1)^2\left[\ln(3x^2-1) - \frac{1}{2}\right] - (3x^2)^2 + \frac{1}{2}\right\}. \qquad (3.23)$$

The singularity at $x = \pm 1/\sqrt{3}$ corresponds to the van der Waals spinodal. In contrast there is no such singularity in the coarse-grained free energy [eq. (3.19)].

The prefactor ratio (2.30) can be also readily obtained. First we rewrite eq. (2.30) using eq. (2.27) and $\phi_{LBM}''(1) = 2$ as:

$$f_0/(f_0)_{LBM} = 2\chi|\tilde{r}^*|. \qquad (3.24)$$

Here χ becomes with eq. (3.13) and the known susceptibility ratio of above and below T_C, [7]

$$\chi = \chi_- = \tfrac{1}{2}y^{\gamma/3}[1 - \epsilon(\ln 2 + 3)/6]|\bar{t}_0|^{-\gamma}. \qquad (3.25)$$

From eqs. (3.18), (3.14), (3.8a), (2.13) and (2.11) we find:

$$|\tilde{r}^*| = |r^*||\bar{t}|^{2\nu}/\xi_0^2 = y^{-\gamma/3}(1+\epsilon/3)|\bar{t}_0|^{2\nu}. \qquad (3.26)$$

Substituting eqs. (3.25) and (3.26) into eq. (3.24) we finally obtain:

$$f_0/(f_0)_{LBM} = 1 - \epsilon(1 + \ln 2)/6. \qquad (3.27)$$

The functions $\phi^*(x), \phi_\infty(x)$, as well as LBM's form $\phi_0(x)$ given by:

$$\phi_0(x) = -\tfrac{1}{2}x^2 + \tfrac{1}{4}x^4 \qquad (3.28)$$

are displayed graphically in fig. 4.

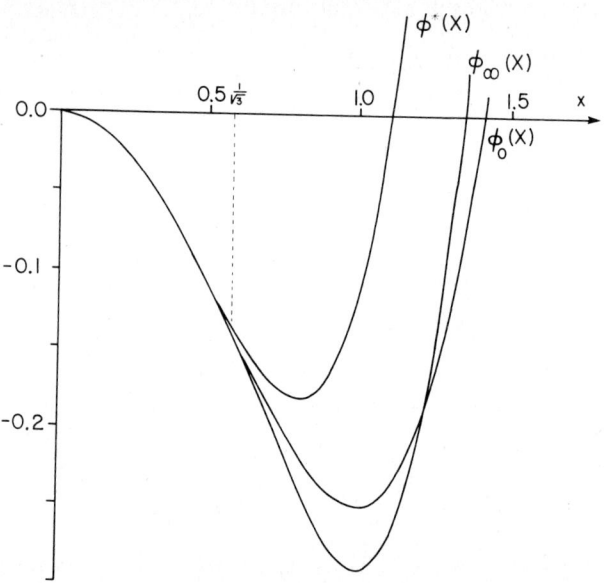

Fig. 4. The scaled free energy functions $\phi^*(x)$, $\phi_\infty(x)$ and $\phi_0(x)$ for $\epsilon = 1$. Note that $\phi_\infty(x)$ is meaningful only for $x > 1/\sqrt{3}$.

4. Discussion and concluding remarks

In the preceding sections we have used the renormalization techniques to investigate the coarse-grained Helmholtz free energy. Besides obtaining explicit forms for the free energy we have also indicated the region where such a free energy loses its meaning due to instability as l exceeds some critical value $l_c^0 = l_c(M=0, T)$ whenever $T < T_C$ (see fig. 1). Thus the free energy is expected to be singular at $l = l_c^0$. In fact a preliminary analysis indicates that the free energy $A_{l_c}(M)$ has a singularity of the form $M^2 \ln M^2$. It is interesting to note that a singular free energy function is also found for the van der Waals system in a similar circumstance [32].

We now turn to the problem of more practical interest, namely, the explicit form of the coarse-grained free energy for $l < l_c^0$. The results of the numerical and approximate analytic solutions are displayed in figs. 2 and 3, respectively. As is expected one common feature of the scaled free energy functions is that the coarse-grained function $\phi^*(x)$ has

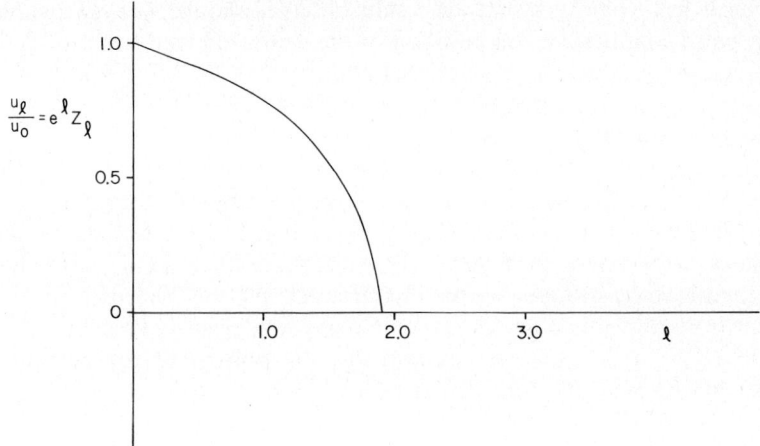

Fig. 5. Graphs of u_l/u_0 calculated by using eq. (A.12) with (A.19) at $t_0 = -0.0537$ and $\epsilon = 1$.

stronger non-linearity as compared to $\phi_0(x)$ or $\phi_\infty(x)$.† At this moment we have not yet fully understood the cause of the sizable difference between the two $\phi^*(x)$ of figs. 3 and 4. We only note that the approximations used are different. If we choose $u_0 = u^*$ in the analysis of sect. 3, we have $u_l = u^*$. On the other hand the same choice in the analysis of sect. 2 gives u_l which decreases appreciably below u^* as l increases toward l^*, as is shown in fig. 5.

Another difference between our coarse-grained free energy and that of LBM lies in the values of f_0. Namely, our values of f_0 extrapolated to $\epsilon = 1$ are about 67% and 72% of that of LBM corresponding to the two approximations of sects. 2 and 3 respectively. The smaller value for f_0 than the LBM value has also been found by using the three dimensional fixed point value for u^* by Kawasaki and Ohta [33].

The present analysis is limited to first order in the ϵ-expansion. Thus the results should not be taken too seriously in quantitative terms.

---

†This can be conjectured also from the fact that the value $\overline{M}(l)$ of M which minimizes $A_l(M)$ decreases first and then increases again as we increase l. This last fact follows by noting

$$d\overline{M}(l)/dl = -[\partial^2 A_l(M)/\partial l \partial M]_{\overline{M}} = \overline{M}/(\partial^2 A_l(M)/\partial M^2)_{\overline{M}}$$

where the denominator is always positive, whereas the sign of the numerator is the same as that of $(e^{-2l} + \tilde{r}_l + 3\tilde{u}_l \overline{M}^2)$ with $\overline{M} > 0$.

Nevertheless we believe that the results obtained should serve as a guide to a more quantitative computation of the coarse-grained free energy.

Acknowledgements

One of us (J.D.G.) wishes to acknowledge the support of a grant from the National Science Foundation and to thank the members of the Physics Department of Kyushu University for their kind hospitality during his sojourn there. We also wish to acknowledge the support of a grant from the United States–Japan Cooperative Science Program. Two of us (K.K. and T.I.) also acknowledge the support of the Scientific Research Fund of the Ministry of Education of Japan.

Appendix

First we summarize the solution of eq. (2.12) correct [32] to $O(\epsilon)$. For this purpose we define the following variables:

$$t_l = \frac{r_l}{(1+r_l)} + \frac{\epsilon}{6} y_l; \tag{A.1}$$

$$y_l \equiv \frac{18}{\epsilon} \frac{u_l}{(1+r_l)^2}; \tag{A.2}$$

$$\phi_l = |1-t_l|^{d/2} e^{\epsilon t_l/6}; \tag{A.3}$$

$$Y_l \equiv (1-y_l/\phi_l) e^{\epsilon t_l y_l/\phi_l}. \tag{A.4}$$

The solution is then expressed in terms of the non-linear scaling fields:

$$S_G(l) = (1+Y_l) t_l Y_l^{-1/3}; \tag{A.5}$$

$$S_W(l) = (1+r_l)^{1/3} t_l y_l^{-1/3}; \tag{A.6}$$

in the form

$$S_G(l) = S_G(0) e^{2l}; \tag{A.7}$$

$$S_W(l) = S_W(0) e^{l/\nu}; \tag{A.8}$$

with

$$\nu^{-1} = 2 - \epsilon/3. \tag{A.9}$$

Equations (A.7) and (A.8) imply the existence of the following renormalization invariant:

$$I \equiv |S_W(l)|^6 / S_G(l)^{6-\epsilon}$$
$$= |t_l|^{\epsilon} |Y_l|^{1/\nu} |1 + r_l|^{-d} y_l^{-2} \qquad (A.10)$$

The critical line corresponds to $t_0 = S_G(0) = 0$. We now obtain expressions for r_l and u_l that enter eq. (A.9). First, eqs. (A.1)–(A.5) and (A.7) lead to:

$$\tilde{r}_l = t_0(1 + r_0) Z_l^{1/3} + r_l^c e^{-2l}, \qquad (A.11)$$

where $Z_l \equiv Y_l / Y_0$ and $r_l^c \equiv -3u_l/(1+r_l)$. r_l^c represents the shift of critical temperature due to fluctuations. From eqs. (A.2), (A.5), (A.6), (A.7) and (A.8) we also obtain:

$$\tilde{u}_l = u_0 Z_l. \qquad (A.12)$$

We now introduce further approximations to facilitate computations. The shift r_l^c can be approximated by:

$$r_l^c \simeq - \frac{3 u_0 e^{\epsilon l} Z_l^0}{1 + e^{2l} t_0 (1 + r_0)(Z_l^0)^{1/3}} \frac{[1 + t_0(1 + r_0)]}{(1 + r_0)}, \qquad (A.13)$$

where the last factor was introduced to match at $l=0$ and Z_l^0 is an approximation to Z_l obtained by replacing I_l by I in eq. (A.19), which amounts to ignoring the shift of T_C. Next we approximate ϕ_l by:

$$\phi_l \simeq |1 - t_l|^{d/2} = \left[|1 + r_l^c| / |1 + r_l| \right]^{d/2}. \qquad (A.14)$$

Then, combining with eq. (A.10) we find:

$$y_l / \phi_l \simeq I^{-1/2} |t_l|^{\epsilon/2} |Y_l|^{1/2\nu} |1 + r_l^c|^{-d/2}, \qquad (A.15)$$

where we have assumed that y_l is non-negative. Thus we can approximate Y_l as:

$$Y_l \simeq 1 - y_l / \phi_l \simeq 1 - I_l^{-1/2} |t_l|^{\epsilon/2} |Y_l|^{1/2\nu}, \qquad (A.16)$$

with

$$I_l \equiv |1 + r_l^c|^d I. \qquad (A.17)$$

This can be rewritten as:

$$|Y_l| = I_l^\nu |t_l|^{-\epsilon\nu}(1-Y_l)(1-Y_l)^{2\nu-1}. \tag{A.18}$$

If the last factor on the right hand side is replaced by 1, we have:

$$|Y_l| \simeq \frac{I_l^\nu}{\pm I_l^\nu + |t_l|^{\epsilon\nu}}, \tag{A.19}$$

where $\pm$ corresponds to positive or negative Y_l. Substituting this into the last factor of eq. (A.18) we obtain the next approximation for Y_l as follows:

$$Y_l \simeq \frac{\pm I_l^\nu}{\pm I_l^\nu + |t_l|^{\epsilon\nu}(\pm I_l^\nu + 1)^{2\nu-1}}. \tag{A.20}$$

Consistency requires the denominator to be positive. The finiteness of $\tilde{r}_\infty$ and $\tilde{u}_\infty$ leads to $r_\infty^c = 0$ and $t_\infty = 1$. Hence:

$$Y_\infty = \frac{\pm I^\nu}{\pm I^\nu + (\pm I^\nu + 1)^{2\nu-1}}. \tag{A.21}$$

On the other hand, for $l=0$, the right hand side of eq. (A.20) does not reduce to Y_0 but to the following when $t_0 = 0$:

$$(1 - y_0 + y_0^{2\nu})^{-1} Y_0. \tag{A.22}$$

The discrepancy is always $O(\epsilon)$.

Since r_l^c contained in I_l is given by eq. (A 13), we only need to know t_l to obtain Y_l and hence $\tilde{r}_l$ and $\tilde{u}_l$. For this purpose we return to eqs. (A.5) and (A.6) which we approximate as:

$$S_G(l) \simeq t_l(1-t_l)^{-1}\left[1 - y_l(1-t_l)^{-2}\right]^{-1/3} \tag{A.23}$$

$$S_W(l) \simeq t_l(1-t_l)^{-1/3} y_l^{-1/3} \tag{A.24}$$

where use has been made of eq. (A.1) without y_l. This then yields:

$$t_l \simeq (1 + P_l^{1/3})^{-1}, \tag{A.25}$$

where

$$P_l \equiv \frac{\{S_G(l)^{-3} + S_W(l)^{-3}\}(1+r_{l_c})^3}{\{1 + (18/\epsilon)u_l(1-(1+r_{l_c})^{-2})\}}. \tag{A.26}$$

Substituting this into eq. (A.20) and multiplying by a numerical factor to ensure that $Z_{l=0} = 1$, we are ready to compute $A_l(M)$, if the initial values r_0 and u_0 are given.

References

[1] J. S. Rowlinson, J. Stat. Phys. **20** (1979) 197.
[2] J. W. Cahn, Acta Metal. **9** (1961) 795; **10** (1962) 179; **14** (1966) 1685.
[3] J. W. Cahn, J. Chem. Phys. **42** (1965) 93.
[4] J. S. Langer, Acta. Metal. **21** (1973) 1649.
[5] J. S. Langer, Physica **73** (1974) 61.
[6] R. B. Heady and J. W. Cahn, J. Chem. Phys. **58** (1973) 896.
[7] J. S. Langer, M. Bar-on and H. D. Miller, Phys. Rev. **A11** (1975) 1417.
[8] K. Kawasaki and T. Ohta, in: Proc. of 1978 Oji Seminar at Kyoto on Nonlinear Nonequilibrium Statistical Physics, Suppl. of Prog. of Theor. Phys. #**64** (1978); K. Kawasaki and T. Ohta, Prog. Theor. Phys. **59** (1978) 362.
[9] W. I. Goldburg, in: Proc. of 1978 Oji Seminar at Kyoto on Nonlinear Nonequilibrium Statistical Physics, Suppl. of Prog. of Theor. Phys., #**64** (1978).
[10] M. Rao, M. H. Kalos, J. L. Lebowitz and J. Marro, Phys. Rev. **B13** (1976) 4328.
[11] A. Sur, J. L. Lebowitz, J. Marro and M. H. Kalos, Phys. Rev. **B 15** (1977) 3014.
[12] S. Fisk and B. Widom, J. Chem. Phys. **50** (1969) 3219.
[13] B. Widom, in: Phase Transitions and Critical Phenomena, Vol. **2**, eds. C. Domb and M. S. Green (Academic Press, London, 1972).
[14] T. Ohta and K. Kawasaki, Prog. Theor. Phys. **58** (1977) 467.
[15] J. Rudnick and D. Jasnow, Phys. Rev. **B17** (1978) 1351.
[16] F. F. Abraham, J. Chem. Phys. **63** (1975) 157.
[17] Y. Singh and F. F. Abraham, J. Chem. Phys. **67** (1977) 537.
[18] A. J. M. Yang, P. D. Fleming and J. H. Gibbs, J. Chem. Phys. **64** (1975) 3732; **65** (1976) 7; **67** (1977) 74.
[19] D. G. Triezenberg and R. Zwanzig, Phys. Rev. Lett. **28** (1972) 1183.
[20] H. T. Davis, J. Chem. Phys. **67** (1977) 3636.
[21] C. A. Leng, J. S. Rowlinson and S. M. Thompson, Proc. Roy. Soc. A **352** (1976) 1; **358** (1977) 267.
[22] A. B. Bhatia and N. H. March, J. Chem. Phys. **68** (1978) 4651.
[23] E. Brezin, J. C. LeGuillou and J. Zinn-Justin, in: Phase Transitions and Critical Phenomena, Vol. **6**, eds. C. Domb and M. S. Green (Academic Press, London, 1976).
[24] A. J. M. Yang, P. D. Fleming and J. H. Gibbs, J. Chem. Phys. **64** (1976) 3732.
[25] Y. Singh and F. F. Abraham, J. Chem. Phys. **67** (1977) 537.
[26] J. F. Nicoll and T. S. Chang, Phys. Rev. **A17** (1978) 2083.
[27] J. F. Nicoll, T. S. Chang and H. E. Stanley, Phys Lett. **57A** (1976) 7.
[28] J. F. Nicoll and T. S. Chang, Phys. Lett. **62A** (1977) 287.
[29] J. Rudnick and D. R. Nelson, Phys. Rev. **B13** (1976) 2208.
[30] F. Wegner and A. Houghton, Phys. Rev. **A8** (1973) 401.
[31] J. F. Nicoll, T. S. Chang and H. E. Stanely, Phys. Rev. Lett. **32** (1974) 1446; ibid **36** (1976) 113.
[32] G. Dee, J. D. Gunton and K. Kawasaki, J. Stat. Phys. (to be published).
[33] K. Kawasaki and T. Ohta, Progr. Theor. Phys. **59** (1978) 362.

CHAPTER 13

Exact Renormalization in Two Dimensional Ising Systems

J. M. J. VAN LEEUWEN

Laboratorium voor Technische Natuurkunde
Technische Hogeschool Delft
The Netherlands

© *North-Holland Publishing Company 1981*   *Perspectives in Statistical Physics*
Ed. H. J. Raveché

Contents

1. Introduction — 227
2. Differential form of real space renormalization — 228
3. Star-triangle transformations — 231
4. The $d=2$ renormalization — 232
5. Results of the renormalization equations — 235
6. Discussion — 237
References — 238

1. Introduction

The advantage of the real space formulation of the renormalization theory for phase transitions is that the definition of the renormalization transformation can be written down in one line. Let $H(s)$ be the Hamiltonian of the starting system with N degrees of freedom (spins) represented by s. The renormalization transformation (RT) maps this $H(s)$ on to a (renormalized) $H'(s')$ involving a system with $N' < N$ spins s'. An RT is built on a weight factor $P(s', s)$ connecting configuration s and s' with the properties:

$$P(s', s) \geqslant 0, \qquad Tr' P(s', s) = 1, \tag{1}$$

where Tr' in the normalization stands for the sum (integral) over all configurations of the s'. The RT is then defined as:

$$H'(s') = \log[Tr P(s', s) \exp H(s)], \tag{2}$$

where Tr sums over all the configurations of s. The normalization (1) of $P(s', s)$ guarantees that the free energies of the system $H(s)$ and $H'(s')$ are the same.

As simple as the definition is, the difficulties are contained in the nature of $H'(s')$. Starting with a nearest neighbor $H(s)$ and a reasonable $P(s', s)$ leads mostly to a $H'(s')$ involving all kinds of interactions besides the nearest neighbor coupling. As it is the primary goal of a renormalization approach to find a fixed point H^* of eq. (2) i.e. a H which leads to the same coupling constants in H', one has to study eq. (2) in a many dimensional Hamiltonian space. The problem of the proliferation of the number of coupling constants is a serious one as has recently been pointed out by Griffiths and Pearce [1].

From the early days of the renormalization approach as introduced by Wilson [2] it was realized by Melville Green that the renormalization theory had to be formulated in an infinite dimensional Hamiltonian space. In fact, long before, he had proposed [3] a generalization of the Ornstein–Zernike theory which also seeks to find a regular (short ranged) direct correlation function by extending it to a space of multivariable correlation functions. In a similar way the RT (2) is hoped to be regular in this infinite dimensional Hamiltonian space. The analogy between the renormalization approach and Green's theory of the correlation functions becomes stronger when the correlation functions are discussed in the real space renormalization context. The transformation for the correlation functions obtains indeed a generalized Ornstein–Zernike form, with the essential difference of a scale transformation in the argument of the correlation functions [4]. This feature allows the power-like behavior of the correlation functions at criticality as a result of regular transformations.

The proliferation of the number of interaction constants is connected to the dimension of the lattice on which the spins are located. In a one-dimensional chain it is easy to derive renormalization equations involving only nearest neighbor interactions.

This contribution discusses a limited number of two-dimensional systems in which one can set up a renormalization scheme, which is also closed at the nearest neighbor level, as has been shown by Hilhorst et al. [5]. This provides a model with a non-trivial critical structure which can be solved exactly by renormalization and in which the ideas and assumptions of the renormalization approach can be checked.

The cases discussed are rather special in the sense that a fortuitous combination of factors cooperate to the success of the method. At the end of this paper a discussion is devoted on the relation of this result to the more general case.

2. Differential form of real space renormalization

The reduction of the number of degrees of freedom N to N' is expressed as:

$$N' = b^{-d} N, \tag{3}$$

with b the linear reduction factor (d=dimension). In the real space renormalization b is usually a number distinctly larger than 1. In this section we explore the possibilities for $b \to 1$. The only way to influence b is the choice of $P(s', s)$ which we translate in a combined system Hamiltonian $H_c(s', s)$:

$$H_c(s', s) = H(s) + \log P(s', s). \tag{4}$$

The advantage of the introduction of $H_c(s', s)$ is that eqs. (1) and (2) can be combined to the symmetric form:

$$H(s) = \log[Tr' \exp H_c(s', s)], \tag{5a}$$

$$H'(s') = \log[Tr \exp H_c(s', s)]. \tag{5b}$$

In the formulation (5) the renormalization is broken up into two steps:

(a) The representation problem: find for a given $H(s)$ an $H_c(s', s)$ such that $H(s)$ can be written as in eq. (5a).
(b) The integration problem: sum out the old degrees of freedom s' in $\exp H_c(s', s)$ such that $H'(s')$ results.

In principle the first step is no problem as the definition (3) of $H_c(s', s)$ shows that any $P(s', s)$ would do. The limitation is however to find a $H_c(s', s)$ which admits the second step. It is therefore more practical to search for an $H_c(s', s)$ leading to similar H and H'.

In this section we investigate the case where H' and H are nearly the same ($b \to 1$), which will be the case when s' and s appear nearly interchangeable in $H_c(s', s)$. Let us illustrate this point in a one-dimensional chain.

In fig. 1 we have drawn a chain of 9 old spins and 8 new spins interacting via a nearest neighbor $H_c(s', s)$, with coupling $p(i)$ between site i' and i and $q(i)$ between i' and $i+1$. Site 1 coincides with $1'$ and site N ($=9$) with N' ($=8'$). The lattice distances a and a' are such that:

$$Na = (N-1)a' = L, \tag{6}$$

where L is the length of the chain. The reduction factor b follows for large N as:

$$b = 1 + a/L, \tag{7}$$

and deviates from 1 only by the infinitesimal quantity a/L (for $N \to \infty$).

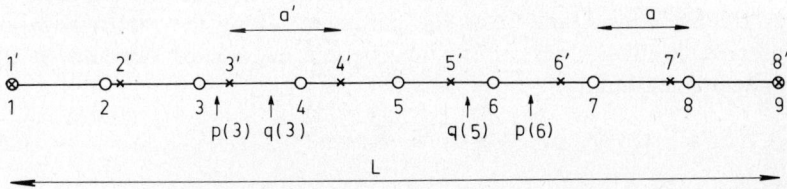

Fig. 1.   Linear chain differential arrangement.

As $H_c(s', s)$ is a nearest neighbor Hamiltonian, the summations in eqs. (5a) and (5b) can be performed easily [6] and result in H and H' which are also nearest neighbor Hamiltonians. If we denote by K_i the coupling between i and $i+1$ and by K'_i the corresponding quantity in the new system we have the expressions:

$$K_i = \tfrac{1}{2}\log\{\cosh[p(i)+q(i)]/\cosh[p(i)-q(i)]\}, \tag{8a}$$

$$K'_i = \tfrac{1}{2}\log\{\cosh[p(i)+q(i-1)]/\cosh[p(i)-q(i-1)]\}. \tag{8b}$$

We note that if we were to make the $p(i)$ and the $q(i)$ to be independent of i, the K_i and K'_i would be everywhere the same. But we have to express the fact that the new system has one degree of freedom less, which is indicated in fig. 1 by the identification of the endpoints in both systems. This can be done by making the configurations of s_1 and s'_1 as well as s_N and s'_{N-1} identical through infinite strong coupling $p(1) = \infty$ and $q(N-1) = \infty$. Consequently the $p(i)$ and $q(i)$ have to vary along the chain. Thus the renormalization is achieved by the spatial variation of the coupling constants in $H_c(s', s)$.

A spatial inhomogeneous $H_c(s', s)$ is an unwanted byproduct of the desire to renormalize in a continuous way ($b \to 1$). In $d=1$ the damage is limited because we still have the option to vary $p(i)$ and $q(i)$ such that K_i and K'_i become independent of i. Starting in the middle of the chain, where $p(i) = q(i)$ by symmetry, one can determine the $p(i)$ and $q(i)$ completely towards both ends under the requirement of constant K_i and K'_i. The boundary condition at either end $p(1) = \infty$ or $q(N-1) = \infty$ then fixes a relation between K' and K. The important point here is that the difference $\delta K = K' - K$ is proportional to the change in $p(i)$ (or $q(i)$) over one lattice distance which in turn is proportional to the smallness parameter a/L. Explicitly one finds (van Saarloos et al. [6]):

$$\delta K = (a/L) \operatorname{tgh} K \log(\operatorname{tgh} K)/[1 - \operatorname{tgh}^2 K]. \tag{9}$$

We present this result in a slightly different form by writing $a/L = \delta t$ and considering "time" t as the parameter along the renormalization process, keeping track of the number of degrees of freedom in the system according to:

$$\delta N = -\mathrm{d} N \delta t \quad \text{or} \quad N(t) = N e^{-\mathrm{d} t}. \tag{10}$$

Then eq. (9) can be cast in the differential form:

$$\partial \chi / \partial t = \chi \log \chi, \tag{11}$$

with $\chi = \mathrm{tgh}\, K$. Obviously $\chi = 0$ (∞-temperature and $\chi = 1$, 0-temperature) are fixed points of this relation.

Extending this idea to higher dimensions one would draw two interpenetrating lattices differing along the edges by one site and trying to couple each site to its immediate neighborhood (in order to allow for the summability of eqs. (5a) and (5b)). Systems where this leads to similar H and H' are those allowing for a star-triangle transformation.

3. Star-triangle transformations

In a star-triangle transformation the interactions of a central spin with the corner spins of a triangle are replaced by equivalent interactions along the edges of the triangle (see fig. 2). The best known case is the Ising model. If we denote by p_i the interactions of the central spin we may write:

$$\sum_{s'=\pm 1} \exp s'(p_1 s_1 + p_2 s_2 + p_3 s_3) = 2\cosh(p_1 s_1 + p_2 s_2 + p_3 s_3)$$

$$= \exp[g + K_1 s_2 s_3 + K_2 s_3 s_1 + K_3 s_1 s_2], \quad (12)$$

where the K_i are given by (i, j, k cyclical):

$$K_i = \tfrac{1}{4}\log\big[\cosh(p_i + p_j + p_k)\cosh(-p_i + p_j + p_k) \\ /\cosh(p_i - p_j + p_k)\cosh(p_i + p_j - p_k)\big]. \quad (13)$$

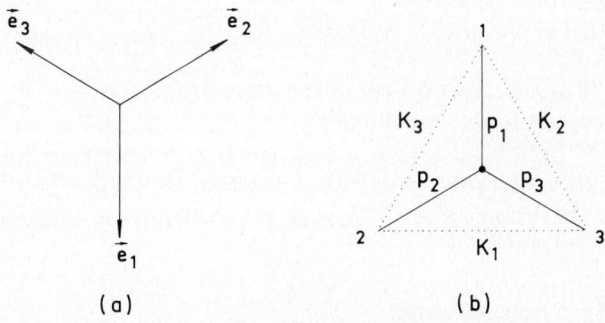

Fig. 2. Basic vectors e_i (a) and star-triangle configuration (b).

The K_i are equivalent to the p_i in the sense that the free energy of the system with spins s', s_1, s_2, s_3 and bonds p_i is the same as the free energy of the system s_1, s_2, s_3 with bonds K_i whatever interactions s_1, s_2, s_3 might have with other spins.

In the same way the q-state Potts-model admits a star-triangle transformation in the limit $q \to 0$. The transformation becomes the same as for linear resistors [7] p_i which can be replaced by K_i and vice versa according to:

$$K_i = p_j p_k / (p_i + p_j + p_k) \qquad (i, j, k, \text{cyclical}). \tag{14}$$

The third example is the Gaussian model [9] in which the spins range from $-\infty$ to $+\infty$ and are weighted by a Gaussian weight $\exp - s^2/2$. In analogy with eq. (12) the integration of the central spin amounts to:

$$(1/\sqrt{\pi}) \int_{-\infty}^{\infty} ds' \exp\left[-\tfrac{1}{2} s'^2 + s'(p_1 s_1 + p_2 s_2 + p_3 s_3)\right] =$$
$$= \exp \tfrac{1}{2}(p_1 s_1 + p_2 s_2 + p_3 s_3)^2, \tag{15}$$

working out the square, yields interactions $p_j p_k$ between s_j and s_k and a self-interaction $\tfrac{1}{2} p_i^2$ for spin s_i. These self-interactions change the width of the Gaussian weight which can be brought back in the standard form by a spin rescaling. Including this effect one arrives (for the hexagonal lattice that we will use later) at the transformation:

$$K_i = p_j p_k / (1 - p_i^2 - p_j^2 - p_k^2). \tag{16}$$

For the next sections we need only to remember that all the examples have a transformation of the form:

$$K_i = K_i(p_1, p_2, p_3), \tag{17}$$

and that the relation of the set of the three p_i and the three K_i is one to one (in contrast to the one-dimensional formula (8) where one K can be represented by several sets of p and q). As a consequence we must be prepared to work with spatial inhomogeneous Hamiltonians when we apply the star-triangle transformation to p_i which vary with position.

4. The $d = 2$ renormalization

In this section we combine the idea of deriving differential renormalization equations through a spatial inhomogeneous $H_c(s', s)$ with the

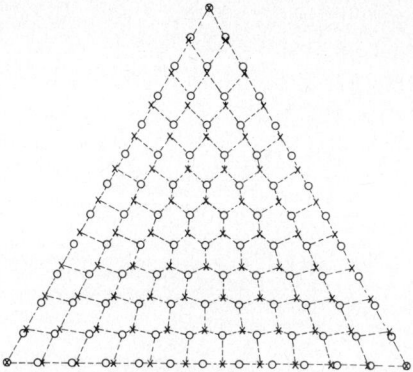

Fig. 3. Two triangular lattices $\bigcirc$ and $\times$ coupled in a hexagonal structure.

star-triangle transformation. In fig. 3 we have drawn two triangular lattices differing by one site along an edge and mutually connected by a (distorted) hexagonal nearest neighbor hamiltonian $H_c(s', s)$. The bonds $p_i(\mathbf{R})$ depend on the position $\mathbf{R}$ in space and by convention we give them the coordinates of the centers of the uptriangles of the old (larger) triangular lattice.

Applying the star-triangle transformation to such an up-triangle the $p_i(\mathbf{R})$ will be replaced by $K_i(\mathbf{R})$ referring to bonds between the 3 sites in the triangle. For convenience we label the bonds in the triangular lattice also by the coordinate of the center of the uptriangle to which they belong. This allows to calculate the $K_i(\mathbf{R})$ from eq. (17) by simply substituting the $p_i(\mathbf{R})$.

Now the calculation of the $K_i'(\mathbf{R})$ is more involved. Firstly we must obey our convention also for the new lattice and label them by the centers of the uptriangles of the new lattice. From fig. 3 we see that the uptriangles of the new lattice are empty and that the star-triangle must be applied to down-triangles. These are occupied by bonds p_i having different coordinates and we must pay attention to this fact as the spatial dependence of the $p_i(\mathbf{R})$ is the motor of the renormalization. In fig. 4 a part of the lattice near the center is drawn and the situation is sketched for the calculation of $\tilde{K}_1'(\mathbf{R})$ (which is not yet $K_1'(\mathbf{R})$ as the distortion of the hexagonal lattice is not yet taken into account). $\tilde{K}_1(\mathbf{R})$ is made up of p_i having the coordinate $\mathbf{R} - a(\mathbf{e}_1 + \mathbf{e}_i)$, where the $\mathbf{e}_i$ are the 3 vectors pointing from an old site (in the center) to its surrounding

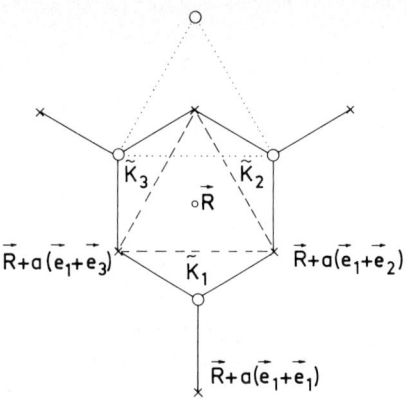

Fig. 4. Contributions to $\tilde{K}_i(R)$.

new sites (see fig. 2). Thus we obtain for $\tilde{K}_i(R)$:

$$\tilde{K}_i(R) = K_i(p_1[R - a_i(e_1 + e_1)], p_2[R - a(e_i + e_2)], \\ \times p_3[R - a(e_i + e_3)]). \tag{18}$$

Once the $\tilde{K}_i(R)$ are found the $K_i'(R)$ result simply by the observation that the new lattice is obtained from the centers of the uptriangles of the old lattice by a dilation from the origin in the center of the big triangle by a factor $L/(L-a)$:

$$K_i'(R) = \tilde{K}_i\left(R\left(1 - \frac{a}{L}\right)\right). \tag{19}$$

The three equations (17) (taken at $p_i(R)$ and $K_i(R)$) (18) and (19) form the renormalization transformation. The only thing to be done is to use explicitly that $a/L = \delta t$ is an infinitesimal quantity. This is achieved by considering the functions as function of $r = R/L$ and expand in δt, which gives the $\delta K_i(r) = K_i'(r) - K_i(r)$ as δt times $\nabla K_i(r)$ and $\nabla p_i(r)$. Re-expressing the ∇p_i in terms of ∇K_i by the inverse of (17) yields the renormalization equations:

$$\frac{\partial K_i(r, t)}{\partial t} = \sum_j D_{ij} \cdot \nabla K_j(r, t), \tag{20}$$

with the D_{ij} given by:

$$D_{ij} = \sum_l \frac{\partial K_i}{\partial p_l}(e_i + e_l)\frac{\partial p_l}{\partial K_j} - r\delta_{ij}. \tag{21}$$

The partial differential equations (20) need boundary conditions, which are most easily formulated in terms of the $p_i(r)$. From the 1-dimensional example we learned already that we have to take $p_1 = \infty$ in the top of the triangle, $p_2 = \infty$ in the left lower corner and $p_3 = \infty$ in the right lower corner*). In addition we have to require for the validity of eq. (20) near the edges of the triangle that the p_i vanish on the edge to which they are perpendicular. With these boundary conditions the renormalization problem is fully specified. Via the star-triangle transformation (17) the boundary conditions in terms of the p_i can be translated in the K_i or alternatively eq. (20) can be transformed in a flow equation for the p_i.

5. Results of the renormalization equations

Equation (20) yields the flow for any system of $K(r)$ fulfilling the boundary conditions. Since the star-triangle transformation is regular, the flow equation themselves do not show any singularity. This basic assumption of the renormalization approach is apparently fulfilled for this case. The remarkable feature of eq. (20) is that the renormalization process takes place strictly in the space of nearest neighbor interactions, albeit at the expense of the introduction of spatially varying interaction parameters.

One of the first questions to be answered is the existence of non-trivial (critical) fixed points. The search for these fixed points is simplified by the existence of an invariant critical subspace. Criticality can be expressed in the three models as:

$$v_1 + v_2 + v_3 = 1, \tag{22}$$

where the v_i are functions of the p_i, reading respectively:

Ising

$$v_i = \left[\sinh 2p_j \sinh 2p_k\right]^{-1} \quad (i, j, k \text{ cyclical}); \tag{23a}$$

---

*More accurately we require the correlations of the old top spin s and the new top spin s' to be perfect: $\langle ss' \rangle = \langle s^2 \rangle = \langle s'^2 \rangle$. This is realized with $p_1 = \infty$ for the Ising model and the $q \to 0$ Potts model. For the Gaussian model this is true for $p_1 = 1$.

Potts $q \to 0$

$$v_i = [p_j p_k]^{-1} \quad (i, j, k \text{ cyclical}); \tag{23b}$$

Gaussian

$$v_i = p_i. \tag{23c}$$

Thus fields $K_i(r)$, fulfilling everywhere the condition (22), flow to new $K_i'(r)$ fulfilling again eq. (22).

It turns out that a fixed point is located in the critical subspace, having the full triangular symmetry of the form:

$$v_i^*(r) = \tfrac{1}{3} - 2(r \cdot e_i). \tag{24}$$

One verifies that this fixed point satisfies the boundary conditions including the requirements for the Gaussian model near the corners of the triangle (see footnote in sect. 4).

It is noteworthy that the fixed point assumes the simple form (24) in the coordinates v_i in which the criticality condition is linear.

The critical propertics are given in terms of the eigenvalues of the linearized transformation. Let $\Psi_i(r)$ be a small deviation,

$$\Psi_i(r) = K_i(r) - K_i^*(r), \tag{25}$$

where the $K_i^*(r)$ are the fixed point values following from eqs. (24), (23) and (17). Then we look for eigen-deviations $\Psi_i(r)$ which develop according to:

$$\partial \Psi_i(r)/\partial t = y \Psi_i(r). \tag{26}$$

Because of the invariance of the critical subspace we can divide the eigenvalues into two groups:

(a) eigenvalues corresponding to flow inside the critical subspace (22);
(b) eigenvalues corresponding to deviations away from criticality.

The first group has been studied by Knops and Hilhorst [8]. They find a rather complex structure leading also to lines of new fixed points. The remarkable thing is that the equations for the flow inside the critical subspace simplifies when expressed in the v_i and obtains the same form for all three models. Yamazaki and Hilhorst [9] call this a fundamental property of the star-triangle transformation without direct physical meaning.

The second group, which corresponds to thermal deviations, turns out to be degenerate with the values [6, 9]:

$y_T = 1$ Ising; (27a)

$y_T = 0$ Potts $q \to 0$; (27b)

$y_T = 2$ Gaussian; (27c)

In the $q \to 0$ Potts model the exponent is marginal and follows in fact from the observation that the fixed point (24) is located on a fixed line in the temperature direction which is generated by a parameter λ according to:

$$v_i^*(r;\lambda) = \lambda v_i^*(r), \quad 0 < \lambda < \infty. \quad (28)$$

The degeneracy of the temperature-like exponent y_T implies that the eigendeviations can be constructed as local deviations from criticality which grow everywhere in the triangle with the same rate y_T. This is an expression of universality as in the limit $a/L \to 0$, one has locally different triangular systems, ranging from the square lattice in the top (the horizontal triangular bond being zero) through an isotropic triangular system in the center to a one-dimensional (zero-temperature, critical) system at the bottom line.

Thus in the nearest neighbor space we have for the temperature dependence a single exponent. It means that the nearest neighbor coupling is an exact scaling field and that no corrections to scaling exist (other than the trivial ones due to combination of regular and singular powers). This confirms the earlier findings from the exact solution [10] of the triangular Ising system and the easier Potts $q \to 0$ and Gaussian solutions.

6. Discussion

In view of the above described renormalization technique we come back on the question of the proliferation of the number of coupling constants. What are the chances to renormalize exactly with a finite number of coupling constants? If the usual renormalization picture holds, it means that exact scaling fields could be constructed within this finite space for which corrections to scaling are absent. All the known exact solutions have trivial corrections to scaling and so it seems that exact renormalization is confined to those cases which admits an exact solution along other lines.

One may turn around this argument and say that the renormalization approach indicates that the singular structure is in general very complicated with infinite many corrections to scaling. To cast such a singular behavior in a form which we accept as an exact solution (a finite number of integrals over known functions as integrand) is out of sight. Thus exact solutions are exceptional from the renormalization viewpoint, which is not a surprising result.

More interesting seems the question whether one could use this renormalization solution as a starting point for a systematic approximation technique. Many candidates qualify such as the inclusion of further interactions in the Ising model or a power series expansion in $\sqrt{q}$ in Potts model. Despite serious efforts in this direction the success is as large as the usefulness of e.g. the exact Ising solution as a starting point for perturbation theory. Moreover the sacrifice we had to make, by going over to spatially dependent interaction constants in order to arrive at the exact renormalization, is a rather high price in the light of perturbative calculations.

Acknowledgement

On this occasion it is with gratitude that the author acknowledges a most stimulating conversation with Melville Green which aroused his interest in the renormalization approach. He is also much indebted to M. Schick and H. J. Hilhorst with whom the main part of the work described here was carried out. In particular the many discussions with H. J. Hilhorst on the implications and the prospects of the differential renormalization method were very beneficial.

References

[1] R. B. Griffiths and P. A. Pearce, Phys. Rev. Lett. **41** (1978) 917.
R. B. Griffiths and P. A. Pearce, J. Stat. Phys. **20** (1979) 5.
[2] K. G. Wilson, Phys. Rev. B **4** (1971) 3174, 3184.
[3] M. S. Green, Journ. of Math. Phys. **9** (1968) 875.
M. S. Green, Phys. Rev. **185** (1969) 176.
J. D. Gunton and M. S. Green, Phys. Rev. A **4** (1971) 1282.
[4] Th. Niemeyer and J. M. J. van Leeuwen, in: Phase Transitions and Critical Phenomena, eds. C. Domb and M. S. Green (Academic Press, London, 1976) Vol. **6**, p. 425.
[5] H. J. Hilhorst, M. Schick and J. M. J. van Leeuwen, Phys. Rev. B **19** (1979) 2749.
[6] W. van Saarloos, J. M. J. van Leeuwen and A. L. Stella, Physica **97A** (1979) 319.
[7] P. W. Kasteleyn and C. W. Fortuin, J. Phys. Soc. Japan **26** (suppl.) (1969) 11.
C. W. Fortuin and P. W. Kasteleyn, Physica **57** (1972) 536.
[8] H. J. F. Knops and H. J. Hilhorst, Phys. Rev. B **19** (1979) 3689.
[9] Y. Yamazaki and H. J. Hilhorst, Physics Letters **70** A (1979) 329.
[10] R. M. F. Houtappel, Physica **16** (1950) 425.
G. F. Newell, Phys. Rev. **79** (1950) 876.

CHAPTER 14

How Close is "Close to the Critical Point"?

J. M. H. LEVELT SENGERS

National Bureau of Standards, Washington, DC 20234

J. V. SENGERS

Institute for Physical Science and Technology
University of Maryland, College Park, MD 20742
and
National Bureau of Standards, Washington, DC 20234

© *North-Holland Publishing Company 1981*

Perspectives in Statistical Physics
Ed. H. J. Ravechè

Contents

1. Introduction — 241
2. Critical-point scaling in fluids — 243
3. Revised and extended scaling — 250
   3.1. Corrections to scaling — 250
   3.2. The Wegner expansion for fluids — 251
   3.3. Parametric formulation of corrections to scaling — 253
   3.4. Application to the analysis of experimental data for fluids — 255
   3.5. Range of asymptotic scaling in fluids — 257
   3.6. Assessment of the Wegner expansion for fluids — 259
4. Crossover; an outlook — 262
5. Appendix. Revised and extended scaling equations for the thermodynamic properties of fluids — 264
   5.1. Reduced thermodynamic quantities — 265
   5.2. Thermodynamic relations — 265
   5.3. Fundamental equations — 265
   5.4. Derived thermodynamic quantities — 265
   5.5. Critical exponents — 266
   5.6. Parametric equations for singular terms — 266
   5.7. Auxiliary functions — 267
   5.8. Critical amplitude ratios — 268
References — 268

1. Introduction

The theoretical ideas concerning the nature of critical-point phase transitions have evolved significantly during the past fifteen years. The theme of this chapter is the application of these ideas to the critical behavior of fluids with emphasis on the gas–liquid critical point. This same theme was also the fiber of our scientific association with Mel Green. Our sketch of the development will, therefore, have a strong component of personal reminiscences of the persistent questioning and growing insights Mel shared with us.

Our association with Mel dates back to the period when he was organizing the 1965 Conference on Critical Phenomena at the National Bureau of Standards [1], motivated by "the expectation or hope that a common thread of explanation would be found among such diverse phenomena as liquid–vapor phase transitions, magnetism, superconductivity, order–disorder transitions" [2]. At this conference Fisher summarized his definitions of the critical power laws [3], later generalized by Widom [4] and by Domb and Hunter [5] to the scaling hypothesis for the asymptotic behavior of systems close to a critical point.

In response to Fisher's theoretical suggestions at the conference, Debye remarked with his characteristic Limburg accent [6]: "I would like that the theoretical people tell me when I am so and so far away from the critical point, then my curve should look so and so", and, indeed, the question: "How large is the range in which the asymptotic scaling laws are valid in fluids?" is at the heart of our story as it will unfold in the next pages.

Actually, since no rigorous theory of critical phenomena in fluids existed, the answer had to come from experiment. We started with the hope that the critical anomalies would dominate in a large range around the critical point. Initially, the analysis of available fluid data seemed to fulfill this hope. Voronel, who had begun with the opposite expectation, reached on the basis of his own specific heat data a similar conclusion concerning the dominance of the critical behavior [7]:

"...I still continued to hope for some time that it would be possible to limit critical phenomena and the relevant singularities to a narrow range of parameters in the vicinity of the critical point.

Permit me now to express the diametrically opposite view, which was reached after all such hopes were systematically destroyed. Now I believe that practically the entire domain of the liquid state is related to the critical point – in fact, the liquid state may be defined as the neighborhood of the critical point ($t = (T_c - T)/T_c \leqslant 1$) – and that the study of critical phenomena should form the basis for research into the liquid state as well as concentrated solutions."

However, new experiments, carried out much closer to the critical point, forced us to give up this idea.

Section 2 of this chapter sketches the evolution of our insights. It begins with our application of the Widom–Griffiths formulation of scaling to fluids [4, 8] and the search for critical-exponent values. It then discusses the ideas about critical-point universality that were formulated by Kadanoff [9, 10], Griffiths [11] and others [12–14] and that were recast by the renormalization-group approach to critical phenomena [15]. This section closes with the message conveyed by newer experiments in fluids carried out much closer to the critical point, such as the interferometric experiments performed by Balzarini et al. [16, 17] and by Hocken and Moldover [18]. This message was that the range of asymptotic scaling in fluids is quite small, so that corrections to scaling will be needed in most of the range accessible to conventional experimentation.

Section 3 of this chapter deals with these corrections to scaling. Ley-Koo and Green applied Wegner's expansion [19] of the Gibbs free energy of spin systems to fluids by choosing the appropriate scaling variables and carefully treating the successive corrections to scaling [20]. In subsequent work, we have carried the method of Ley-Koo and Green to practical application using a parametric representation of the scaling functions. These results are discussed and we present complete expressions for the thermodynamic properties of fluids near the critical point including the first Wegner correction, using a potential close to that of Ley-Koo and Green.

From fits to fluid data performed by ourselves and others, we deduce estimates for the range of validity of asymptotic scaling in fluids and fluid mixtures. Of particular importance is Ley-Koo and Green's analysis of the coexistence curve diameter of sulfur hexafluoride [21]. This work forcefully illustrates the applicability of Wegner's expansion as well as its inherent weakness: poor convergence.

Ley-Koo and Green's analysis forms the bridge to sect. 4. The small size of the asymptotic range and the poor convergence of Wegner's expansion force us to consider a new direction, that of a "crossover" from a regime of Ising-like scaled behavior to another regime in which the critical anomalies due to large fluctuations can be ignored. This was the problem Mel, amongst others, worked on during the past few years [22], and it would undoubtedly have been the direction of his further collaboration with us if his death had not intervened.

2. Critical-point scaling in fluids

As mentioned earlier our close interaction with Mel Green on the critical behavior of fluids dates back to the time of the conference which Mel organized in 1965. At this conference, a venerable German professor, Ernst Schmidt, appeared with a set of pictures of density gradients

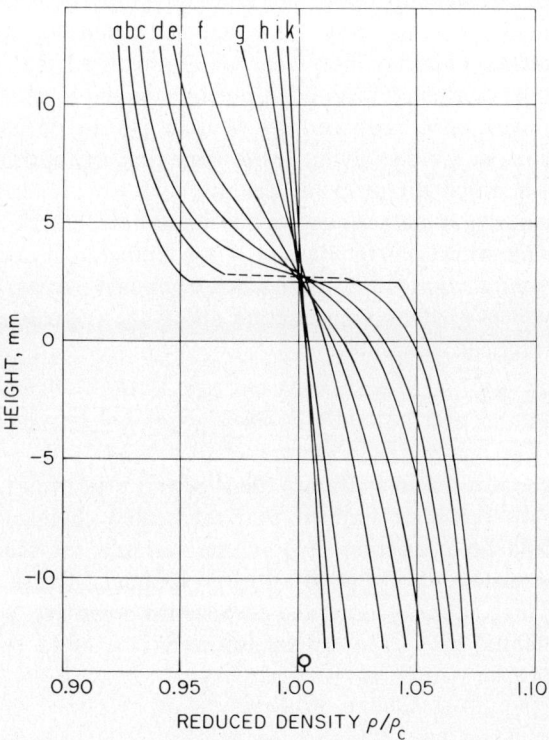

Fig. 1. Reduced density ρ/ρ_c versus height for N_2O near the critical point as measured by Schmidt and Straub [23, 24]. Mean density: $\rho/\rho_c = 1.005$. Temperatures: (a) $T-T_c = -0.002$ K; (b) $T-T_c = 0.003$ K; (c) $T-T_c = 0.010$ K; (d) $T-T_c = 0.020$ K; (e) $T-T_c = 0.032$ K; (f) $T-T_c = 0.038$ K; (g) $T-T_c = 0.063$ K; (h) $T-T_c = 0.119$ K; (i) $T-T_c = 0.167$ K; (k) $T-T_c = 0.218$ K.

observed experimentally in CO_2, N_2O and $CClF_3$ using refractive index measurements [23]. Within the last few hundredths of a degree from the critical point, the compressibility of the fluid has become so large that gravity produces a noticeable density gradient. In Schmidt's pictures, the density varied by about 10% over a height of 2 cm (fig. 1). It was clear to us that these pictures contained detailed information about the close vicinity of the critical point, information that cannot be obtained in conventional equation-of-state experiments where gravitational stratification causes loss of accuracy [25, 26].

When Matilde Vicentini-Missoni came to the National Bureau of Standards as a guest worker in 1966, Mel asked her and one of us (Anneke Levelt Sengers) to analyze these pictures. Matilde and Anneke spent a fair amount of time numerically integrating the density versus height curves, converting these curves to pressure–density isotherms with the purpose of comparing them with the results of conventional PVT experiments. One day, however, Matilde realized that the density versus height isotherms corresponded precisely to the form in which the scaling laws had been proposed by Widom [4]. In Mel's Statistical Physics Section, we were studying at the same time a preprint of a paper by Griffiths in which the proposed scaling laws were further discussed [27]. The quantity to be scaled was the chemical potential difference $\mu(\rho, T) - \mu_0(T)$, where ρ is the density, T the temperature and $\mu_0(T)$ the chemical potential along the coexistence boundary below the critical temperature or its analytic continuation above the critical temperature. Introducing reduced variables

$$t = \frac{(T - T_c)}{T_c}, \quad \Delta\rho = \frac{(\rho - \rho_c)}{\rho_c}, \quad \Delta\mu = \frac{\rho_c}{P_c}[\mu(\rho, T) - \mu_0(T)], \qquad (1)$$

where T_c, ρ_c, P_c are the temperature, density and pressure at the critical point, Widom's postulate implied that the scaled chemical potential, $\Delta\mu/|t|^{\beta\delta}$ would be a function of only one variable, the scaled density $\Delta\rho/|t|^{\beta}$. The exponents β and δ are the exponents of the asymptotic power laws for the density along the coexistence boundary, ρ_{cxc}, and the chemical potential along the critical isotherm (see table 1). Thus, the scaling law can be written as:

$$\frac{\Delta\mu}{|t|^{\beta\delta}} = G\left(\frac{\Delta\rho}{|t|^{\beta}}\right). \qquad (2)$$

The height versus density curves from Schmidt's experiments represent precisely the chemical potential as a function of density [25]. So

Table 1. Critical power laws

Path	Power law
$\rho=\rho_c$, $T>T_c$	$\tilde{C}_v = \dfrac{A^+}{\alpha}|\Delta \tilde{T}|^{-\alpha}$
$\rho=\rho_c$, $T<T_c$	$\tilde{C}_v = \dfrac{A^-}{\alpha}|\Delta \tilde{T}|^{-\alpha'}$
$\rho=\rho_{cxc}$, $T<T_c$	$\Delta\tilde{\rho} = \pm B |\Delta \tilde{T}|^{\beta}$
$\rho=\rho_c$, $T>T_c$	$\left(\dfrac{\partial \tilde{\rho}}{\partial \tilde{\mu}}\right)_T = \Gamma^+ |\Delta \tilde{T}|^{-\gamma}$
$\rho=\rho_{cxc}$, $T<T_c$	$\left(\dfrac{\partial \tilde{\rho}}{\partial \tilde{\mu}}\right)_T = \Gamma^- |\Delta \tilde{T}|^{-\gamma'}$
$T=T_c$	$\Delta\tilde{\mu} = \pm D |\Delta\tilde{\rho}|^{\delta}$

Exponent relations: $2-\alpha=\beta(\delta+1)$, $\gamma=\beta(\delta-1)$, $\alpha=\alpha'$, $\gamma=\gamma'$.
Universal amplitude ratios: A^+/A^-, Γ^+/Γ^-, $\Gamma^+ D B^{\delta-1}$, $A^+\Gamma^+/B^2$.

Matilde and Anneke reversed their gears and started converting conventional PVT data for CO_2, Xe and SF_6 to chemical potential data as a function of density. We remember the day when the scaling hypothesis was tested from the swelling collection of optical and conventional chemical potential data. Using a door-size sheet of logarithmic graph paper, Matilde and Anneke did the plotting, while Mel looked spellbound over their shoulders. When the isotherms for the five fluids in a range of a few millidegrees to 30°C from the critical temperature started tracing out the scaling function $G(x)$ for fluids, a thrill of discovery was experienced that each scientist hopes to encounter in his research. The plot [28] is reproduced in fig. 2. The scaled dependent variable $\Delta\mu/|t|^{\beta\delta}$ spans about five decades. The two branches in the plot correspond to super- and subcritical behavior. In order to obtain optimum coincidence of the isotherms, exponent values $\beta=0.35$ and $\delta=5$ were used. Had we realized that, in addition, two system-dependent scaling constants were allowed for each fluid, an even more striking plot could have been produced. In the subsequent years, we greatly refined our method of analysis, performing careful error estimates for the individual experiments and resorting to non-linear least-squares techniques, thus sharpening our estimates for the critical exponents of fluids [29–32].

In the mean time, however, Kadanoff [9, 10], Griffiths [11] and others [12–14] had introduced the concept of critical-point universality, an idea strongly affirmed by the evolving renormalization-group approach [15]. Critical behavior is dominated by fluctuations which are departures

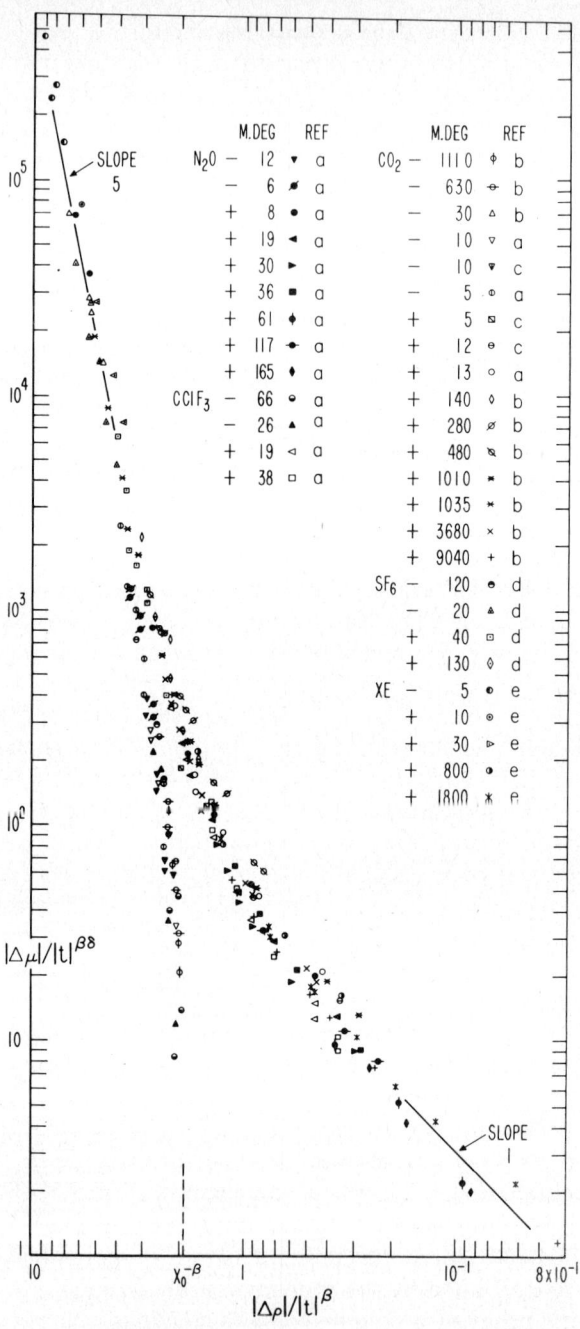

Fig. 2.

of the order parameter from its average value that may extend over ranges in space much larger than the distances between neighboring molecules. Hence, it is expected that details of the molecular interaction are irrelevant to the nature of the critical behavior. This expectation is supported by model calculations [33]. Only a few general features determine the type of critical behavior that a system will display, or the "universality class" to which it belongs: dimensionality of the system, number of components of the order parameter and whether the intermolecular forces are short- or long-range [15, 34]. Considerations of this type inevitably put fluids in the same universality class as the 3-dimensional Ising model or, more precisely, in the universality class of Ising-like systems for which the critical behavior is described by the so-called 1-component ϕ^4 model [15]. The critical exponents for this universality class have been calculated using series expansion techniques for the 3-dimensional Ising model [35–37] and field theoretical methods for the 1-component ϕ^4 model [38, 39]. The two methods yield numerically similar results, but unresolved discrepancies do exist [40–42]. The theoretical exponent values are summarized in table 2. Our "best" critical exponent values for fluids, however, differed systematically from those of the 3-dimensional Ising model or of Ising-like systems. For fluids near the gas–liquid critical point, the critical exponent β, with few exceptions, tended to fall between 0.34 and 0.36 [30–32, 43, 44] to be compared with the series expansion result $\beta = 0.312 \pm 0.005$ for the Ising model and the renormalization-group value $\beta = 0.325 \pm 0.001$ for Ising-like systems.

The issue came to a head at the Renormalization-Group Conference organized by Mel at Temple University in 1973 [45]. Based on his experience with scaling near the superfluid phase transition in liquid helium, Ahlers suggested that corrections to scaling be incorporated in the analysis of fluid data [46, 47]. Results of several experiments performed closer to the critical point, in particular those of Balzarini

Fig. 2. First test of the scaling hypothesis for equations of state data of fluids [28]. Data sources: (a) E. H. W. Schmidt, in: Critical Phenomena, Proceedings of a Conference, eds. M. S. Green and J. V. Sengers, NBS Miscellaneous Publ. 273 (U. S. Gov't Printing Office, Washington, D.C., 1966), p. 13; (b) A. Michels et al., Proc. Roy. Soc. (London) **A153** (1935) 201, 214; **A160** (1937) 358; (c) H. L. Lorentzen, Acta Chem. Scand. **7** (1953) 1335; **9** (1955) 1724; Proceedings International Symposium on Statistical Mechanics and Thermodynamics, ed. J. Meixner (North-Holland, Amsterdam, 1965), p. 262; (d) R. H. Wentorf, J. Chem. Phys. **24** (1956) 607; (e) H. W. Habgood and W. G. Schneider, Can. J. Chem. **32** (1954) 98; M. A. Weinberger and W. G. Schneider, Can. J. Chem. **30** (1952) 422.

Table 2. Theoretical critical exponent values

	Ising model series expansions [35–37]	Renormalization group theory [38, 39]	Adopted in this paper
α	0.125 ± 0.010	0.109 ± 0.004	0.1085
β	0.312 ± 0.005	0.325 ± 0.001	0.325
γ	1.250 ± 0.003	1.241 ± 0.002	1.2415
δ	5.05 ± 0.15	4.82 ± 0.02	4.82
Δ_1	0.50 ± 0.08	0.496 ± 0.004	0.50

and Ohrn [16] and of Carr et al. [48] were also discussed at the meeting; these experiments seemed to indicate a decrease in the value of β as the critical point was approached more closely. Other experiments, in particular those of Weber [49], did not reveal such a change [50].

Clearly, a careful experimental study of the region close to the critical point was called for. Robert Hocken who had received his training with

Fig. 3. A seven-layered thermostat, built by Hocken and capable of 20 microdegree control at its core. Laser light traversing the fluid in an optical cell inside the thermostat forms a diffraction pattern due to the fluid's density gradients. (Figure copied from ref. [52]).

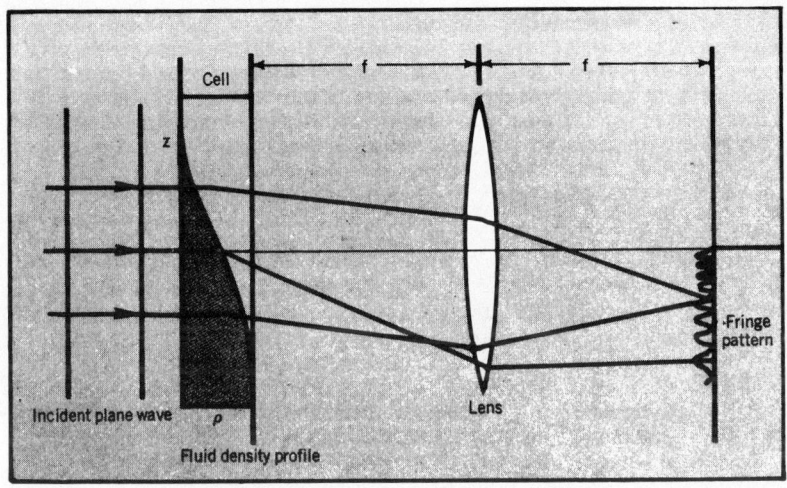

Fig. 4. Schematic representation of the formation of a Fraunhofer pattern. The light rays are bent over an angle proportional to the local density gradient and experience a phase shift proportional to the local density. Hence, the interference pattern yields the density gradient as a function of density [51] (figure copied from ref. [52]).

Wilcox, developer of an interferometric method for studying density gradients near the critical point of fluids [51], joined the Equation of State Section of NBS as a postdoctoral research associate in 1973. He developed a thermostat with high temperature stability, an optical cell and an interferometer for measuring profiles within a few hundredths of a degree from the critical temperature with high accuracy (figs. 3, 4). With Michael Moldover he measured these profiles for CO_2, SF_6 and Xe and demonstrated that the critical exponents do approach Ising-like values [18]. Specifically, the values they obtained for the exponent β ranged from 0.321 to 0.328. Similarly small values for β have been subsequently reported by other investigators [53–56].

The experiments of Hocken and Moldover made a profound impression on Mel. He realized immediately that these experiments implied the necessity of including correction-to-scaling terms in a description of the behavior of fluids in all but the last fraction of a degree from the critical temperature. The form of these correction-to-scaling terms and their incorporation in an analysis of fluid data are the subject of the following section.

3. Revised and extended scaling

3.1. Corrections to scaling

In the late sixties, Mel left the National Bureau of Standards and joined the Physics Department at Temple University. However, he remained a frequent visitor to NBS and, at that time, was very interested in the form of the corrections to scaling beyond the asymptotic range. With Martin Cooper and one of us (J.M.H.L.S.), he proposed a form for these corrections [57]. It was arrived at by a marriage of the ideas of Schofield et al. [58] concerning parametric representations and the ideas of Griffiths and Wheeler [59] concerning the preferred direction in the space of independent field variables. Schofield's parametric representation was expanded to include higher-order terms without confluent singularities. A demand for invariance of the form of the potential in all directions except the "preferred" direction, that of the coexistence curve in μ-T space, then led to the introduction of a "gap" exponent.

This attempt to formulate corrections to scaling failed for two reasons. First, there was no hope of extracting reliable information on corrections to scaling from experimental data as long as uncertainty existed about the critical-exponent values to be used in the asymptotic term. Secondly, as a renormalization-group analysis by Wegner later showed [19], confluent singularities are present in the corrections to scaling. Wegner used the renormalization-group notion of irrelevant scaling fields to obtain an expansion of the thermodynamic potential P in terms of the two relevant scaling fields u_t and u_h. For a magnetic system, considered by Wegner, P is the Gibbs free energy, u_h the magnetic field H and u_t an analytic function of temperature which close to the critical point reduces to $(T-T_c)/T_c$. The Wegner expansion has the form [19].

$$P(u_t, u_h) = P_{\text{reg}}(u_t, u_h) + ak_0 |u_t|^{2-\alpha} g_0\left(\frac{u_h}{|u_t|^{\beta\delta}}\right)$$

$$+ ak_1 |u_t|^{2-\alpha+\Delta_1} g_1\left(\frac{u_h}{|u_t|^{\beta\delta}}\right) + \dots. \tag{3}$$

Here P_{reg} is an analytic function of the scaling fields u_t and u_h; α, β and δ are the usual critical exponents and Δ_1 the first correction to scaling exponent. The critical exponents α, β, δ, Δ_1 and the scaling functions $g_0, g_1 \dots$ are assumed to be universal, i.e. identical for all systems belonging to the same universality class. The constant a is a non-universal scale factor for the potential; the subsequent terms in the expansion are multiplied by additional non-universal scale factors $k_0, k_1 \ldots$.

3.2. The Wegner expansion for fluids

Shortly after Hocken and Moldover had obtained their new experimental results, Mel arrived at NBS one day with a manuscript prepared by his student Marcos Ley-Koo who had worked out in detail the consequences of Wegner's expansion for the description of fluid critical behavior [20]. Mel sat down with us and explained the steps that led from Wegner's expansion eq. (3) for the Gibbs free energy of a magnetic system to a potential useful for fluids. The potential is such that both independent variables are intensive variables or fields that are equal in coexisting phases. Possibilities are $P(\mu, T)$ or $\mu(P, T)$; the choice that follows most naturally from the analogy between Ising model and lattice gas is the potential P as a function of μ and T (or, more precisely, P/T as a function of μ/T and $1/T$).

Mel always delighted in explaining the properties of this potential. Earlier he had asked his student, Frank Cook, to construct a model which is shown in fig. 5 [60]. Two-phase equilibrium occurs where two parts of the surface meet with a discontinuous change in slope. This change in slope indicates that the derivative properties, energy and number density, are unequal in coexisting phases. The "preferred direction" of Griffiths and Wheeler is that of the two-phase curve or seam. The seam ends at the critical point where the two coexisting phases become identical. At this point, however, the curvature of the surface becomes infinite, reflecting the anomalies in compressibility and specific heat.

The scaling variables u_h, u_t are constructed as two analytic functions of the physical variables $\mu/T, 1/T$. The first variable, u_h, is measured from the coexistence curve and its analytic continuation as projected onto the $(\mu/T, 1/T)$ plane. The second variable, u_t, is measured from an axis intersecting the coexistence curve. In the lattice gas, the axis is the locus $T = T_c$, but in fluids, which lack the lattice-gas symmetry, the $u_t = 0$ axis may depart from the critical isotherm, so that u_t is a "mixture" of μ/T and $1/T$. The effect of this mixing of μ and T, referred to as revised scaling, has been studied in non-symmetric models and leads to subtle anomalies such as a weak divergence of the slope of the diameter of the coexisting densities at the critical point [61–63].

In terms of the reduced variables:

$$\tilde{P} = \frac{P}{T} \frac{T_c}{P_c}, \quad \tilde{\mu} = \frac{\mu}{T} \frac{\rho_c T_c}{P_c}, \quad \tilde{T} = -\frac{T_c}{T}, \tag{4}$$

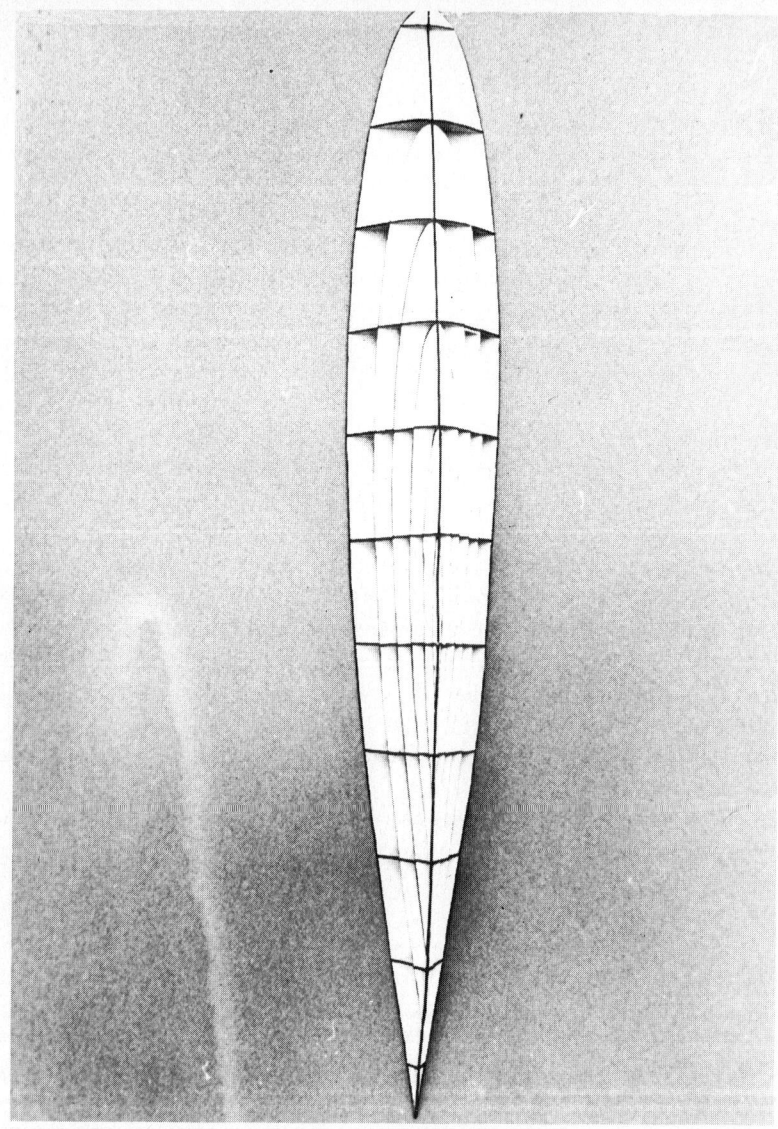

Fig. 5. Cook and Green's model of the P-μ-T surface of a fluid in the vicinity of the critical point viewed from the direction of the P-axis [60].

the potential $\tilde{P}(\tilde{\mu}, \tilde{T})$ satisfies the differential relation:

$$d\tilde{P} = \tilde{U} d\tilde{T} + \tilde{\rho} d\tilde{\mu}, \tag{5}$$

where $\tilde{U} = U/VP_c$ is the reduced energy per unit volume and $\tilde{\rho} = \rho/\rho_c$ the reduced number density. Introducing the quantities:

$$\Delta \tilde{T} = \tilde{T} + 1 = (T - T_c)/T,$$
$$\Delta \tilde{\mu} = \tilde{\mu} - \tilde{\mu}_0(\tilde{T}), \tag{6}$$

where

$$\tilde{\mu}_0(T) = \tilde{\mu}_c + \sum_{j=1} \tilde{\mu}_j (\Delta \tilde{T})^j, \tag{7}$$

we expand the scaling fields u_h and u_t in a Taylor series around the critical point:

$$u_\mu \equiv u_h = \sum_{ij} a_{ij} (\Delta \tilde{\mu})^i (\Delta \tilde{T})^j,$$
$$u_t = \sum_{ij} b_{ij} (\Delta \tilde{\mu})^i (\Delta \tilde{T})^j, \tag{8}$$

with $a_{00} = b_{00} = 0$. Ley-Koo and Green [20] substituted the expansion (8) into the expansion (3) for the potential $\tilde{P}$ and derived the explicit expansion that resulted for a number of thermodynamic properties along special curves: the critical isotherm, the critical isochore and the coexistence curve[†].

Sandra Greer [53] at NBS, was the first to apply Ley-Koo and Green's formulation of Wegner's expansion to an analysis of coexistence data obtained by herself [53] and by Gopal et al. [64] for binary liquid mixtures. Ley-Koo and Green [21] analyzed the coexistence curve data obtained by Weiner et al. [65] for SF_6. The results of these and other analyses using the Wegner expansion will be discussed in sect. 3.5.

3.3. Parametric formulation of corrections to scaling

By the mid-seventies, the usefulness of a parametric representation of critical scaling functions, introduced by Schofield et al. [58], had been amply demonstrated. With some simplifying assumptions, namely those leading to the linear and cubic models, the thermodynamic properties could all be formulated in closed parametric form. For a survey the reader is referred to a review published elsewhere [66]. The integrability and the absence of spurious anomalies in the one-phase region are very desirable features that have thus far only been obtained by using

---

[†] Ley-Koo and Green actually used $\tilde{\mu} - \tilde{\mu}_c$, rather than $\tilde{\mu} - \tilde{\mu}_0$, in the expansion (8).

parametric variables. It seemed, therefore, worthwhile to cast Wegner's expansion in parametric form. Together with Frances Balfour and Michael Moldover we have formulated a parametric representation that includes the first Wegner correction term [67].

For this purpose we retain only the lowest-order terms in the expansion (8) of the scaling fields:

$$u_\mu = \Delta\tilde{\mu}, \quad u_t = \Delta\tilde{T} + c\Delta\tilde{\mu}, \tag{9}$$

where c is a system-dependent mixing parameter which equals zero in the case of the symmetric lattice gas. The potential $\tilde{P}$ is given by eq. (3):

$$\tilde{P} = \tilde{P}_{\text{reg}} + \Delta\tilde{P}, \tag{10}$$

where the analytic background $\tilde{P}_{\text{reg}}$ can be expanded in a Taylor series around the critical point. We have used the approximation:

$$\tilde{P}_{\text{reg}} = \tilde{P}_0(\tilde{T}) + \Delta\tilde{\mu} + \tilde{P}_{11}\Delta\tilde{\mu}\Delta\tilde{T}, \tag{11}$$

with

$$\tilde{P}_0(\tilde{T}) = \sum_{j=1} \tilde{P}_j(\Delta\tilde{T})^j. \tag{12}$$

In practice, two or three terms are required in the expansions (7) and (12) for $\tilde{\mu}_0(\tilde{T})$ and $\tilde{P}_0(\tilde{T})$.

In order to cast the singular part $\Delta\tilde{P}(u_\mu, u_t)$ into a parametric form, we relate the scaling fields u_μ and u_t in the usual way to parametric variables r and θ,

$$\begin{aligned} u_\mu &= \Delta\tilde{\mu} = ar^{\beta\delta}\theta(1-\theta^2), \\ u_t &= \Delta\tilde{T} + c\Delta\tilde{\mu} = r(1-b^2\theta^2). \end{aligned} \tag{13}$$

In lowest-order scaling, the linear model of Schofield et al. [58] is obtained by assuming that the density difference $\Delta\tilde{\rho} = (\rho - \rho_c)/\rho_c$ be linear in θ. Since $\tilde{\rho} = (\partial\tilde{P}/\partial\tilde{\mu})_{\tilde{T}}$, this linear-model condition determines the form of the leading scaling function g_0 in eq. (3). We have made a similar linear-model assumption for the first correction-to-scaling function g_1, leading to:

$$\Delta\tilde{P} = r^{2-\alpha}ak_0 p_0(\theta) + r^{2-\alpha+\Delta_1}ak_1 p_1(\theta). \tag{14}$$

Substituting eqs. (11) and (14) into eq. (10), one obtains for the density

$$\tilde{\rho} = (\partial \tilde{P}/\partial \tilde{\mu})_{\tilde{T}},$$

$$\tilde{\rho} = 1 + \tilde{P}_{11} \Delta \tilde{T} + r^{\beta} k_0 \theta + r^{\beta + \Delta_1} k_1 \theta$$
$$+ c \left[r^{1-\alpha} a k_0 s_0(\theta) + r^{1-\alpha+\Delta_1} a k_1 s_1(\theta) \right]. \tag{15}$$

Here $p_i(\theta)$ and $s_i(\theta)$ are polynomials in θ^2 presented in the Appendix. The resulting parametric expressions for a number of other thermodynamic properties are also presented in the Appendix.

3.4. Application to the analysis of experimental data for fluids

The adjustable parameters in the potential (10) with eqs. (11)–(14) fall into five groups: the critical parameters P_c, ρ_c, T_c; the critical exponents $\alpha, \beta, \gamma, \delta, \Delta_1$; the "background" constants $\tilde{P}_{11}, \tilde{P}_j, \tilde{\mu}_j$; the linear-model parameters a, k_0, b^2; the correction-to-scaling amplitude k_1 and the mixing parameter c. Our procedure has been to fix the values of the critical exponents at the values predicted by renormalization-group methods as shown in table 2. The "background" constants $\tilde{P}_{11}$ and $\tilde{P}_j$ can be determined from experimental PVT data, but a determination of the constants $\tilde{\mu}_j$ requires caloric information. Out of the linear model parameters a, k_0, k_1 and b^2, the parameter b^2 should be universal. In fitting the linear-model equation to experimental PVT data, we have left the linear-model parameters including b^2 adjustable which permits a test of universality of critical amplitude ratios, to be discussed below.

The transformation to parametric variables is achieved by simultaneously solving the parametric equations for the temperature and density as described by Moldover [68], using educated first guesses for the adjustable parameters. In fitting PVT data, the pressure is then predicted at the given densities and temperatures, the weighted sum of squares is calculated and minimized by varying the adjustable parameters using a non-linear least squares procedure. The procedure was implemented by Frances Balfour [69]. It was applied to the critical-region data of steam [67, 70] and of ethylene [69, 71]. For both fluids the equation-of-state data were fitted to within the accuracy claimed by the experimenters, covering a range of approximately ±30% around the critical density and a range from −1% to +10% around the critical temperature. Supplementary caloric data were used to determine the constants $\tilde{\mu}_j$. The parameter sets thus obtained for the two fluids are given in table 3.

The hypothesis of critical-point universality implies that certain ratios of the critical power-law amplitudes listed in table 1 should assume

Table 3. Parameters in two-term Wegner expansion for steam and ethylene [69–71]

	Steam	Ethylene
Critical exponents (fixed)	$\beta = 0.325$ $\delta = 4.82$ $\Delta_1 = 0.50$	$\beta = 0.325$ $\delta = 4.82$ $\Delta_1 = 0.50$
Fitted parameters	$T_c = 647.00$ K $\rho_c = 322.68$ kg/m^3 $b^2 = 1.3757$ $a = 23.848$ $k_0 = 1.4643$ $k_1 = 0.2253$ $c = -0.01978$ $\tilde{P}_1 = 6.8426$ $\tilde{P}_2 = -26.8381$ $\tilde{P}_3 = 4.513$ $\tilde{\mu}_c = -11.29$ $\tilde{\mu}_1 = -22.67$ $\tilde{\mu}_2 = -19.14$ $\tilde{\mu}_3 = 0$	$T_c = 282.345$ K $\rho_c = 214.16$ kg/m^3 $b^2 = 1.6747$ $a = 23.981$ $k_0 = 1.3762$ $k_1 = 0.7662$ $c = -0.0078$ $\tilde{P}_1 = 5.3439$ $\tilde{P}_2 = -16.9686$ $\tilde{P}_3 = 0$ $\tilde{\mu}_c = -36.48$ $\tilde{\mu}_1 = -27.35$ $\tilde{\mu}_2 = -12.47$ $\tilde{\mu}_3 = -11.60$
Derived parameters	$P_c = 22.0276$ MPa $\tilde{P}_{11} = 0.511$	$P_c = 5.0403$ MPa $\tilde{P}_{11} = -0.1818$

universal values [72]. By leaving the parameter b^2 free in our fits, a test of this universality hypothesis was possible (see eqs. (A.31)–(A.34) in the Appendix). Values of the universal critical amplitude ratios obtained for steam and ethylene are compared in table 4 with the values predicted theoretically for Ising-like systems. The good agreement supports the hope that our use of the Wegner expansion has correctly identified the range of asymptotic scaling in these fluids.

Table 4. Critical amplitude ratios

	Theory[a]	Theory[b]	Steam [70]	Ethylene [71]
A^+/A^-	0.51	0.55	0.53	0.50
Γ^+/Γ^-	5.07	4.80	4.89	5.31
$\Gamma^+ D B^{\delta-1}$	1.75	1.6	1.69	1.76
$A^+\Gamma^+/B^2$	0.059	0.066	0.057	0.051

[a] From series expansions for Ising model [72]
[b] From ε-expansion of renormalization-group equations [72]

3.5. Range of asymptotic scaling in fluids

In this section we estimate the range of validity of the asymptotic scaling laws for a variety of fluids near the gas–liquid critical point and binary liquids near the critical mixing point. Our sources of information are analyses including corrections to scaling, performed by a number of investigators, including ourselves, for a variety of fluid properties. In particular, we have used Ley-Koo and Green's [21] analysis of the coexistence curve data for sulfur hexafluoride, Greer's analysis of the coexistence curve data for the isobutyric acid-water and carbon disulfide-nitromethane mixtures [53], the coexistence curve data for polystyrene–cyclohexane measured and analyzed by Nakata et al. [73], the analyses of Balfour et al. of the PVT data of steam [70] and ethylene [71] referred to in sect. 3.4, the specific heat data for the triethylamine–water mixture measured and analyzed by Thoen et al. [74], the coexistence curve data for xenon measured and analyzed by Hayes and Carr [54], the coexistence curve and compressibility data for He^3 measured and analyzed by Pittman et al. [75] and the specific heat data for carbon dioxide measured by Lipa et al. [76].

We summarize the information available for these fluids in a succinct way by the following procedure. Consider the three properties coexistence curve, compressibility and specific heat, and let the size of the asymptotic range be determined by requiring that the first Wegner correction contributes not more than 1% to the property in question. In most cases we accepted the fits as given by the authors*. Since the amplitude of the first correction term is not universally constant, we should expect variations in the size of the asymptotic range from fluid to fluid; additional variations are introduced because of the spread in exponent values used by the different authors. Nevertheless, a general picture emerges. The range of validity of the asymptotic power law for the coexistence curve is at most of the order of 10^{-3} in reduced temperature $\Delta \tilde{T}$, sometimes even smaller (fig. 6). For the susceptibility, $(\partial \tilde{\rho}/\partial \tilde{\mu})_T$, of H_2O and C_2H_4 we find a mean value of 10^{-3} in $\Delta \tilde{T}$ for the asymptotic range of the supercritical data, but a much smaller range for He^3 and for the data below T_c (fig. 7). This is consistent with the fact that the exponent equality $\gamma = \gamma'$ (see table 1) has not been demonstrated in practice. It also explains why apparent critical exponents for He^3 have been persistently different from those determined for other

---

*For the purpose of estimating the size of the asymptotic range, the difference between t as defined in eq. (1) and $\Delta \tilde{T}$ as defined in eq. (6) is irrelevant.

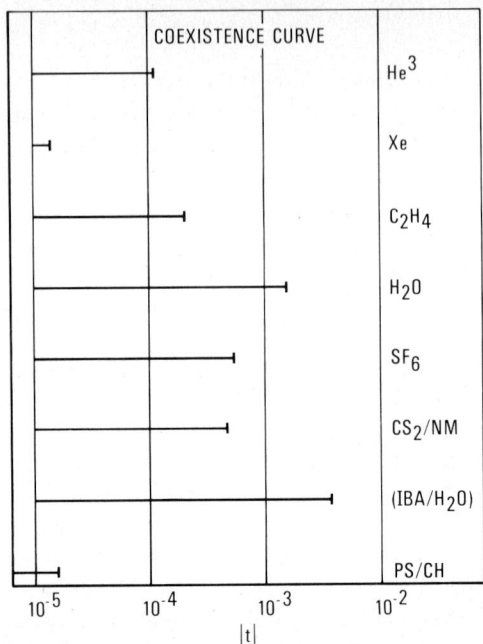

Fig. 6. Range of the asymptotic power law for the coexistence curve of fluids near the critical point. He^3: ref. [75]; Xe: ref. [54]; C_2H_4: ref. [71]; H_2O: ref. [70]; SF_6: ref. [21]; CS_2 – nitromethane and isobutyric acid – H_2O: ref. [53]; polystyrene–cyclohexane: ref. [73]. In IBA/H_2O the amplitude of the first Wegner correction is quite uncertain.

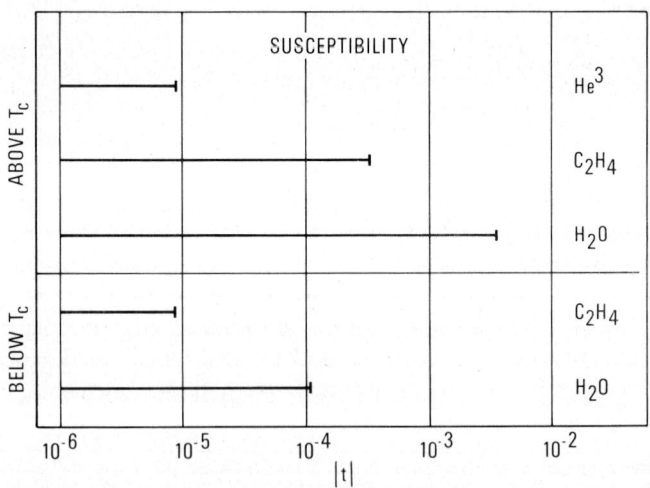

Fig. 7. Range of the asymptotic power law for the susceptibility $(\partial\rho/\partial\mu)_T$ of fluids near the critical point. He^3: ref. [75]; C_2H_4: ref. [71]; H_2O: ref. [70].

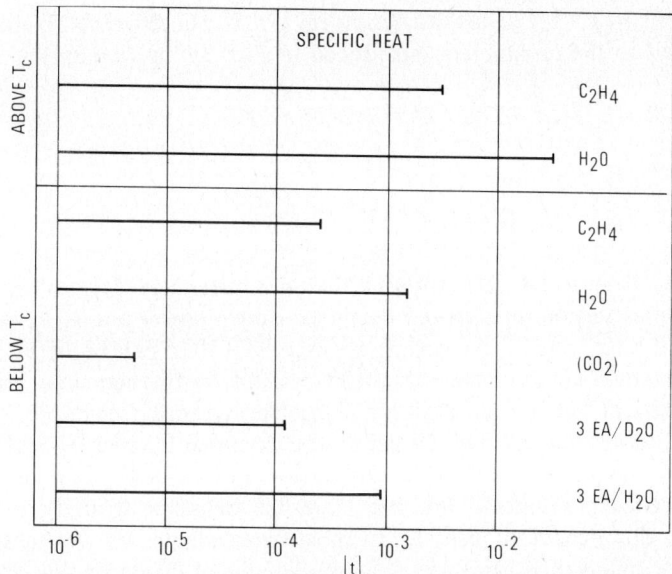

Fig. 8. Range of the asymptotic power law for the specific heat of fluids near the critical point. C_2H_4: ref. [71]; H_2O: ref. [70]; CO_2: ref. [76] and this work, triethylamine – $H_2O(D_2O)$: ref. [74]. In CO_2 the amplitude of the first Wegner correction is quite uncertain.

fluids in equally large regions [30]. For the specific heat we find again a small asymptotic range, less than 1°C for fluids with a critical temperature near room temperature. This explains why fits of the form $C_v/T = A^\pm |t|^{-\alpha} + B^\pm$, applied to specific heat data over ranges of several degrees, result in different analytic backgrounds above (B^+) and below (B^-) the critical temperature [74].

3.6. Assessment of the Wegner expansion for fluids

There is no better illustration of both the power and the weakness of the Wegner expansion than Ley-Koo and Green's analysis of the coexistence curve of sulfur hexafluoride [21]. The Wegner expansion implies that the densities ρ_l, ρ_g at the liquid and gas sides of the coexistence boundary are to be represented by:

$$\Delta\rho_{cxc} \equiv (\rho_l - \rho_g)/2 = \rho_c \left[B_0 |t|^\beta + B_1 |t|^{\beta + \Delta_1} + \ldots \right], \qquad (16)$$

$$\rho_d \equiv (\rho_l + \rho_g)/2 \equiv \rho_c + D_0 |t|^{1-\alpha} + D_1 t + D_2 |t|^{1-\alpha+\Delta_1} \ldots, \qquad (17)$$

where $t=(T-T_c)/T_c$ as defined in eq. (1). The coefficients B_i and D_i are related to the parameters introduced in sect. 3.3 by ($\theta = \pm 1$):

$$B_0 = k_0/(b^2-1)^\beta, \quad B_1 = k_1/(b^2-1)^{\beta+\Delta_1},$$
$$D_0/\rho_c = cak_0 s_0(1)/(b^2-1)^\alpha, \quad D_1/\rho_c = \tilde{P}_{11},$$
$$D_2/\rho_c = cak_1 s_1(1)/(b^2-1)^{\alpha+\Delta_1}. \tag{18}$$

Ley-Koo and Green showed that the difference $\Delta\rho_{cxc}$ of coexisting densities can be represented by the one-term power law $B_0|t|^\beta$ only for $|t| \leqslant 7 \times 10^{-4}$; with two terms, a range $|t| \leqslant 10^{-2}$ can be fitted and the exponent β assumes the value 0.327 ± 0.005, in fine agreement with the theoretical value predicted for Ising-like systems (see table 2). With three terms, however, the fit cannot be extended beyond $|t| = 2.1 \times 10^{-2}$. The strength of the Wegner expansion is its ability to reconcile the theoretically predicted limiting Ising-like behavior with the departure from this behavior observed in most experiments. Its weakness is the poor convergence, because of the small value of the correction-to-scaling exponent Δ_1. Many terms, each with an adjustable amplitude, are needed to describe a coexistence curve that is deceptively easily described by a simple power law with an apparent exponent $\beta \simeq 0.35$ in an appreciable temperature range [30, 43, 75]. Ley-Koo and Green's analysis of the coexistence curve diameter ρ_d brings this point home even more forcefully. As originally noticed by Cailletet and Mathias [77], for most fluids and for most practical purposes, the diameter is a straight line over a range as large as $|t| \leqslant 0.1$. The gas-liquid asymmetry leads us to expect, however, that the slope of the diameter should asymptotically exhibit a weak divergence, manifesting itself in some curvature of the diameter near the critical point [62, 63]. In the experiment of Weiner et al. this curvature was seen [65]. Ley-Koo and Green showed that the asymptotic behavior, $\rho_d = \rho_c + D_0|t|^{1-\alpha}$, applied in a range no larger than $|t| \leqslant 8 \times 10^{-4}$. Adding the linear term $D_1 t$, they could fit the data out to $|t| \leqslant 3 \times 10^{-3}$ (see fig. 9). However, a curious fact emerged: the coefficient D_1 of the linear term had a sign opposite to that of the slope of the rectilinear diameter observed over a large temperature range. Similarly, the values found in table 3 for $\tilde{P}_{11}$ of H_2O and C_2H_4 differ considerably from the values -1.60 and -0.777, respectively, observed for the slope of the rectilinear diameter for these fluids. In other words, the analytic backgrounds in the scaling laws do not seem to have an obvious relation to the behavior of the property in question far away from the critical point. In order to reproduce this global behavior, many Wegner terms will have to combine with the analytic background.

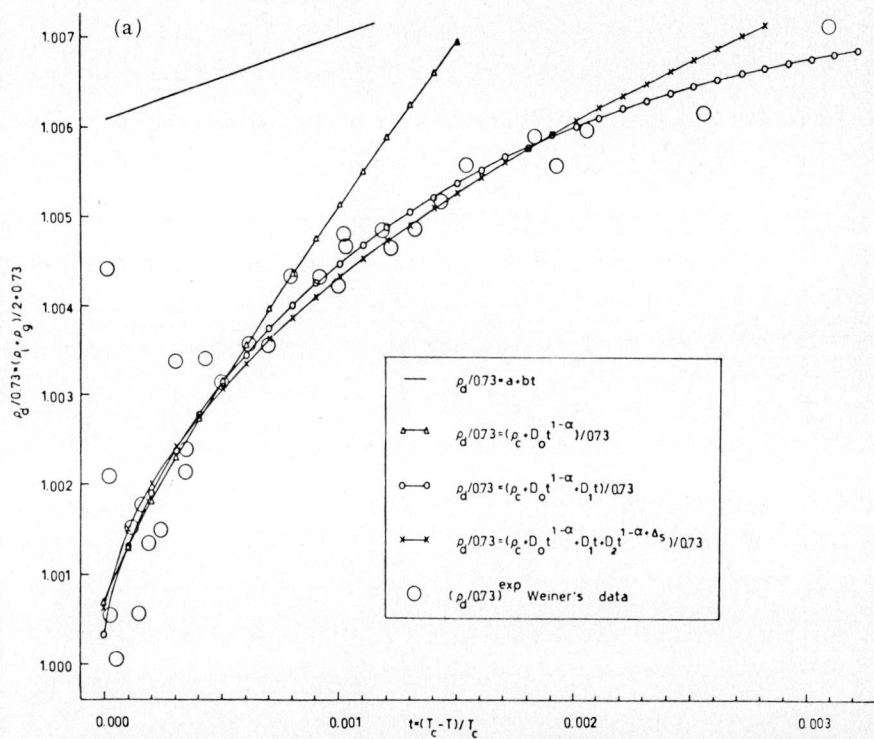

Fig. 9. Comparison of the experimental data for the diameter ρ_d of SF_6 with the Wegner expansion in the temperature range $0 < |t| < 0.0035$ as reported by Ley-Koo and Green [21].

Figure 10, taken from the paper of Ley-Koo and Green [21], illustrates dramatically the problem we are facing. The empirical law of the rectilinear diameter holds with great precision for a temperature range exceeding $|t| = 0.05$ in the case of SF_6. The $|t|^{1-\alpha}$ term is needed to describe the curvature near the critical point. In order to mimic what is, away from the critical temperature, a straight line, four terms, including one Wegner correction, will not be sufficient beyond $|t| = 0.02$.

The same difficulty is experienced when specific heat data near the critical point of fluids are fitted to the expression $C_v = A^\pm |t|^{-\alpha} + B^\pm$. The value found for the constant background $B^+ = B^-$ does not bear any obvious relation to the value of C_v far away from the critical temperature; it is not even uncommon for the constant to be negative [29, 60, 76, 78]. For the binary liquid triethylamine-water near the critical mixing point, Beysens and Bourgou identified the coefficient D_1 of the linear background term in eq. (17) for the concentration diameter

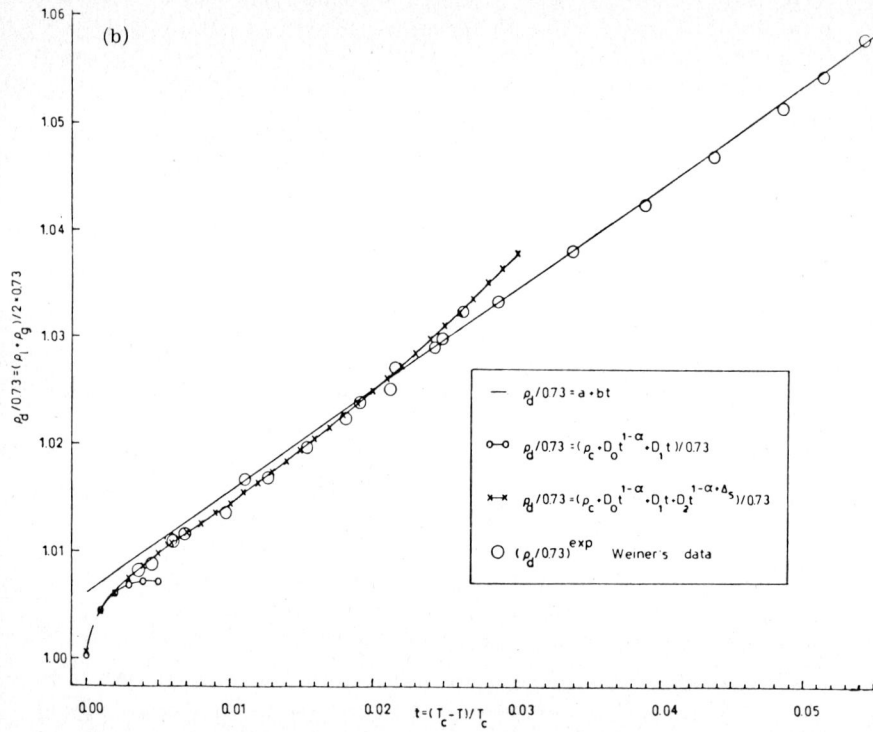

Fig. 10. Comparison of the experimental data for the diameter ρ_d of SF_6 with the Wegner expansion in the temperature range $0 < |t| < 0.055$ as reported by Ley-Koo and Green [21].

with the value deduced from the ideal behavior away from the critical point [79]. However, when the specific heat data for the same system are fitted to $C_{p,x} = A_0^\pm |t|^{-\alpha} + A_1^\pm |t|^{-\alpha + \Delta_1} + B^\pm$, the background $B^+ = B^-$ is very different from the background the specific heat data appear to approach away from the critical temperature [74]; it seems, again, that many Wegner correction terms would be needed to produce the value of the specific heat away from the critical temperature. Hence, we expect that the practical applicability of the Wegner expansion will be rather limited.

4. Crossover; an outlook

At the end of their paper on the coexistence curve of SF_6, Ley-Koo and Green suggested that, instead of the poorly-converging Wegner expansion a more promising approach might be a crossover description from non-classical behavior close to the critical point to mean-field behavior further away from the critical point [21]. They obtained a mean-field description by fitting the data for the coexisting densities to the expansion [80, 81]:

$$\rho^{\pm}/\rho_c = 1 \pm b_1|t|^{1/2} + b_2 t \pm b_3|t|^{3/2} + \ldots, \tag{19}$$

where $\rho^+ = \rho_l$ and $\rho^- = \rho_g$. Leaving the critical temperature as an adjustable parameter, they found a reasonable fit to the difference $(\rho^+ - \rho^-)/\rho_c$ in the range $0.02 \leq |t| \leq 0.055$ with an effective mean-field critical temperature 2 K above the actual critical temperature [21].

This result encouraged Mel to find a theoretical foundation for the crossover phenomenon. In his contribution to a volume honoring Elliott Montroll, he described his quest for a Van der Waals-type fixed point using the renormalization-group approach [22]. He located such a fixed point in the limit of weak long-range forces, but found it to be unstable for dimensionality smaller than 4. Hence, under a renormalization transformation, the critical Hamiltonian of a real fluid may tend towards the mean-field point initially, but the trajectory will ultimately change direction and head to the Ising-like fixed point.

Nicoll, Chang and Stanley [82, 83], Nelson and Rudnick [84] and de Pasquale and Tombesi [85] consider the crossover to take place not to a Van der Waals type fixed point, but rather to the well-established Gaussian fixed points. These authors give an explicit expression for the crossover function to first order in $\varepsilon = 4 - d$, where d is the dimensionality. This expression contains only one non-universal parameter. The global equation of state, obtained in magnetic language, has the feature that it scales with non-classical exponents for small $|t|$, but yields mean-field exponents for large $|t|$. Expansion around the critical point gives the Wegner corrections due to the first irrelevant scaling field with amplitudes that are all interrelated, because the crossover function contains only one non-universal constant. Chang et al. tried to fit their crossover function to the coexistence curve data of xenon determined by Garland and Thoen, but were dissatisfied with the result [86]. It should be remarked that the xenon data extended to $|t|=0.03$ or 10 K below the critical temperature, a fairly large range. De Pasquale et al. proposed a parametric representation of the crossover function [85] and reported success in fitting this function to Greer's coexistence curve data for the isobutyric acid–water system [87]. The range of this data set, however, extends only to 2 K below the critical temperature, where an expansion with one Wegner correction term does equally well.

The crossover functions discussed above have the advantages of simple structure and few adjustable parameters. However, they have the drawback of crossing over at $t \to \infty$ to the limiting classical scaling behavior rather than to the full classical behavior which itself would require corrections to the mean-field asymptotic power laws. To give an example, Van der Waals' equation as analyzed by Van Laar [80], requires a 1% correction to the pure $\beta = \frac{1}{2}$ power-law behavior of the coexistence curve at $|t| = 0.025$, while Ley-Koo and Green's analysis of the coexistence curve of SF_6 with a mean-field expansion indicates the presence of a 1% correction already at $|t| = 0.005$. These limits are well within the range of the xenon data analyzed by Chang et al. [86] so that a crossover to a $\beta = \frac{1}{2}$ behavior at $t = \infty$ will not do justice to the classical behavior that may prevail away from the critical point in a real fluid.

Concluding, we have found that the Wegner expansion describes the departure from asymptotic scaling in fluids adequately, but that its range of applicability is limited. The alternative, crossover from Ising-like to mean-field behavior, is being developed by various investigators, and has already resulted in proposals for an explicit form for the crossover functions. The functions proposed thus far, however, cross over only to the limiting critical behavior of the classical equation and higher order contributions to the classical behavior need to be incorporated. Hence, much work remains to be done before Voronel's goal, a cohesive description of the entire fluid range radiating out from the critical point, will be attained.

Acknowledgments

This article is dedicated to the memory of our friend and colleague Melville S. Green. We are indebted to Frances W. Balfour for her participation in this project. We also acknowledge stimulating discussions with Jeffrey F. Nicoll.

The research at the University of Maryland was supported by the National Science Foundation under grant DMR 79-10819.

5. Appendix. Revised and extended scaling equations for the thermodynamic properties of fluids

5.1. Reduced thermodynamic quantities

$$\tilde{T} = -\frac{T_c}{T}, \quad \tilde{\mu} = \frac{\mu}{T} \frac{\rho_c T_c}{P_c}, \quad \tilde{P} = \frac{P}{T} \frac{T_c}{P_c},$$

$$\tilde{\rho} = \frac{\rho}{\rho_c}, \quad \tilde{U} = \frac{U}{V}\frac{1}{P_c}, \quad \tilde{S} = \frac{S}{V}\frac{T_c}{P_c},$$

$$\tilde{A} = \frac{A}{VT}\frac{T_c}{P_c}, \quad \tilde{H} = \frac{H}{VT}\frac{T_c}{P_c}, \quad \tilde{\chi}_T = \left(\frac{\partial \tilde{\rho}}{\partial \tilde{\mu}}\right)_T, \qquad \text{(A.1)}$$

$$\tilde{C}_v = \frac{C_v}{V}\frac{T_c}{P_c}, \quad \tilde{C}_p = \frac{C_p}{V}\frac{T_c}{P_c}.$$

5.2. Thermodynamic relations

$$\begin{aligned}
d\tilde{P} &= \tilde{U}d\tilde{T} + \tilde{\rho}d\tilde{\mu}, \\
d\tilde{A} &= -\tilde{U}d\tilde{T} + \tilde{\mu}d\tilde{\rho}, \\
d\tilde{H} &= -\tilde{T}d\tilde{U} + \tilde{\rho}d\tilde{\mu}, \\
d\tilde{S} &= -\tilde{T}d\tilde{U} - \tilde{\mu}d\tilde{\rho},
\end{aligned} \qquad \text{(A.2)}$$

with

$$\begin{aligned}
\tilde{A} &= \tilde{\rho}\tilde{\mu} - \tilde{P}, \\
\tilde{H} &= \tilde{P} - \tilde{T}\tilde{U}, \\
\tilde{S} &= \tilde{H} - \tilde{\rho}\tilde{\mu} = -\tilde{T}\tilde{U} - \tilde{A}.
\end{aligned} \qquad \text{(A.3)}$$

5.3. Fundamental equations

$$\Delta \tilde{T} = \tilde{T} + 1, \qquad \text{(A.4a)}$$
$$\Delta \tilde{\mu} = \tilde{\mu} - \tilde{\mu}_0(\tilde{T}), \qquad \text{(A.4b)}$$
$$\tilde{P} = \tilde{P}_0(\tilde{T}) + \Delta\tilde{\mu} + \tilde{P}_{11}\Delta\tilde{\mu}\Delta\tilde{T} + \Delta\tilde{P}, \qquad \text{(A.5)}$$

with

$$\tilde{\mu}_0(T) = \tilde{\mu}_c + \sum_{j=1}^{3} \tilde{\mu}_j (\Delta \tilde{T})^j, \qquad \text{(A.6a)}$$

$$\tilde{P}_0(\tilde{T}) = 1 + \sum_{j=1}^{3} \tilde{P}_j (\Delta \tilde{T})^j. \qquad \text{(A.6b)}$$

5.4. Derived thermodynamic quantities

$$\tilde{\rho} = 1 + \tilde{P}_{11}\Delta\tilde{T} + \left(\frac{\partial \Delta \tilde{P}}{\partial \Delta \tilde{\mu}}\right)_{\Delta \tilde{T}}, \qquad \text{(A.7)}$$

$$\tilde{U} = \frac{d\tilde{P}_0}{d\tilde{T}} - \tilde{\rho}\frac{d\tilde{\mu}_0}{d\tilde{T}} + \tilde{P}_{11}\Delta\tilde{\mu} + \left(\frac{\partial\Delta\tilde{P}}{\partial\Delta\tilde{T}}\right)_{\Delta\tilde{\mu}},\tag{A.8}$$

$$\tilde{\chi}_T = \left(\frac{\partial^2\Delta\tilde{P}}{\partial\Delta\tilde{\mu}^2}\right)_{\Delta\tilde{T}},\tag{A.9}$$

$$\left(\frac{\partial\tilde{P}}{\partial\tilde{T}}\right)_{\tilde{\rho}} = \frac{d\tilde{P}_0}{d\tilde{T}} + \tilde{P}_{11}\left[\Delta\tilde{\mu} - \frac{\tilde{\rho}}{\tilde{\chi}_T}\right] + \left(\frac{\partial\Delta\tilde{P}}{\partial\Delta\tilde{T}}\right)_{\Delta\tilde{\mu}} - \frac{\tilde{\rho}}{\tilde{\chi}_T}\frac{\partial^2\Delta\tilde{P}}{\partial\Delta\tilde{\mu}\Delta\tilde{T}},\tag{A.10}$$

$$\frac{\tilde{C}_v}{\tilde{T}^2} = \frac{d^2\tilde{P}_0}{d\tilde{T}^2} - \tilde{\rho}\frac{d^2\tilde{\mu}_0}{d\tilde{T}^2} - \frac{\tilde{P}_{11}^2}{\tilde{\chi}_T} + \left(\frac{\partial^2\Delta\tilde{P}}{\partial\Delta\tilde{T}^2}\right)_{\Delta\tilde{\mu}}$$

$$- \frac{2\tilde{P}_{11}}{\tilde{\chi}_T}\frac{\partial^2\Delta\tilde{P}}{\partial\Delta\tilde{\mu}\Delta\tilde{T}} - \frac{1}{\tilde{\chi}_T}\left(\frac{\partial^2\Delta\tilde{P}}{\partial\Delta\tilde{\mu}\Delta\tilde{T}}\right)^2,\tag{A.11}$$

$$\tilde{C}_p = \tilde{C}_v + \frac{\tilde{\chi}_T}{\tilde{\rho}^2}\left[\tilde{P} - \tilde{T}\left(\frac{\partial\tilde{P}}{\partial\tilde{T}}\right)_{\tilde{\rho}}\right]^2.\tag{A.12}$$

5.5. Critical exponents

$$\begin{aligned}&\alpha_0 = \alpha, \quad \alpha_1 = \alpha - \Delta_1,\\ &\beta_0 = \beta, \quad \beta_1 = \beta + \Delta_1,\\ &\gamma_0 = \gamma, \quad \gamma_1 = \gamma - \Delta_1,\end{aligned}\tag{A.13}$$

with

$$2 - \alpha = \beta(\delta + 1), \quad \gamma = \beta(\delta - 1).\tag{A.14}$$

5.6. Parametric equations for singular terms

$$\Delta\tilde{\mu} = r^{\beta\delta}a\theta(1 - \theta^2),\tag{A.15}$$

$$\Delta\tilde{T} = r(1 - b^2\theta^2) - c\Delta\tilde{\mu},\tag{A.16}$$

$$\Delta\tilde{P} = \sum_{i=0}^{1} r^{2-\alpha_i}ak_i p_i(\theta),\tag{A.17}$$

$$\left(\frac{\partial\Delta\tilde{P}}{\partial\Delta\tilde{\mu}}\right)_{\Delta\tilde{T}} = \sum_{i=0}^{1}\left[r^{\beta_i}k_i\theta + cr^{1-\alpha_i}ak_i s_i(\theta)\right],\tag{A.18}$$

$$\left(\frac{\partial \Delta \tilde{P}}{\partial \Delta \tilde{T}}\right)_{\Delta\tilde{\mu}} = \sum_{i=0}^{1} r^{1-\alpha_i} ak_i s_i(\theta), \tag{A.19}$$

$$\left(\frac{\partial^2 \Delta \tilde{P}}{\partial \Delta \tilde{\mu}^2}\right)_{\Delta\tilde{T}} = \sum_{i=0}^{1} \left[r^{-\gamma_i} \frac{k_i}{a} u_i(\theta) + 2cr^{\beta_i - 1} k_i v_i(\theta) + c^2 r^{-\alpha_i} ak_i w_i(\theta) \right], \tag{A.20}$$

$$\frac{\partial^2 \Delta \tilde{P}}{\partial \Delta \tilde{\mu} \Delta \tilde{T}} = \sum_{i=0}^{1} \left[r^{\beta_i - 1} k_i v_i(\theta) + cr^{-\alpha_i} ak_i w_i(\theta) \right], \tag{A.21}$$

$$\left(\frac{\partial^2 \Delta \tilde{P}}{\partial \Delta \tilde{T}^2}\right)_{\Delta\tilde{\mu}} = \sum_{i=0}^{1} r^{-\alpha_i} ak_i w_i(\theta). \tag{A.22}$$

5.7. Auxiliary functions

$$p_i(\theta) = p_{0i} + p_{2i}\theta^2 + p_{4i}\theta^4, \tag{A.23}$$
$$s_i(\theta) = s_{0i} + s_{2i}\theta^2, \quad s'_i(\theta) = 2s_{2i}\theta, \tag{A.24}$$
$$q(\theta) = 1 + \{b^2(2\beta\delta - 1) - 3\}\theta^2 - b^2(2\beta\delta - 3)\theta^4, \tag{A.25}$$
$$u_i(\theta) = \left[1 - b^2(1 - 2\beta_i)\theta^2\right]/q(\theta), \tag{A.26}$$
$$v_i(\theta) = \left[\beta_i(1 - 3\theta^2)\theta - \beta\delta(1 - \theta^2)\theta\right]/q(\theta), \tag{A.27}$$
$$w_i(\theta) = \left[(1 - \alpha_i)(1 - 3\theta^2)s_i(\theta) - \beta\delta(1 - \theta^2)\theta s'_i(\theta)\right]/q(\theta), \tag{A.28}$$

with

$$p_{0i} = + \frac{\beta\delta - 3\beta_i - b^2\alpha_i\gamma_i}{2b^4(2-\alpha_i)(1-\alpha_i)\alpha_i},$$

$$p_{2i} = - \frac{\beta\delta - 3\beta_i - b^2\alpha_i(2\beta\delta - 1)}{2b^2(1-\alpha_i)\alpha_i},$$

$$p_{4i} = + \frac{2\beta\delta - 3}{2\alpha_i}, \tag{A.29}$$

$$s_{0i} = (2 - \alpha_i)p_{0i},$$

$$s_{2i} = - \frac{\beta\delta - 3\beta_i}{2b^2\alpha_i}. \tag{A.30}$$

5.8. Critical amplitude ratios

$$\frac{A^+}{A^-} = \frac{(b^2-1)^{2-\alpha} p_{00}}{p_{00}+p_{20}+p_{40}}, \tag{A.31}$$

$$\frac{\Gamma^+}{\Gamma^-} = \frac{2}{(b^2-1)^{\gamma-1}\{1-b^2(1-2\beta)\}}, \tag{A.32}$$

$$\Gamma^+ D B^{\delta-1} = \frac{b^{\delta-3}}{(b^2-1)^{\gamma-1}}, \tag{A.32}$$

$$\frac{A^+ \Gamma^+}{B^+} = (2-\alpha)(1-\alpha)\alpha(b^2-1)^{2\beta} p_{00}. \tag{A.34}$$

References

[1] M. S. Green and J. V. Sengers, eds., Critical Phenomena, Proceedings of a Conference held in Washington, D.C., NBS Miscellaneous Publ. **273** (U.S. Government Printing Office, Washington, D.C., 1966).
[2] Ref. [1] p. vii.
[3] M. E. Fisher, ref. [1] p. 21.
[4] B. Widom, J. Chem. Phys. **43** (1965) 3898.
[5] C. Domb and D. L. Hunter, Proc. Phys. Soc. **86** (1965) 1147.
[6] Ref. [1] p. 130.
[7] A. V. Voronel, in: Critical Phenomena and Phase Transitions, Vol. **5B**, eds. C. Domb and M. S. Green (Academic Press, New York, 1976) p. 344.
[8] R. B. Griffiths, Phys. Rev. **158** (1967) 176.
[9] L. P. Kadanoff, Physics **2** (1966) 263.
[10] L. P. Kadanoff, in: Critical Phenomena, Proceedings of the International School of Physics, Enrico Fermi, Course LI, ed. M. S. Green (Academic Press, New York, 1971), p. 100.
[11] R. B. Griffiths, Phys. Rev. Lett. **24** (1970) 1479.
[12] M. E. Fisher, Phys. Rev. Lett. **16** (1966) 11.
[13] P. G. Watson, J. Phys. C2 (1969) 1883; **2** (1969) 2158.
[14] D. Jasnow and M. Wortis, Phys. Rev. **176** (1968) 739.
[15] K. G. Wilson and J. Kogut, Phys. Rep. **12C** (1974) 75.
[16] D. Balzarini and K. Ohrn, Phys. Rev. Lett. **29** (1972) 840.
[17] D. Balzarini, Can. J. Phys. **52** (1974) 499.
[18] R. Hocken and M. R. Moldover, Phys. Rev. Lett. **37** (1976) 29.
[19] F. Wegner, Phys. Rev. **B5** (1972) 4529.

[20] M. Ley-Koo, Consequences of the Renormalization Group for the Thermodynamics of Fluids near the Critical Point, Thesis (Department of Physics, Temple University, Philadelphia, PA, 1976); M. Ley-Koo and M. S. Green, Phys. Rev. (to be published).
[21] M. Ley-Koo and M. S. Green, Phys. Rev. A **16** (1977) 2483.
[22] M. S. Green, in: Statistical Mechanics and Statistical Methods in Theory and Application", ed. U. Landman (Plenum Publ. Corp., 1977) p. 73.
[23] E. H. W. Schmidt, ref. [1] p. 13.
[24] J. Straub, Chem. Ing.-Techn. **5/6** (1967) 291.
[25] J. M. H. Levelt Sengers, in: Experimental Thermodynamics, Vol. II, eds. B. Le Neindre and B. Vodar (Butterworths, London, 1975) p. 657.
[26] M. R. Moldover, J. V. Sengers, R. W. Gammon and R. J. Hocken, Revs. Mod. Phys. **51** (1979) 79.
[27] R. B. Griffiths, Phys. Rev. **158** (1967) 176.
[28] M. S. Green, M. Vicentini-Missoni and J. M. H. Levelt Sengers, Phys. Rev. Lett. **18** (1967) 1113.
[29] M. Vicentini-Missoni, J. M. H. Levelt Sengers and M. S. Green, J. Res. Natl. Bur. Stand. **73A** (1969) 563.
[30] J. M. H. Levelt Sengers, Physica **73** (1974) 73.
[31] J. M. H. Levelt Sengers and J. V. Sengers, Phys. Rev. A **12** (1975) 2622.
[32] J. M. H. Levelt Sengers, W. L. Greer and J. V. Sengers, J. Phys. Chem. Ref. Data **5** (1976) 1.
[33] C. Domb and M. S. Green, eds. Phase Transitions and Critical Phenomena, Vol. 3 (Academic Press, New York, 1974).
[34] L. P. Kadanoff, in: Phase Transitions and Critical Phenomena, Vol. 5A, eds. C. Domb and M. S. Green (Academic Press, New York, 1976) p. 1.
[35] C. Domb, ref. [33] p. 357.
[36] W. J. Camp, D. M. Saul, J. P. Van Dyke and M. Wortis, Phys. Rev. B **14** (1976) 3990.
[37] D. S. Gaunt and M. F. Sykes, J. Phys. A **12** (1979) L25.
[38] G. A. Baker, B. G. Nickel, M. S. Green and D. I. Meiron, Phys. Rev. Lett. **36** (1976) 1351;
G. A. Baker, B. G. Nickel and D. I. Meiron, Phys. Rev. B **17** (1978) 1365.
[39] J. C. Le Guillou and J. Zinn-Justin, Phys. Rev. Lett. **39** (1977) 95.
[40] G. A. Baker, Phys. Rev. B **15** (1977) 1552.
[41] G. A. Baker and J. M. Kincaid, Phys. Rev. Lett. **42** (1979) 1431.
[42] J. Zinn-Justin, J. Physique **40** (1979) 969.
[43] J. M. H. Levelt Sengers, J. Straub and M. Vicentini-Missoni, J. Chem. Phys. **54** (1971) 5034.
[44] B. Wallace and H. Meyer, Phys. Rev. A **2** (1970) 1563.
[45] J. D. Gunton and M. S. Green, eds., Renormalization Group in Critical Phenomena and Quantum Field Theory: Proceedings of a Conference (Department of Physics, Temple University, Philadelphia, PA, 1973).
[46] G. Ahlers, ref. [45] p. 137; Phys. Rev. A **8** (1973) 530.
[47] D. S. Greywall and G. Ahlers, Phys. Rev. A **7** (1973) 2145.
[48] L. M. Stacey, B. Pass and H. Y. Carr, Phys. Rev. Lett. **23** (1969) 1424.
[49] L. A. Weber, Phys. Rev. A **2** (1970) 2379.
[50] J. M. H. Levelt Sengers, ref. [45] p. 112.
[51] L. R. Wilcox and D. Balzarini, J. Chem. Phys. **48** (1968) 753;
W. T. Estler, R. Hocken, T. Charlton and L. R. Wilcox, Phys. Rev. A **12** (1975) 2118.
[52] A. Levelt Sengers, R. J. Hocken and J. V. Sengers, Physics Today **30** (12) (1977) 42.
[53] S. C. Greer, Phys. Rev. A **14** (1976) 1770.

[54] C. E. Hayes and H. Y. Carr, Phys. Rev. Lett. **39** (1977) 1558.
[55] D. T. Jacobs, D. L. Anthony, R. C. Mockler and W. J. O'Sullivan, Chem. Phys. **20** (1977) 219.
[56] D. Beysens, J. Chem. Phys. **71** (1979) 2557.
[57] M. S. Green, M. J. Cooper and J. M. H. Levelt Sengers, Phys. Rev. Lett. **26** (1971) 491.
[58] P. Schofield, Phys. Rev. Lett. **22** (1969) 606;
P. Schofield, J. D. Litster and J. T. Ho, Phys. Rev. Lett **23** (1969) 1098.
[59] R. B. Griffiths and J. C. Wheeler, Phys. Rev. A **2** (1970) 1047.
[60] F. J. Cook, Theoretical Study and Data Analysis of Thermodynamic Properties in the Neighborhood of the Critical Point, Ph.D. thesis (Department of Physics, Temple University, Philadelphia, Pa., 1972).
[61] B. Widom and J. S. Rowlinson, J. Chem. Phys **52** (1970) 1670.
[62] N. D. Mermin and J. J. Rehr, Phys. Rev. Lett. **26** (1971) 1155; Phys. Rev. A **4** (1971) 2408;
J. J. Rehr and N. D. Mermin, Phys. Rev. A **8** (1973) 472.
[63] J. A. Zollweg and G. W. Mulholland, J. Chem. Phys. **57** (1972) 1021;
G. W. Mulholland, J. A. Zollweg and J. M. H. Levelt Sengers, J. Chem. Phys. **62** (1975) 2535.
[64] E. S. R. Gopal, R. Ramachandra and P. Chandra Sekhar, Pramana **1** (1973) 260;
E. S. R. Gopal, R. Ramachandra and P. Chandra Sekhar, K. Govindarajun and S. V. Subramanyam, Phys. Rev. Lett. **32** (1974) 284.
[65] J. Weiner, K. H. Langley and N. C. Ford, Phys. Rev. Lett. **32** (1974) 879.
[66] J. V. Sengers and J. M. H. Levelt Sengers, in: Progress in Liquid Physics, ed. C. A. Croxton (Wiley, New York, 1978) p. 103.
[67] F. W. Balfour, J. V. Sengers, M. R. Moldover and J. M. H. Levelt Sengers, in: Proceedings 7th Symposium on Thermophysical Properties, ed. A. Cezairliyan, (American Society of Mechanical Engineers, New York, 1977) p. 786; Phys. Lett. **65A** (1978) 223.
[68] M. R. Moldover, J. Res. Natl. Bur. Stand. **83** (1978) 329.
[69] F. W. Balfour, Ph.D. thesis (Department of Chemistry, University of Maryland, College Park, MD, 1980).
[70] F. W. Balfour, J. V. Sengers and J. M. H. Levelt Sengers, in: Water and Steam: Their Properties and Current Industrial Applications, eds. J. Straub and K. Scheffler (Pergamon Press, Oxford, New York, 1980) p. 128.
[71] J. R. Hastings, J. M. H. Levelt Sengers, and F. W. Balfour, J. Chem. Thermod. (in press).
[72] A. Aharony and P. C. Hohenberg, Phys. Rev. B **13** (1976) 3081.
[73] M. Nakata, T. Dobashi, H. Kuwahara, M. Kaneko and B. Chu, Phys. Rev. A **18** (1978) 2683.
[74] J. Thoen, E. Bloemen and W. Van Dael, J. Chem. Phys. **68** (1978) 735.
[75] C. Pittman, T. Doiron and H. Meyer, Phys. Rev. B **20** (1979) 3678.
[76] J. A. Lipa, C. Edwards and M. J. Buckingham, Phys. Rev. A **15** (1977) 778.
[77] L. Cailletet and E. Mathias, Compt. Rend. **102** (1886) 1202; **104** (1887) 1563.
[78] M. R. Moldover, Phys. Rev. **182** (1969) 342.
[79] D. Beysens and A. Bourgou, Phys. Rev. A **19** (1979) 2407.
[80] J. J. Van Laar, Comm. Koninklijke Akademie van Wetenschappen, Amsterdam **14** (II), 1091 (1911/12); Die Zustandsgleichung von Gasen und Flüssigkeiten (Leopold Vos, Leipzig, 1924).
[81] J. M. H. Levelt Sengers, Ind. Eng. Chem. Fundam. **9** (1970) 470.

[82] J. F. Nicoll, T. S. Chang and H. E. Stanely, Phys. Rev. Lett. **32** (1974) 1446; Phys. Rev. B **12** (1975) 458.
[83] J. F. Nicoll and T. S. Chang, Phys. Rev. A **17** (1978) 2083.
[84] D. R. Nelson and J. Rudnick, Phys. Rev. Lett. **35** (1975) 178; J. Rudnick and D. R. Nelson, Phys. Rev. B **13** (1976) 2208.
[85] F. de Pasquale and P. Tombesi, J. Phys. A **10** (1977) 399.
[86] T. S. Chang, C. W. Garland and J. Thoen, Phys. Rev. A **16** (1977) 446.
[87] F. de Pasquale, P. Tartaglia and P. Tombesi, J. Phys. A **11** (1978) 2033.

CHAPTER 15

*The Interfaces between Fluid Phases**

B. WIDOM

*Department of Chemistry, Cornell University
Ithaca, New York 14853, U.S.A.*

*This work was supported by the National Science Foundation and the Cornell University Materials Science Center.

© *North-Holland Publishing Company 1981*

*Perspectives in Statistical Physics
Ed. H. J. Raveché*

Contents

1. Introduction — 275
2. Non-critical interface near a critical end point — 276
3. The Cahn transition — 280
4. Two-component lattice gas — 285
References — 291

It is an honor to have been asked to take part in this tribute to Mel Green. We all know of the deep interest Mel had in the theory of phase transitions and critical phenomena, and of his highly original and seminal contributions to the subject. I believe he would have been keenly interested in the recent ideas about the interfaces between coexisting phases, and am most gratified that this account of them is being dedicated to his memory.

1. Introduction

In the van der Waals theory [1] of the liquid-vapor interface we have a free-energy density F, shown in fig. 1 as a function of the density ρ at fixed temperature. It has two equal minima occurring at the densities ρ_g and ρ_l of the coexisting vapor and liquid phases. The density profile and the tension of the liquid–vapor interface both follow from $F(\rho)$ by a dynamical analogy in which $-F$ is the potential energy of a particle of coordinate ρ; see fig. 2. The distance z in the direction perpendicular to the plane of the interface is the time in the dynamical analogy. Then the density profile $\rho(z)$, in the analogy, is the coordinate of the particle as a function of the time when the particle moves subject to the potential $-F(\rho)$ with a total energy equal to the common value of the potential at its two maxima (0, by convention). Figure 2 shows such a profile $\rho(z)$, with the density varying from its value ρ_g deep in the bulk vapor phase ($z \to -\infty$), to its value ρ_l deep in the bulk liquid phase ($z \to \infty$). The surface tension σ is the action in the dynamical analogy,

$$\sigma = \sqrt{2m} \int_{\rho_g}^{\rho_l} \sqrt{F(\rho)}\, d\rho. \tag{1}$$

The van der Waals theory [1] includes a prescription for calculating the parameter m (the mass in the dynamical analogy) from the intermolecular forces.

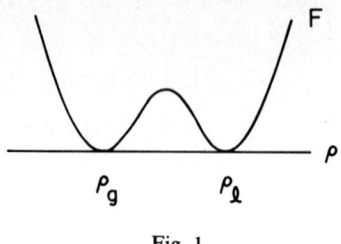

Fig. 1.

Three independent extensions and applications of this theory are in the sections that follow.

2. Non-critical interface near a critical end point

We consider three phases in equilibrium, as in fig. 3. When β and γ are at the critical point of their phase separation while still in equilibrium with the distinct phase α, the system is said to be at a critical end point. We are interested in the nature of the interfaces between phases in and near such thermodynamic states [2]. The proximity of the critical end point is indicated in fig. 3 by showing the $\beta\gamma$ interface as a dotted line.

A common example of a critical end point is the critical solution point of a two-component liquid mixture in the presence of an equilibrium vapor phase (which is then the phase α). We have also

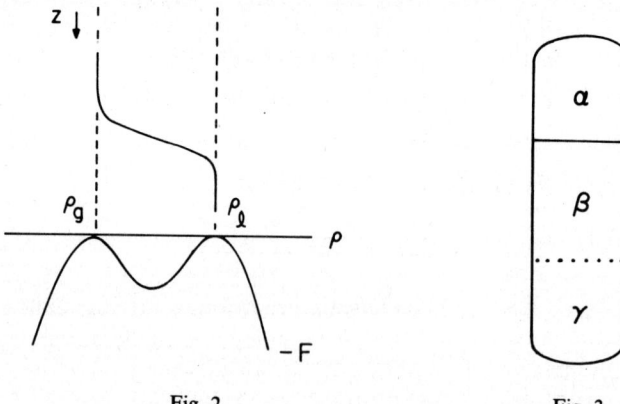

Fig. 2.   Fig. 3.

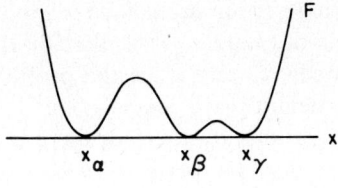

Fig. 4.

studied critical end points in four-component mixtures in which all three phases are liquid [3, 4, 5].

The $\beta\gamma$ critical end point is an ordinary critical point, and the $\beta\gamma$ interface is unaffected by the presence of the α phase. Both the structure (i.e., the composition profile) and the tension $\sigma_{\beta\gamma}$ of the $\beta\gamma$ interface are then well known from earlier studies of critical interfaces [6] going back to van der Waals [1]. The new developments [2, 4, 7] concern the non-critical interfaces $\alpha\beta$ and $\alpha\gamma$.

In fig. 4 we show the free-energy density F as a function of some composition variable x for a three-phase equilibrium, with x_α, x_β and x_γ the values of x in the bulk α, β and γ phases. Griffiths [8] has given such an F. Extending the van der Waals theory, we find the composition profiles $x(z)$ of the $\alpha\beta$, $\beta\gamma$ and $\alpha\gamma$ interfaces by the same dynamical analogy as before, treating $-F$ as the potential of a particle of coordinate x, and the depth z as the time. That is shown in fig. 5. Between x_α

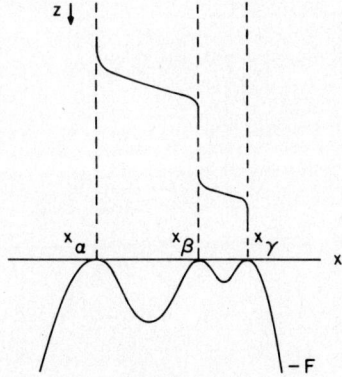

Fig. 5.

and x_β, the $x(z)$ shown in the figure is the composition profile of the $\alpha\beta$ interface, and between x_β and x_γ it is that of the $\beta\gamma$ interface. The composite curve between x_α and x_γ is the profile of the $\alpha\gamma$ interface, showing a step of indeterminate macroscopic height at $x=x_\beta$. That means that in the equilibrium $\alpha\gamma$ interface there is necessarily a layer of β phase, and that the β phase perfectly wets [9], or spreads at, that interface. Identifying the interfacial tensions with the actions in the dynamical analogy, as in the original van der Waals theory, we conclude:

$$\sigma_{\alpha\gamma} = \sigma_{\alpha\beta} + \sigma_{\beta\gamma}, \tag{2}$$

which is Antonow's rule [10], the condition that the β phase spread at the $\alpha\gamma$ interface.

Figure 6 shows $-F$ at the critical end point, and the composition profile of the interface between the critical $\beta\gamma$ phase and the non-critical α phase. The composition approaches its limiting value in the bulk α phase exponentially rapidly, but approaches its limiting value in the bulk $\beta\gamma$ phase much more slowly, roughly proportionally to $1/z$ as $z \to \infty$, reflecting the finiteness of the coherence length of composition fluctuations in a non-critical phase and the divergence of that length in a critical phase. That distinct asymmetry of the composition profile [2, 11] should be detectable in the optical properties of the interface.

We have measured [4] the three interfacial tensions $\sigma_{\alpha\beta}$, $\sigma_{\beta\gamma}$ and $\sigma_{\alpha\gamma}$ in each of several three-liquid-phase mixtures of the four components benzene, ethanol, water and ammonium sulfate. The mixtures spanned the three-phase region from the $\alpha\beta$ critical end point to the $\beta\gamma$ critical end point, all at room temperature. The experimental results are in good

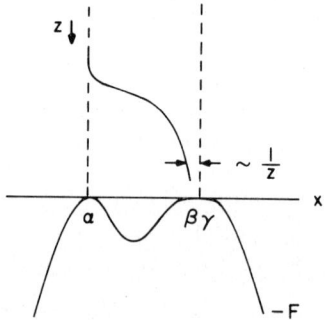

Fig. 6.

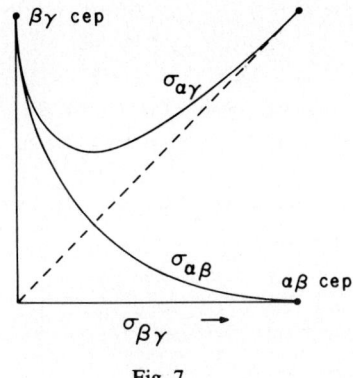

Fig. 7.

qualitative accord with the theoretical predictions [4], which are shown schematically in fig. 7. The figure shows how $\sigma_{\alpha\beta}$ and $\sigma_{\alpha\gamma}$ vary with $\sigma_{\beta\gamma}$, from the $\beta\gamma$ critical end point, at which $\sigma_{\beta\gamma}=0$, to the $\alpha\beta$ critical end point, at which $\sigma_{\alpha\beta}=0$. The relation (2) is implicit in the figure; the experimental results [4] are consistent with that relation.

It would have been possible, in principle, to continue the $\sigma_{\alpha\beta}$ and $\sigma_{\alpha\gamma}$ measurements beyond the $\beta\gamma$ critical end point into the two-phase region, where those two tensions would have become the single tension $\sigma_{\alpha,\beta\gamma}$ of the interface between the still distinct α phase and the now single $\beta\gamma$ phase. We did not do that in the system we studied earlier [4], but we did the equivalent of that in two-component liquid mixtures, where we measured [7] the liquid–vapor surface tensions $\sigma_{\alpha\beta}$ and $\sigma_{\alpha\gamma}$ of the two liquid phases β and γ below an upper critical solution temperature, and the surface tension $\sigma_{\alpha,\beta\gamma}$ of the single liquid phase, $\beta\gamma$, of critical composition, above that temperature. The theoretical prediction [2, 12] is shown in fig. 8. The curves of $\sigma_{\alpha\gamma}$ and $\sigma_{\alpha\beta}$ are related to each other as in fig. 7, but the abscissa is now the temperature rather than the interfacial tension $\sigma_{\beta\gamma}$, which is not defined for $T>T_c$. The critical solution point at $T=T_c$ in fig. 8 is a $\beta\gamma$ critical end point like that in fig. 7. Our first experimental results [7] are not in complete accord with fig. 8; they show an anomaly at $T=T_c$ roughly describable as a discontinuity in the surface tensions that is about 1% as great as the surface tensions themselves, whereas the theory as summarized in fig. 8 requires the tensions and their first derivatives with respect to temperature to be continuous. The observed anomaly may not be real, and we propose to measure those surface tensions again by other methods [13].

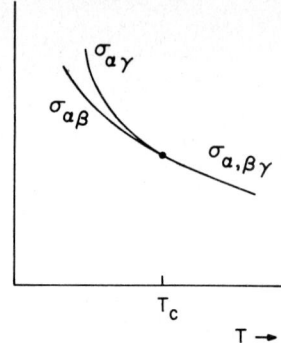

Fig. 8.

3. The Cahn transition

We saw that if in three-phase equilibrium the free-energy density F is represented as a function of a single composition variable x, as in figs. 4 and 5, then the intermediate phase β will be found to wet (spread at) the $\alpha\gamma$ interface, and the three tensions will be found to satisfy Antonow's rule, eq. (2). But there are many known examples of a phase β not spreading, but forming a lens at the interface between two other phases α and γ with which it is in equilibrium; so that there is then a line along which the three phases meet, and at which all three of the dihedral angles (contact angles) between the planes that separate pairs of phases are positive. That is a condition of imperfect wetting, and occurs when eq. (2) is replaced by the inequality

$$\sigma_{\alpha\gamma} < \sigma_{\alpha\beta} + \sigma_{\beta\gamma}. \tag{3}$$

(The inequality cannot go in the other direction; (2) and (3) are the only possibilities in three-phase equilibrium [14, 15].)

How can imperfect wetting, or (3), be reconciled with the van der Waals theory? Cahn [16] made the important remark that to account for (3) within the framework of a theory like that of van der Waals the free-energy density F must be taken to be a function of two (or more) independent composition or density variables [17], not merely of one as in our earlier theory.

Let us take F to be a function of two such densities, say x and y. In fig. 9 we show contours of constant $-F$ in the x, y plane. For three-phase equilibrium F has three equal minima, so that $-F$, which is the

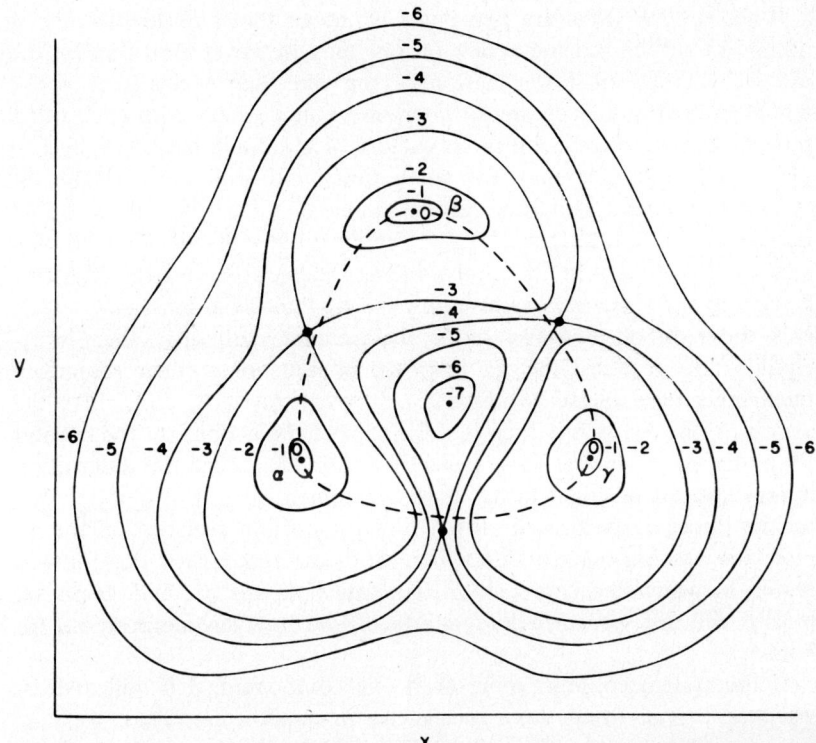

Fig. 9.

potential energy in the dynamical analogy, has three equal maxima. We see two paths from α to γ, one an indirect path via β, and the other a more direct path not via β. The indirect path always exists; the direct path from α to γ exists only if α, β, γ are not too nearly collinear (with β between α and γ) in the x, y plane. As α, β, γ become more nearly collinear, with β the intermediate phase, the saddle point in the potential surface between α and γ and the bowl near the middle of the $\alpha\beta\gamma$ triangle annihilate each other, and that, roughly, is when a direct $\alpha\gamma$ path, distinct from the indirect path via β, ceases to exist. Near a tricritical point, where the three phases all become identical, the three points α, β, γ do become asymptotically collinear [8]. Then the only path from α to γ is that via β, the β phase perfectly wets the $\alpha\gamma$ interface, and eq. (2) holds [19], as we verified experimentally [4].

When we say there are two paths we mean that both are locally of minimum action, but the action on one may be lower than that on the other. The one of lower action is the one that yields the stable, equilibrium structure of the $\alpha\gamma$ interface: x and y vary with each other through the interface as on the trajectory of minimum action in the x, y plane. The interface may for some time, and with some degree of stability, assume a structure corresponding to a local but not absolute minimum in the interfacial tension (action), but it would then be only metastable, and would ultimately, and spontaneously, undergo transformation to the structure of absolutely lowest interfacial tension.

If the path of lower action is the indirect path $\alpha\beta\gamma$, then in its equilibrium structure the $\alpha\gamma$ interface consists of a macroscopically thick layer of β phase; β spreads at the $\alpha\gamma$ interface; and, since the action on the $\alpha\gamma$ path is then just the sum of the actions on the $\alpha\beta$ and $\beta\gamma$ paths, eq. (2) holds. This is exactly as before, though the description is now two-dimensional rather than one-dimensional. But if the action on the direct $\alpha\gamma$ path is the lower, the equilibrium structure of the $\alpha\gamma$ interface is as on that path; nowhere in the $\alpha\gamma$ interface is there now a region in which the composition is identical to that of bulk β phase; bulk β will not spread at that interface; and now (by supposition) (3) holds.

If the system contains more than one component, it is still at least univariant even when three phases are in equilibrium. Then we may vary some thermodynamic field (temperature, pressure, one of the chemical potentials, or some combination of all of those) to change continuously the two competing trajectories and the action on each. We may then reach a point of transition on one side of which the indirect $\alpha\beta\gamma$ path, say, is the path of lesser action, and on the other side of which the direct $\alpha\gamma$ path is. That is the transition predicted by Cahn [9], and recently studied in very beautiful experiments by Moldover and Cahn [20]. The transition has also been seen theoretically in computer calculations [21] of trajectories on model potential surfaces like that in fig. 9. It is a first-order phase transition of the $\alpha\gamma$ interface (first order, because the relevant free energy, which is the interfacial tension, is a continuous function of the varying field while the rate at which the tension varies with the field is discontinuous at the transition). At the transition point itself two distinct interfacial structures of equal free energy (equal interfacial tension) are in competition. The alternative structures might be made to coexist for some time as islands of one in the other, if the one-dimensional boundaries between the distinct surface phases were not themselves of too high free energy or could somehow be stabilized.

Cahn [9] predicted an even more general phase transition of the $\alpha\gamma$ interface, not requiring a third phase β in equilibrium with α and γ, but occurring in the two-phase system. It is then not quite a transition between perfect and imperfect wetting of an interface by a third phase, but is very much like it, and becomes the wetting transition itself in the limit in which the third phase comes to be present as a bulk phase. Suppose in fig. 9 that the peak in $-F$ is slightly lower at β (the minimum in F slightly higher at β) than at α and γ, so that only α and γ are present as bulk phases, not β. There may still be two competing $\alpha\gamma$ trajectories as in fig. 9, but now the indirect one will in general only pass near, rather than exactly over, the β peak. There can still be a transition between the two trajectories as before; but now, in the interfacial structure that corresponds to the indirect trajectory, which goes via β, the β layer, while thick, is not macroscopically thick, and its composition, while nearly that of bulk β, is not exactly that. The closer β is to being in equilibrium as a bulk phase the more nearly macroscopic is that layer of β at the $\alpha\gamma$ interface; its thickness is predicted to diverge as the logarithm of the reciprocal of the distance from the point of three-phase equilibrium [9, 22, 23].

A two-component system of three phases is univariant and is thus invariant at its Cahn transition. With only two phases, the two-component system is bivariant, and univariant at its interfacial transition. There is thus a whole locus of these transitions in the two-component, two-phase system, and the transition in the three-phase system is just a terminus of that locus. That is Cahn's picture [9]. But Cahn also observes that as one follows that transition line in the direction away from its terminus at the three-phase region, one may expect to come to a critical point of that first-order interfacial phase transition. That is seen most simply in fig. 9: moving away from the point of three-phase equilibrium means, in fig. 9, letting the β peak in $-F$ sink progressively lower; whereupon the two alternative trajectories come nearer to each other and ultimately merge into one. Until that moment of merging, if we have indeed been following the locus of Cahn transitions, the interfacial tensions associated with the two alternative and competing structures of the $\alpha\gamma$ interface would at every moment be equal, while the structures themselves were different; but then as the critical point was approached the two structures would become more alike and finally identical.

If there were yet one more thermodynamic degree of freedom in the system—i.e., if the three-phase system were at least bivariant, which would require at least three components—the critical point of the interfacial phase transition could in principle be made to occur in the

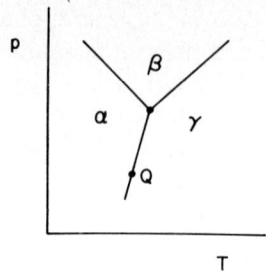

Fig. 10.

three-phase system, with β as well as α and γ present as bulk phases. The picture would be as just described—the direct and indirect paths approaching each other and then merging into one, while the actions on the two paths are at every moment equal—but now the extra degree of freedom would be used to ensure that the β peak was at the same height as the α and γ peaks, so that the indirect path from α to γ went via bulk β.

The more general Cahn transition, which occurs in the interface between two bulk phases even when no third bulk phase is in equilibrium with them, could in principle be found even in a one-component system. (I emphasize *in principle*; it might be hard to find an example.) Figure 10 shows a hypothetical phase diagram in the pressure–temperature plane, with an $\alpha\beta\gamma$ triple point and a point of Cahn transition, marked Q, on the $\alpha\gamma$ equilibrium line. In this hypothetical case, as we followed that line through Q, the $\alpha\gamma$ interface would undergo a discontinuous change of structure. Such a transition would perhaps be more likely to occur if, at the triple point, the β phase perfectly wet the $\alpha\gamma$ interface, for then we could imagine a transition from a β-like structure of the $\alpha\gamma$ interface to a non-β-like structure as we passed through Q in the direction away from the triple point.

So far, the only Cahn transitions that have with certainty been seen in fluid interfaces are in three-phase systems [20], where they are transitions between perfect and imperfect wetting of an $\alpha\gamma$ interface by a bulk β phase that is in thermodynamic equilibrium with α and γ. It is the presence of the third phase that makes the phenomenon macroscopically visible as a transition between spreading and non-spreading, with concomitant changes in the three-phase contact angles [20]. The phenomenon would manifest itself also as a change from the equality (2) to the inequality (3), and as a discontinuity in the derivative of $\sigma_{\alpha\gamma}$ with respect to the varying thermodynamic field. The more general Cahn transition, which occurs when there are only two phases, might have no immediately visible, macroscopic manifestation; but it, too, would lead to a discontinuity in the derivative of $\sigma_{\alpha\gamma}$, and it would in addition be detectable as a discontinuous change in the microscopic structure—hence, in the optical properties—of the interface.

4. Two-component lattice gas

We saw in sect. 3 that it may sometimes be useful, or even necessary, to formulate the van der Waals theory with two or more densities varying independently through the interface. Such a formulation and the dynamical analogy associated with it are illuminated by a two-component lattice-gas model to be described shortly.

But first, as an introduction to this topic, we consider the conventional one-component lattice gas and its liquid–vapor interface as treated in mean-field approximation by Ono and Kondo [24]. The volume containing the fluid is divided into cells, each of which may be at most singly occupied. Molecules in neighboring cells interact with an attractive potential energy $-\varepsilon$ ($\varepsilon>0$). There are no other interactions. The interface is horizontal; distance in the vertical direction is indexed by the discrete, dimensionless variable z. Each cell at height z has c' neighbors at $z+1$ and at $z-1$, and $c-2c'$ neighbors in the same layer with itself, at z, so the total coordination number is c. Let $\zeta(\rho)$ be the activity of the fluid as a function of ρ, the mean number of molecules per cell ($0 \leq \rho \leq 1$). This $\zeta(\rho)$ is the analytic function given by the mean-field approximation; it is the activity of a hypothetical fluid constrained to be uniform at the density ρ. It is given explicitly, in units of the reciprocal of the cell volume, by [24, 25]:

$$\zeta(\rho) = [\rho/(1-\rho)]e^{-c\theta\rho}, \tag{4}$$

where $\theta = \varepsilon/kT$. The equilibrium activity obtained from $\zeta(\rho)$ by the Maxwell construction is $\exp(-\frac{1}{2}c\theta)$, which we shall simply call ζ. Figure 11 then shows $\ln[\zeta(\rho)/\zeta]$. The shaded areas are equal; ρ_g and ρ_l are the densities of the coexistent phases. This $\ln[\zeta(\rho)/\zeta]$ is related to the $F(\rho)$ in fig. 1 by:

$$kT\ln[\zeta(\rho)/\zeta] = dF(\rho)/d\rho. \tag{5}$$

In mean-field approximation the density profile $\rho(z)$ satisfies the func-

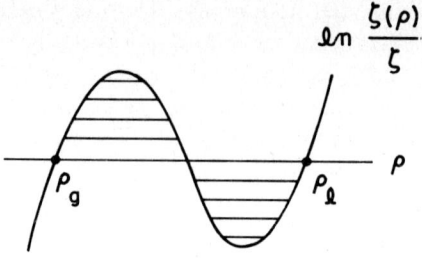

Fig. 11.

tional equation [24, 25]:

$$c'\theta\Delta^2\rho = \ln[\zeta(\rho)/\zeta], \qquad (6)$$

where Δ^2 is the second-difference operator. This is the equation-of-motion of a particle of coordinate ρ, mass $c'\varepsilon$, and potential energy $-F(\rho)$. With the right-hand side of eq. (6) as shown in fig. 11, the solution of eq. (6) that has the properties $\rho(-\infty)=\rho_g$, $\rho(\infty)=\rho_l$ is as shown in fig. 12. That is the density profile of the liquid–vapor interface of the lattice gas in mean-field approximation.

Now we shall find the density profiles $\rho_A(z)$ and $\rho_B(z)$ in a two-component lattice gas with components A and B. Here ρ_A and ρ_B are, respectively, the mean numbers of A's and of B's per cell. Again, no cell may be more than singly occupied, so that $0 \leqslant \rho_A + \rho_B \leqslant 1$; and unlike molecules may not occupy neighboring cells. There are no other interactions. As before, each cell at z has c' neighbors at $z+1$ and at $z-1$, and $c-2c'$ neighbors in the same layer with itself at z.

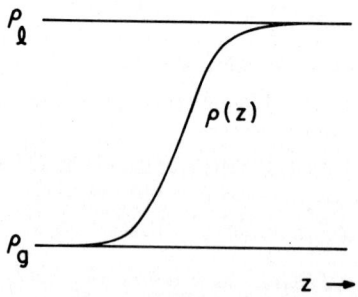

Fig. 12.

The activities of the two species are $\zeta_A(\rho_A, \rho_B)$ and $\zeta_B(\rho_A, \rho_B)$. The mean-field approximation is in this case the Bethe–Guggenheim approximation, in which, in a homogeneous phase of this model fluid, the analytic ζ_A and ζ_B (in units of the reciprocal of the cell volume) are [26]:

$$\zeta_A(\rho_A, \rho_B) = \frac{\rho_A}{(1-\rho_A-\rho_B)(1-q_B)^c},$$

$$\zeta_B(\rho_A, \rho_B) = \frac{\rho_B}{(1-\rho_A-\rho_B)(1-q_A)^c}, \tag{7}$$

where q_B and q_A are, respectively, the probabilities that a given neighbor of an empty cell be occupied by a molecule of type B or by a molecule of type A. That ζ_A thus depends on q_B and ζ_B on q_A is due to the repulsion between neighboring A's and B's. In this same approximation q_B and q_A are related to ρ_A and ρ_B by [26]:

$$\rho_A = \frac{q_A(1-q_B)}{1-2q_Aq_B}, \quad \rho_B = \frac{q_B(1-q_A)}{1-2q_Aq_B}. \tag{8}$$

At sufficiently high total density there is separation into an A-rich phase and a B-rich phase. The coexistence curve in the ρ_A, ρ_B plane, the critical point of the phase separation, and a few representative tie-lines, are shown schematically in fig. 13 [26].

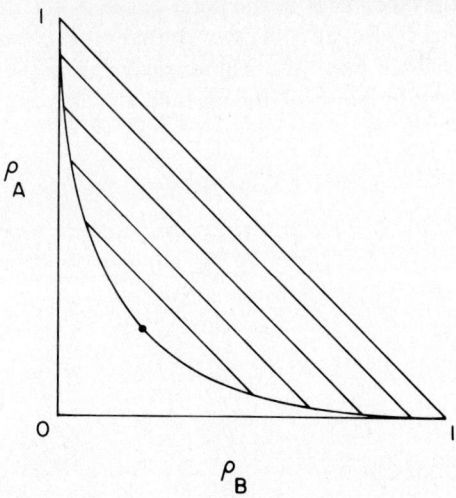

Fig. 13.

When there are two phases in equilibrium, ζ_A and ζ_B have a common value [26], call it ζ. In the Bethe–Guggenheim approximation, $q_A(z)$ and $q_B(z)$ are then found [27] to satisfy the coupled difference equations:

$$c'\Delta^2 \ln(1-q_B) = \ln[\zeta_A(\rho_A, \rho_B)/\zeta],$$
$$c'\Delta^2 \ln(1-q_A) = \ln[\zeta_B(\rho_A, \rho_B)/\zeta], \qquad (9)$$

where $\zeta_A(\rho_A, \rho_B)$ and $\zeta_B(\rho_A, \rho_B)$ are given by eq. (7). If we assume that the gradients are small, as they are near the critical point (where the thickness of the interface, being equal to the coherence length of composition fluctuations, diverges), then q_A and q_B will be related to ρ_A and ρ_B as in the homogeneous fluid, hence by eq. (8). Then from eq. (7), (8), and (9),

$$c'\Delta^2 \ln(1-q_B) = \ln\left[q_A/(1-q_A-q_B)(1-q_B)^{c-1}\zeta\right],$$
$$c'\Delta^2 \ln(1-q_A) = \ln\left[q_B/(1-q_A-q_B)(1-q_A)^{c-1}\zeta\right]. \qquad (10)$$

These coupled difference equations may be solved for $q_A(z)$ and $q_B(z)$, and then the density profiles $\rho_A(z)$ and $\rho_B(z)$ may be obtained from eq. (8). The $\rho_A(z)$ and $\rho_B(z)$ thus calculated [27] for $c=7$, $c'=1$, $\zeta=1$ are shown in fig. 14.

By eliminating z between $\rho_A(z)$ and $\rho_B(z)$ we obtain the trajectory in the ρ_A, ρ_B plane, which shows how ρ_A and ρ_B vary with each other through the interface. That is the solid curve in fig. 15, again for $c=7$, $c'=1, \zeta=1$. The figure also shows, dotted, the coexistence curve for $c=7$, with its critical point, C. The dashed line is the $\zeta=1$ tieline. The point S is the saddle point of the surface $F = F(\rho_A, \rho_B)$, the free-energy

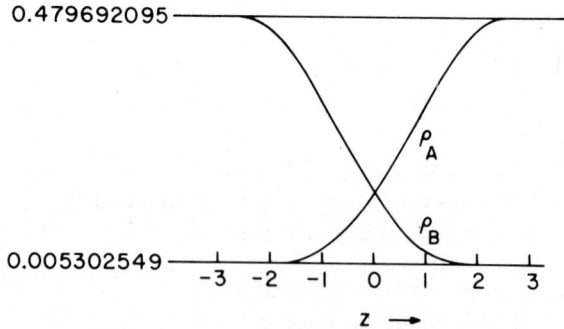

Fig. 14.

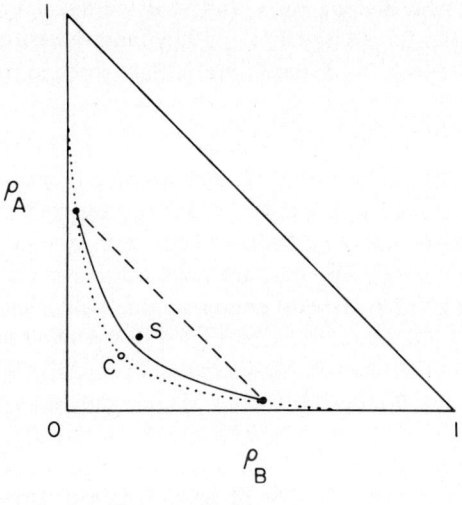

Fig. 15.

density, which is related to the activities by:

$$\partial F(\rho_A, \rho_B)/\partial \rho_A = kT \ln[\zeta_A(\rho_A, \rho_B)/\zeta],$$
$$\partial F(\rho_A, \rho_B)/\partial \rho_B = kT \ln[\zeta_B(\rho_A, \rho_B)/\zeta]. \tag{11}$$

The saddle point shown is for $c = 7$, $\zeta = 1$.

Note that the trajectory is concave toward the saddle point. That is the opposite of what we would expect if there were a dynamical analogy in which $-F$ was the potential energy and in which the trajectory was that of a particle of positive mass moving subject to that potential; for then, due to centrifugal effects, the trajectory would be convex toward the saddle point, as in fig. 9, for example. We shall see that in the present problem there is indeed a dynamical analogy, but now to a system that is described by generalized coordinates and in which the relation of the kinetic energy to the velocities is of an uncommon and surprising form.

At low densities, we see from eq. (8) that $q_A \sim \rho_A$ and $q_B \sim \rho_B$, so that $\ln(1 - q_B) \sim -\rho_B$ and $\ln(1 - q_A) \sim -\rho_A$; while near the critical point, where the gradients are small, the second difference Δ^2 is practically the same as the second derivative d^2/dz^2. In the Bethe–Guggenheim approximation the densities at the critical point are [26] $\rho_A = \rho_B = (c-1)/(c^2 - 2)$, which are indeed small when the coordination number is large. Thus, the coupled eqs. (9), with (11), are roughly of the form:

$$-c'kT d^2\rho_B/dz^2 = \partial F/\partial \rho_A,$$
$$-c'kT d^2\rho_A/dz^2 = \partial F/\partial \rho_B; \tag{12}$$

or, if we replace ρ_A and ρ_B by x and y, if we again think of F as the negative of a potential energy $V(x, y)$, say, and if we use a dot to mean derivative with respect to z (the "time"), then these are of the form:

$$m\ddot{x} = -\partial V/\partial y, \quad m\ddot{y} = -\partial V/\partial x, \tag{13}$$

with a negative "mass" $m = -c'kT$. The x-component of acceleration is proportional to the y-component of the force and vice versa.

It is the repulsion between unlike neighbors, and the lack of interaction between like neighbors, that are responsible for this result; for that makes the activity of A depend primarily on the density of B, and vice versa. The same phenomenon is seen also in the two-component primitive version of the penetrable-sphere model [27, 28], where, in the limit of small gradients, the densities ρ_A and ρ_B again satisfy eq. (13). (The negative mass m in that case is $-kT/2(s+2)$, with s the dimensionality.) In that model, too, unlike molecules repel infinitely strongly while like molecules do not interact, so the physical origin of eq. (13) is the same in the two models. The ρ_A, ρ_B trajectories in both models have the peculiarity of being concave toward the saddle points of the respective $F(\rho_A, \rho_B)[= -V(x, y)]$ surfaces [27, 28].

The equations-of-motion (13) now have as a first integral

$$V + m\dot{x}\dot{y} = E, \tag{14}$$

with some constant energy E (which, in our application to the interface, is again equal to the value of V at its maxima). The kinetic energy K, the Lagrangian L, the generalized momenta p_x and p_y conjugate, respectively, to x and y, and the Hamiltonian H, are.

$$K = m\dot{x}\dot{y},$$
$$L = m\dot{x}\dot{y} - V,$$
$$p_x = m\dot{y}, \quad p_y = m\dot{x}, \tag{15}$$
$$H = V + p_x p_y/m.$$

The x-component of momentum is here the mass times the y-component of velocity, and vice versa. The equations-of-motion (13) are the Euler–Lagrange equations for the minimization of the action

$$\sigma = \int_{-\infty}^{\infty} (K - V)\,dz. \tag{16}$$

In the interface problem $K - V$ is the excess free-energy density due to the inhomogeneity; the minimized σ is therefore again the equilibrium interfacial tension, just as in the dynamical analogy to the original van der Waals theory. In our model fluid the repulsions between unlike molecules are such as to make ρ_A decrease with increasing ρ_B in the interface, and vice versa, as seen in figs. 14 and 15. In the mechanical system, analogously, $\dot{x}$ and $\dot{y}$ are of opposite sign; so, with $m < 0$, the kinetic energy $m\dot{x}\dot{y}$ is always positive: the density gradients in the interface still make positive contributions to the interfacial tension, as in the earlier theories.

References

[1] J. D. van der Waals, Z. Phys. Chem. **13** (1894) 657;
J. S. Rowlinson, J. Stat. Phys. **20** (1979) 197.
[2] B. Widom, J. Chem. Phys. **67** (1977) 872.
[3] J. C. Lang, Jr. and B. Widom, Physica **81A** (1975) 190.
[4] J. C. Lang, Jr., P. K. Lim and B. Widom, J. Phys. Chem. **80** (1976) 1719.
[5] P. Bocko, Fluid Phase Equil. **4** (1980) 137; Physica, in press.
[6] B. Widom, in: Phase Transitions and Critical Phenomena, Vol. **2**, Chap. 3, eds. C. Domb and M. S. Green (Academic Press, 1972).
[7] M. P. Khosla and B. Widom, J. Coll. Interf. Sci. **76** (1980) 375.
[8] R. B. Griffiths, J. Chem. Phys. **60** (1974) 195.
[9] J. W. Cahn, J. Chem. Phys. **66** (1977) 3667.
[10] G. N. Antonow, J. Chim. Phys. **5** (1907) 372; Kolloid-Z. **59** (1932) 7; **64** (1933) 336;
N. K. Adam, The Physics and Chemistry of Surfaces, 3rd ed. (Oxford University Press, London, 1941) pp. 7, 214–215.
[11] A. I. Rusanov, in: Progress in Surface and Membrane Science, Vol. **4**, eds. J. F. Danielli, M. D. Rosenberg and D. A. Cadenhead (Academic Press, New York, London, 1971) pp. 57–114, particularly the concluding paragraph.
[12] F. Ramos-Gómez and B. Widom, Physica, in press.
[13] N. Nagarajan, W. W. Webb and B. Widom, in progress.
[14] J. W. Gibbs, The Scientific Papers of J. Willard Gibbs, Vol. **1** (Dover, 1961) p. 258.
[15] B. Widom, J. Chem. Phys. **62** (1975) 1332.
[16] J. W. Cahn, privately communicated.
[17] J. W. Cahn and J. E. Hilliard, J. Chem. Phys. **28** (1958) 258, footnote || on p. 260.
[18] B. Widom, in: Statistical Mechanics and Statistical Methods in Theory and Application, ed. U. Landman (Plenum, New York, 1977) pp. 33–71.
[19] B. Widom, Phys. Rev. Lett. **34** (1975) 999.
[20] M. R. Moldover and J. W. Cahn, Science **207** (1980) 1073.
[21] J. F. Currie and A. R. Bishop, Can. J. Phys. **57** (1979) 890;
J. F. Currie, Ph.D. thesis, Cornell University (1977).
[22] J. Lajzerowicz, preprint "Parois de Domaines Ferroélectriques et de Macles pour un Transition du Premier Ordre," from the Laboratoire de Spectrométrie Physique, Université Scientifique et Médicale de Grenoble (1976).
[23] B. Widom, J. Chem. Phys. **68** (1978) 3878.
[24] S. Ono and S. Kondo, in: Handbuch der Physik, ed. S. Flügge, Vol. **X**, Struktur der Flüssigkeiten (Springer-Verlag, 1960).
[25] B. Widom, J. Stat. Phys. **19** (1978) 563.
[26] J. C. Wheeler and B. Widom, J. Chem. Phys. **52** (1970) 5334.
[27] J. S. Rowlinson and B. Widom, The Molecular Theory of Capillarity (Oxford University Press) in preparation.
[28] S. J. Hemingway, J. S. Rowlinson and E. S. Severin, to be published.

PART III

Foundations

CHAPTER 16

*On Higher Order WKB Approximations
for the Calculation of Energy Levels*[†]

F. T. HIOE
E. W. MONTROLL
and
M. YAMAWAKI

*Institute for Fundamental Studies
Department of Physics and Astronomy
University of Rochester
Rochester, New York 14627
U.S.A.*

[†]Research partially supported by the Physics Branch of the Army Research Office, Durham, North Carolina.

© *North-Holland Publishing Company 1981*

*Perspectives in Statistical Physics
Ed. H. J. Raveché*

Contents

1. Introduction — 297
2. On the estimation of characteristic values of second order differential equations — 298
3. Remarks on Method of Maslov — 304
4. Energy levels of purely quartic and mixed quartic and harmonic oscillators — 306
5. Density of states in multidimensional systems — 315
Appendix A. Expressions required for derivation of λ_4 and λ_6 — 318
Appendix B. Alternative forms for λ_4 — 321
References — 322

1. Introduction

Melville Green found a certain fascination in critical phenomenon and in his later work with the renormalization group and critical exponents. Critical exponents appear as strange powers not easily identifiable in most approximation methods. In this article dedicated to his memory, we now show how strange powers are evident in the WKB type of approximations and in the Titchmarsh generalization to many variable problems.

The Schrödinger equation first appeared in print in March 1926. The approximation method for its solution known as the WKB method was published independently by Wentzel [1], Kramers [2] and Brillouin [3]. Unbeknown to WK and B, similar ideas had been considered by Jeffreys [4] in 1923, before the birth of the S equation, and indeed seem to have occurred to Liouville [5] and Green [6] as early as 1837. The work of WKB was especially important in 1926 since their first approximation to the solution of the S equation yielded the Bohr–Wilson–Sommerfeld condition $\int p \, dq = (n + \tfrac{1}{2})h$.

Among the many interesting subsequent researches on the WKB method the works of Langer [7], Erdélyi [8], Olver [9] and the Frömans [10] are especially useful as well as those mentioned in the following which have a more direct bearing on the development of this report.

While the original WKB method gave a first approximation to the solution of the one-dimensional Schrödinger equation, the systematics of extending the solution to higher orders was developed in an elegant paper by Dunham. We will start our analysis with Titchmarsh's [11] slight variation of Dunham's [12] work. Our main thrust here is to derive higher order formulae for the estimation of energy levels. Maslov [13] has developed a rather different style for obtaining extensions to higher order. This will be discussed briefly in sect. 3.

Section 4 contains an application of the general formulas derived in sect. 2 to quartic anharmonic oscillators. Section 5 is concerned with densities of states of multidimensional systems. A more extensive systematic investigation of this topic will be presented elsewhere.

2. On the estimation of characteristic values of second order differential equations

Consider the second order differential equation

$$-\varepsilon^2 u''(x) + v(x)u(x) = \lambda u(x), \quad v(x) \to \infty \quad \text{as } |x| \to \infty \tag{1}$$

where ε is a parameter introduced for bookkeeping purposes. Our aim here is to find the characteristic values, λ, of this equation when $\varepsilon = 1$ and when $\varepsilon^2 = \frac{1}{2}$. The latter case is common in discussions of the Schrödinger equation. Certain series expansions in powers of ε will be introduced which may or may not converge for our special values of ε. Generally when $v(x)$ is a polynomial these expansions will be asymptotic expansions in inverse powers of the "quantum number" n which will appear combined with ε. Let

$$u(x) = \exp(\mathrm{i}/\varepsilon) \int^x \chi(x)\,\mathrm{d}x. \tag{2}$$

Then $\chi(x)$ satisfies the first order non-linear differential, Riccati equation

$$-\mathrm{i}\varepsilon \chi' = (\lambda - v) - \chi^2. \tag{3}$$

Now construct the series expansion

$$\chi = \chi_0 + (\varepsilon/\mathrm{i})\chi_1 + (\varepsilon/\mathrm{i})^2 \chi_2 + \ldots \tag{4}$$

with χ_0 defined to be

$$\chi_0 \equiv (\lambda - v)^{1/2}. \tag{5}$$

Then upon substitution of eqs. (4) and (5) into eq. (3) we obtain, equating coefficients of like powers of ε of both sides of the resulting equation,

$$\dot{\chi}_0 = -2\chi_0\chi_1, \quad \dot{\chi}_1 = -2\chi_0\chi_2 - \chi_1^2, \tag{6a}$$

$$\dot{\chi}_2 = -2\chi_0\chi_3 - 2\chi_1\chi_2, \tag{6b}$$

$$\dot{\chi}_{j-1} = -\sum_{k=0}^{j} \chi_{j-k}\chi_k. \tag{6c}$$

Since

$$\dot{\chi}_0 = -\tfrac{1}{2}v'/\chi_0, \tag{6d}$$

the first few functions of the set $\{\chi_j\}$ are:

$$\chi_1 = -2 d\log\chi_0/dx$$
$$= \tfrac{1}{4} v'(x)/[\lambda - v(x)] \equiv \tfrac{1}{4} v'(x)/\chi_0^2, \tag{7a}$$

$$\chi_2 = -\frac{1}{8\chi_0^3}\left[v'' + \frac{5(v')^2}{4\chi_0^2}\right], \tag{7b}$$

$$\chi_3 = \frac{1}{16\chi_0^4}\left[v''' + \frac{9v'v''}{2\chi_0^2} + \frac{15(v')^3}{4\chi_0^4}\right], \tag{7c}$$

$$\chi_4 = -\frac{1}{32\chi_0^5}\left[v'''' + \frac{7v'v'''}{\chi_0^2} + \frac{19(v'')^2}{4\chi_0^2}\right.$$
$$\left. + \frac{221(v')^2 v''}{8\chi_0^4} + \frac{1105(v')^4}{64\chi_0^6}\right]. \tag{7d}$$

Now, let $u(x) \equiv u_n(x)$ be a solution of eq. (1) with n real zeros. Then, upon introduction of the complex variable $z = x + iy$,

$$n = \frac{1}{2\pi i}\oint \frac{u'_n(z)}{u_n(z)} dz, \tag{8}$$

where the integral is evaluated around a counter clockwise contour which contains the n real zeros and none other. By combining eqs. (4) and (8) we obtain

$$n = (2\pi\varepsilon)^{-1}\oint\{\chi_0 + (\varepsilon/i)\chi_1 + (\varepsilon/i)^2\chi_2 + \ldots\} dz. \tag{9}$$

Let $v(x)$ be a real continuous concave up function of x such that (i) $v(\pm\infty) = \infty$ and (ii) the equation $v(x) = \lambda$ has two real solutions x_1 and x_2 for all real $\lambda > v(0)$. Then expanding around x_l with $l = 1, 2$,

$$-v(z) = -(z - x_l) v'(x_l) - \tfrac{1}{2}(z - x_l)^2 v''(x_l) \ldots,$$

we find that

$$\frac{1}{2\pi i}\oint \chi_1(z) dz = \frac{1}{2\pi i}\oint \frac{1}{4}\left(\frac{v'}{\lambda - v}\right) dz$$
$$= -\frac{1}{2},$$

since each of the poles at x_1 and x_2 contributes the residue $-\frac{1}{4}$. Hence [12, 11]

$$\varepsilon(n+\tfrac{1}{2}) = \frac{1}{2\pi}\oint \chi_0 \, dz + \sum_{j=2}^{\infty} (\varepsilon/i)^j \frac{1}{2\pi}\oint \chi_j(z)\, dz. \tag{10}$$

Equation (10) is a basic equation for the determination of the nth characteristic value, $\lambda^{(n)}$, of the differential equation (1). Since χ_0, by its definition (5) contains λ explicitly, so also do $\chi_2, \chi_3 \ldots$ through the recurrence formulae (6). Hence λ appears in all the terms in the series (10). It is therefore required to find that value of $\lambda^{(n)}$, which, when substituted into the right hand side of eq. (10) yields $\varepsilon(n+\tfrac{1}{2})$. Most of the remainder of this report is addressed to that question.

In order to proceed further let us examine the integrals of χ_2, χ_3, χ_4 and χ_6. From eq. (6a):

$$\oint \chi_2 \, dz = -\tfrac{1}{2}\oint (\dot{\chi}_1/\chi_0)\, dz - \tfrac{1}{2}\oint (\chi_1^2/\chi_0)\, dz$$

$$= \tfrac{1}{2}\oint (\chi_1^2/\chi_0)\, dz = \frac{1}{32}\oint \frac{(v')^2 \, dz}{(\lambda - v)^{5/2}}, \tag{11}$$

after integration by parts and remembering that $\dot{\chi}_0 = -2\chi_0 \chi_1$. Also

$$\oint \chi_3 \, dz = \tfrac{1}{2}\oint (2\chi_0 \chi_3/\chi_0)\, dz$$

$$= \tfrac{1}{2}\oint [(2\chi_1 \chi_2 + \dot{\chi}_2)/\chi_0]\, dz$$

$$= -\oint (\chi_1 \chi_2/\chi_0)\, dz - \tfrac{1}{2}\oint (\chi_2/\chi_0^2)(d\chi_0/dz)\, dz = 0. \tag{12}$$

It has been shown by Bender, Olaussen and Wang [14] that when $v(x)$ is a polynomial, all odd order integrals of order 3 or greater vanish.

We find in a similar manner:

$$\oint \chi_4 \, dz = \tfrac{1}{2}\oint [(\dot{\chi}_3/\chi_0) + (2\chi_1 \chi_3/\chi_0) + (\chi_2^2/\chi_0)]\, dz$$

$$= -\tfrac{1}{2}\oint (\chi_2^2/\chi_0)\, dz, \tag{13}$$

where χ_2 is given by eq. (7b), and:

$$\oint \chi_6 \, dz = -\tfrac{1}{2}\oint (\chi_3^2/\chi_0)\, dz - \oint (\chi_4 \chi_2/\chi_0)\, dz. \tag{14}$$

Hence, by combining eqs. (10)–(14) we find:

$$\varepsilon\left(n+\tfrac{1}{2}\right) = \frac{1}{2\pi}\oint \chi_0 \, dz - (\varepsilon^2/4\pi)\oint (\chi_1^2/\chi_0)\, dz$$

$$- (\varepsilon^4/4\pi)\oint(\chi_2^2/\chi_0)\, dz$$

$$+ (\varepsilon^6/4\pi)\left(\oint dz\, \chi_3^2/\chi_0 \right.$$

$$\left. + 2\oint dz\, \chi_2 \chi_4/\chi_0\right) + O(\varepsilon^8). \tag{15}$$

Our aim now is to determine $\lambda^{(n)}$ from eq. (15), remembering that $\chi_0, \chi_1, \ldots$ are all functions of $\lambda^{(n)}$. Since we eventually wish to set $\varepsilon^2 = 1$ or $\tfrac{1}{2}$ there is no difficulty in setting the ε on the left side of eq. (15) equal to a new parameter α, regarding eq. (15) as a two parameter equation with parameters α and ε, and finally in the last stages of our calculation set both α and ε equal to 1 or $(\tfrac{1}{2})^{1/2}$.

We now develop an ε expansion for the characteristic value

$$\lambda^{(n)} = \lambda_0 + \varepsilon\lambda_1 + \varepsilon^2\lambda_2 + \ldots \tag{16}$$

Then, upon postulating λ_0 to be the real solution of

$$\alpha\left(n+\tfrac{1}{2}\right) = (1/2\pi)\oint(\lambda_0 - v)^{1/2} dz, \tag{17}$$

i.e. the WKB approximated to $\lambda^{(n)}$, we find that:

$$\oint(\lambda - v)^{1/2}\, dz = \oint(\lambda_0 - v + \varepsilon\lambda_1 + \varepsilon^2\lambda_2 + \ldots)^{1/2}\, dz$$

$$= \oint(\lambda_0 - v)^{1/2}\left[1 + \frac{\varepsilon\lambda_1}{(\lambda_0 - v)} + \ldots\right]^{1/2} dz$$

$$= \oint(\lambda_0 - v)^{1/2}\left\{1 + \frac{1}{2}\frac{\varepsilon\lambda_1}{(\lambda_0 - v)} + \frac{1}{2}\frac{\varepsilon^2\lambda_2}{(\lambda_0 - v)}\right.$$

$$\left. - \frac{1}{8}\frac{\varepsilon^2\lambda_1^2}{(\lambda_0 - v)^2} + \ldots\right\} dz. \tag{18}$$

Since

$$\oint(\chi_1^2/\chi_0)\, dz = \frac{1}{16}\oint\frac{(v')^2\, dz}{(\lambda - v)^{5/2}} = \frac{1}{16}\oint\frac{(v')^2\, dz}{(\lambda_0 - v)^{5/2}} + O(\varepsilon). \tag{19}$$

We find that to order ε^2, eq. (15) becomes:

$$0 = \varepsilon \lambda_1 \oint \chi_0^{-1} dz$$
$$+ \varepsilon^2 \oint \left[\lambda_2 \chi_0^{-1} - \tfrac{1}{4}\lambda_1^2 \chi_0^{-3} - \tfrac{1}{16}(v')^2(\lambda_0-v)^{-5/2} \right] dz + O(\varepsilon^4). \quad (20)$$

Upon setting the coefficient of each power of ε to be zero we find:

$$\lambda_1 = 0, \quad (21a)$$

$$\lambda_2 = \tfrac{1}{16} \left[\oint (v')^2 (\lambda_0-v)^{-5/2} dz \oint / (\lambda_0-v)^{-1/2} dz \right]. \quad (21b)$$

It can also be shown that $\lambda_3 = \lambda_5 = 0$. Then by employing the extensions of eqs. (18) and (19) to higher order, and the corresponding expansions of $\oint \chi_4 dz$ and $\oint \chi_6 dz$, all listed in appendix A we find:

$$\lambda_4 S_1 = \tfrac{1}{4} \lambda_2^2 S_3 - \frac{5\lambda_2}{64\pi} \oint \frac{(v')^2 dz}{(\lambda_0-v)^{7/2}}$$

$$+ \frac{1}{128\pi} \oint (\lambda_0-v)^{-1/2} \left\{ \frac{v''}{(\lambda_0-v)^{3/2}} + \frac{5(v')^2}{4(\lambda_0-v)^{5/2}} \right\}^2 dz \quad (22)$$

where

$$S_n = \frac{1}{2\pi} \oint (\lambda_0 - v)^{-n/2} dz. \quad (23)$$

The λ_6 expression is quite long.

$$\lambda_6 S_1 = \tfrac{1}{2}\lambda_2 \lambda_4 S_3 - \tfrac{1}{8}\lambda_2^3 S_5 - \frac{5\lambda_4}{64\pi} \oint w^7 (v')^2 dz + \frac{35\lambda_2^2}{256\pi} \oint w^9 (v')^2 dz$$

$$- \frac{7\lambda_2}{256\pi} \oint w^9 (v'')^2 dz - \frac{45\lambda_2}{512\pi} \oint w^{11} (v')^2 v'' dz$$

$$- \frac{275\lambda_2}{4096\pi} \oint w^{13} (v')^4 dz$$

$$- \frac{1}{2\pi} \left(\oint \frac{\chi_3^2 dz}{\chi_0} + 2\oint \frac{\chi_4 \chi_2}{\chi_0} dz \right), \quad (24)$$

where the functions χ_2, χ_3 and χ_4 are defined by eq. (7) and

$$w \equiv (\lambda_0 - v)^{-1/2}. \quad (25)$$

Certain relations are helpful for the derivation of higher order integrals from more primitive ones. For example:

$$S_3 = -2\,\mathrm{d}S_1/\mathrm{d}\lambda_0, \tag{26a}$$

$$S_5 = (4/3)\,\mathrm{d}^2 S_1/\mathrm{d}\lambda_0^2 \quad \text{etc.} \tag{26b}$$

If:

$$I_1 = \frac{1}{4\pi}\oint \frac{(v'')^2\,\mathrm{d}z}{(\lambda_0-v)^{1/2}}, \quad \text{then} \tag{27a}$$

$$\frac{1}{4\pi}\oint \frac{(v'')^2\,\mathrm{d}z}{(\lambda_0-v)^{7/2}} = -\frac{8}{15}\frac{\mathrm{d}^3 I_1}{\mathrm{d}\lambda_0^3}, \quad \text{etc.} \tag{27b}$$

If:

$$I_2 = \frac{1}{4\pi}\oint \frac{(v')^4\,\mathrm{d}z}{(\lambda_0-v)^{1/2}}, \quad \text{then} \tag{28a}$$

$$\frac{1}{4\pi}\oint \frac{(v')^4\,\mathrm{d}z}{(\lambda_0-v)^{11/2}} = -\frac{32}{15\times 63}\frac{\mathrm{d}^5 I_2}{\mathrm{d}\lambda_0^5}. \tag{28b}$$

Expressions for certain derivatives of various order terms in $\lambda = \lambda_0 + \varepsilon^2 \lambda_2 + \ldots$ are sometimes useful. We start with the definition (17) of λ_0,

$$\alpha\left(n+\frac{1}{2}\right) = \frac{1}{2\pi}\oint (\lambda_0-v)^{+1/2}\,\mathrm{d}z.$$

Differentiation with respect to n once and twice yields respectively:

$$\frac{1}{\alpha}\frac{\mathrm{d}\lambda_0}{\mathrm{d}n} = \left[\frac{1}{4\pi}\oint (\lambda_0-v)^{-1/2}\,\mathrm{d}z\right]^{-1}, \tag{29a}$$

$$\alpha\frac{\mathrm{d}^2\lambda_0/\mathrm{d}n^2}{(\mathrm{d}\lambda_0/\mathrm{d}n)^3} = \frac{1}{8\pi}\oint (\lambda_0-v)^{-3/2}\,\mathrm{d}z. \tag{29b}$$

From the definition (21b) of λ_2, one finds:

$$\lambda_2 = \frac{1}{16\alpha}\frac{\mathrm{d}\lambda_0}{\mathrm{d}n}\left[\frac{1}{4\pi}\oint (v')^2(\lambda_0-v)^{-5/2}\,\mathrm{d}z\right]. \tag{30a}$$

Further differentiation with respect to n yields:

$$\frac{5}{32}\left\{\frac{1}{4\pi}\oint (v')^2(\lambda_0-v)^{-7/2}\,\mathrm{d}z\right\} = \lambda_2\frac{(\mathrm{d}^2\lambda_0/\mathrm{d}n^2)}{(\mathrm{d}\lambda_0/\mathrm{d}n)^3} - \frac{(\mathrm{d}\lambda_2/\mathrm{d}n)}{(\mathrm{d}\lambda_0/\mathrm{d}n)^2}. \tag{30b}$$

3. Remarks on Method of Maslov

Maslov has derived the first few terms of the expansion (16) by a somewhat different method (chapt. 9 of ref. [13]). He starts with the harmonic oscillator equation:

$$-\varepsilon^2 W_n + (x^2 - \mu_n) W_n = 0, \quad \int_{-\infty}^{\infty} W_n^2(x)\,dx = 1, \tag{31a}$$

$$\mu_n = \pi\varepsilon(n + \tfrac{1}{2}). \tag{31b}$$

The nth characteristic function of the equation of interest (1) has the same number of nodes as $W_n(x)$ but the zeros of the solution of eq. (1) are somewhat displaced from those of $W_n(x)$, and as a function of x, $u(x)$ varies somewhat from $W_n(x)$. On this basis an attempt is made to distort $W_n(x)$ to give the first order approximation to $u_n(x)$ by the product:

$$F(x) = z(x) W_n(y(x)) \phi(x), \tag{32a}$$

where the distortion mapping function $y(x)$ satisfies:

$$y' = \{[\lambda_0 - v(x)]/(u - y^2)\}^{1/2}, \tag{32b}$$

with λ_0 being the solution of the Bohr–Wilson–Sommerfeld equation (17):

$$\pi\varepsilon(n + \tfrac{1}{2}) = \int_{x_1}^{x_2} [\lambda_0 - v(x)]^{1/2}\,dx, \tag{33a}$$

$$v(x_1) = v(x_2) = \lambda_0. \tag{33b}$$

Also, $z(x) = (y')^{-1/2}$ and $\phi(x) = 1$ in the interval:

$$x_1 - \delta \leqslant x \leqslant x_2 + \delta \tag{34a}$$

and vanishes outside the interval:

$$(x_1 - 2\delta, x_2 + 2\delta). \tag{34b}$$

The following differential equation is found for $F(x)$:

$$-\varepsilon^2 F_n'' + [v(x) - \lambda_n^{(0)}] F_n(x) - \varepsilon^2 \frac{z''}{z} F_n(x) = \varepsilon^2 [\phi'' z W_n + 2\phi'(zW_n)']. \tag{35}$$

Then, through appropriate manipulations of combinations of $F_n(x)$ and the nth characteristic function $\psi_n(x)$ of eq. (1) he shows that:

$$\lambda^{(n)} = \lambda_0 + \varepsilon^2 \lambda_2 + O(\varepsilon^2), \tag{36a}$$

where

$$\lambda_2 = \frac{1}{24T} \left[\frac{d^2}{d\lambda^2} \int_{x_1}^{x_2} \frac{[v'(x)]^2 dx}{(\lambda - v(x))^{1/2}} \right]_{\lambda = \lambda_0}, \tag{36b}$$

and

$$T = \frac{1}{2} \int_{x_1}^{x_2} [\lambda_0 - v(x)]^{1/2} dx. \tag{37}$$

Our expression (30a) is essentially of this form, since:

$$\frac{4}{3} \frac{d^2}{d\lambda_0^2} \oint (v')^2 (\lambda_0 - v)^{-1/2} dz = \oint (v')^2 (\lambda_0 - v)^{-5/2} dz. \tag{38}$$

Maslov then tries to replace $F(x)$ by a new function $R(x)$ whose differential equation is closer to eq. (1) than eq. (35). For this purpose he defines $R(x)$ by:

$$R_n(x) = \phi(x) [z + \varepsilon^2 z_1] W_n(y + \varepsilon^2 y_1), \tag{39}$$

where y_1 and z_1 are appropriately constructed functions to accomplish the original aim. When the function R_n and ψ_n are manipulated together properly he finds that:

$$\lambda^{(n)} = \lambda_0 + \varepsilon^2 \lambda_2 + \varepsilon^4 \lambda_4 + O(\varepsilon^4), \tag{40}$$

where λ_2 is defined by eq. (36) and λ_4 by:

$$\lambda_4 = \lambda_2 \frac{d\lambda_2}{dn} \bigg/ \frac{d\lambda_0}{dn} - \frac{\lambda_2}{2} \frac{d^2\lambda_0}{dn^2} \bigg/ (d\lambda_0/dn)^2$$

$$- \frac{1}{48} \frac{d}{dn} \left[\frac{1}{5} \frac{d^2}{d\lambda^2} \int_{x_1}^{x_2} \frac{(v'')^2 dx}{(\lambda - v)^{1/2}} - \frac{1}{36} \frac{d^4}{d\lambda^4} \int_{x_1}^{x_2} \frac{(v')^4 dx}{(\lambda - v)^{1/2}} \right]_{\lambda = \lambda_0}.$$

$$\tag{41}$$

While this form is somewhat different from eq. (22), we show in appendix B how eq. (22) can be transformed into it.

4. Energy levels of purely quartic and mixed quartic and harmonic oscillators

We begin this section with a discussion of the purely quartic oscillator with

$$v(x) = \gamma x^4. \tag{42}$$

While one need consider only the case $\gamma = 1$ since eq. (1) with this form for v can be transformed into a similar equation with $\gamma = 1$, we will find it useful in some algebraic manipulations to have the extra free parameter γ. Several basic integrals are needed for the calculation of various terms listed in the last section. First by a standard branch point integration:

$$I_1(\mu) = \oint (\mu - z^4)^{1/2} dz$$

$$= \mu^{3/4} \oint (1 - w^4)^{1/2} dw$$

$$= 2\mu^{3/4} \int_{-1}^{1} (1 - w^4)^{1/2} dw = \frac{2}{3} \frac{\left[\Gamma(\tfrac{1}{4})\right]^2}{(2\pi)^{1/2}} \mu^{3/4}. \tag{43}$$

The integration path circles the branch points $w = \pm 1$ and a branch cut has been made along the real axis connecting these two points. In a similar manner, if we define:

$$S_{n,m} = \oint z^{2n} (\mu - z^4)^{m/2} dz, \tag{44}$$

then

$$S_{n,-1} = \pi^{1/2} \mu^{(2n-1)/4} \frac{\Gamma(\tfrac{1}{4}[2n+1])}{\Gamma(\tfrac{1}{4}[2n+3])}, \tag{45}$$

so that:

$$S_{n,-3} = -2S'_{n,-1}; \quad S_{n,-5} = (2^2/1\times 3) S''_{n,-1};$$

$$S_{n,-7} = -(2^3/1\times 3\times 5) S'''_{n,-1}, \text{ etc.} \tag{46}$$

The standard WKB approximation to the nth characteristic value λ of eq. (1), with $v\equiv\gamma x^4$ is obtained from eq. (17):

$$\alpha\left(n+\frac{1}{2}\right)=\frac{1}{2\pi}\oint(\lambda_0-\gamma z^4)^{1/2}dz$$

$$=(\gamma^{1/2}/2\pi)I_1(\lambda_0/\gamma)=\frac{2}{3}\frac{[\Gamma(\frac{1}{4})]^2}{(2\pi)^{3/2}}\gamma^{-1/4}\lambda_0^{3/4}, \qquad (47)$$

so that

$$\lambda_0=(2\pi)^2\gamma^{1/3}\left[\frac{3\alpha(n+\frac{1}{4})}{2[\Gamma(\frac{1}{4})]^2}\right]^{4/3}. \qquad (48)$$

The first correction term λ_2, is obtained from eq. (21b) which, for eq. (42) has the form:

$$\lambda_2=\frac{S_{3,-5}(\lambda_0/\gamma)}{S_{0,-1}(\lambda_0/\gamma)}=(\gamma/\lambda_0)^{1/2}\{\Gamma(\tfrac{3}{4})/\Gamma(\tfrac{1}{4})\}^2$$

$$=2(\gamma/\lambda_0)^{1/2}\pi^2\{\Gamma(\tfrac{1}{4})\}^{-4}, \qquad (49)$$

so that to second order in ε:

$$\lambda=\lambda_0[1+\varepsilon^2(\lambda_2/\lambda_0)+\ldots]$$

$$=(2\pi)^2\gamma^{1/3}\left[\frac{3\alpha(n+\frac{1}{2})}{2[\Gamma(\frac{1}{4})]^2}\right]^{4/3}\left\{1+[\varepsilon/\alpha(n+\tfrac{1}{2})]^2/9\pi+\ldots\right\}. \qquad (50)$$

In the case of a quantum mechanical quartic oscillator $\varepsilon=\alpha=(\tfrac{1}{2})^{1/2}$ and λ corresponds to the nth energy level of the oscillator, E_n. Hence

$$E_n\sim\pi^2\gamma^{1/3}\left[\frac{3(n+\tfrac{1}{2})}{[\Gamma(\tfrac{1}{4})]^2}\right]^{4/3}\left\{1+\frac{1}{9\pi(n+\tfrac{1}{2})^2}+\ldots\right\}. \qquad (51)$$

The first order correction term to the standard WKB result was derived empirically from numerically calculated energy levels in reference [15] (eq. IV.15a). The term corresponding to that in the curly bracket of eq. (51) was found to be:

$$\left\{1+0.02650(n+\tfrac{1}{2})^{-2}+\ldots\right\}^{4/3}=\left\{1+0.03533(n+\tfrac{1}{2})^{-2}+\ldots\right\}. \qquad (52)$$

The coefficient 0.035333... agrees with that of eq. (51):

$$1/9\pi = 0.0353677$$

to three significant figures.

Further terms in the expansion (16) follow from eqs. (22), (24), (7b), (7c) and (7d), again employing eqs. (44) and (45):

$$\lambda_4/\lambda_0 = -\frac{(n+\tfrac{1}{2})^{-4}}{\alpha^4(9\pi)^2}\left\{\frac{5}{8} + \frac{11}{6}\frac{[\Gamma(\tfrac{1}{4})]^8}{(4\pi)^4}\right\}, \tag{53a}$$

$$\lambda_6/\lambda_0 = \frac{11(n+\tfrac{1}{2})^{-6}}{\alpha^6(9\pi)^3}\left\{\frac{1}{12} - \frac{31[\Gamma(\tfrac{1}{4})]^8}{10(4\pi)^4}\right\}. \tag{53b}$$

Again, in the quantum mechanical case $\alpha = \varepsilon = \sqrt{\tfrac{1}{2}}$ and

$$\begin{aligned}E_n \sim \pi^2\gamma^{1/3}&\left\{\frac{3(n+\tfrac{1}{2})}{[\Gamma(\tfrac{1}{4})]^2}\right\}^{4/3}\left\{1 + \frac{1}{9\pi(n+\tfrac{1}{2})^2}\right.\\ &+ \frac{(n+\tfrac{1}{2})^{-4}}{(9\pi)^2}\left[\frac{5}{8} + \frac{11}{6}\frac{[\Gamma(\tfrac{1}{4})]^8}{(4\pi)^4}\right] + \frac{11(n+\tfrac{1}{2})^{-6}}{(9\pi)^3}\\ &\left.\times\left[\frac{1}{12} - \frac{31[\Gamma(\tfrac{1}{2})]^8}{10(4\pi)^4}\right] + \ldots\right\}\end{aligned} \tag{54a}$$

$$= 1.37650740\gamma^{1/3}(n+\tfrac{1}{2})^{4/3}\left\{1 + 0.035367(n+\tfrac{1}{2})^{-2}\right.$$
$$\left. - 0.003527(n+\tfrac{1}{2})^{-4} - 0.001765(n+\tfrac{1}{2})^{-6} + \ldots\right\}. \tag{54b}$$

It is remarkable that while we originally set out to derive a series expansion in ε, we have ended with one in inverse powers of $(n+\tfrac{1}{2})^{-1}$.

In table 1, we compare the exact values of E_n in the last column for $n = 0, 1, \ldots, 10$ with those calculated from eq. (54b) up to the second, fourth and sixth orders respectively. It is seen that the accuracy of our series expansion results increases rapidly with increasing n. The expansion in eq. (54b), being an expansion in $(n+\tfrac{1}{2})^{-1}$, is clearly not appropriate for the ground state energy. It should also be remembered that for any fixed value of n, the series we obtained is asymptotic in

Table 1. Comparison of the results obtained from eq. (54b) ($\gamma=1$) with the exact, results for the purely quartic oscillator.

n	E_n (0th)	E_n (2nd)	E_n (4th)	E_n (6th)	E_n (exact)
0	0.546 267 33	0.623 548 34	0.592 716 32	0.530 984 39	0.667 986 26
1	2.363 561 4	2.400 714 3	2.399 067 3	2.398 701 0	2.393 644 0
2	4.670 519 9	4.696 949 7	4.696 527 9	4.696 494 1	4.696 795 4
3	7.314 802 6	7.335 921 6	7.335 749 7	7.335 742 7	7.335 730 0
4	10.226 536	10.244 398	10.244 310	10.244 308	10.244 309
5	13.363 764	13.379 389	13.379 337	13.379 337	13.379 337
6	16.697 945	16.711 923	16.711 890	16.711 890	16.711 890
7	20.208 166	20.220 872	20.220 850	20.220 849	20.220 850
8	23.878 321	23.890 010	23.889 994	23.889 994	23.889 994
9	27.695 552	27.706 406	27.706 394	27.706 393	27.706 394
10	31.649 313	31.659 466	31.659 457	31.659 456	31.659 456

nature. We shall study this series by other summation methods and shall present the result in another paper.

An alternative derivation of eq. (54) would follow directly from eq. (15). In the case of $v(x) \equiv \gamma x^4$, eq. (15) becomes:

$$2\pi(\varepsilon\gamma^{-1/2})(n+\tfrac{1}{2}) = I_1 - \tfrac{1}{2}(\varepsilon\gamma^{-1/2})^2 S_{3,-5}$$
$$- \tfrac{9}{8}(\varepsilon\gamma^{-1/2})^4 \left(S_{2,-7} + \tfrac{10}{3} S_{4,-9} + \tfrac{25}{9} S_{6,-11} \right) + \ldots \quad (55)$$

where the integrals I_1 and $S_{n,m}$ are defined by eqs. (43) and (44) with the argument μ taken to be $\mu = \gamma/\lambda$. Upon integration, using eqs. (44) and (46), one finds:

$$N = x\left[a_0 + a_1 x^{-2} + a_2 x^{-4} + a_3 x^{-6} + \ldots \right], \quad (56)$$

with

$$N \equiv (n+\tfrac{1}{2})\pi, \quad x \equiv (\lambda^{3/4}/\varepsilon\gamma^{1/4}), \quad (57a)$$

$$a_0 = \left[\Gamma(\tfrac{1}{4})\right]^2 / 3(2\pi)^{1/2} = 1.748\,038\,369, \quad (57b)$$

$$a_1 = (2\pi)^{3/2}/8\left[\Gamma(\tfrac{1}{4})\right]^2 = 0.149\,767\,529, \quad (57c)$$

$$a_2 = 11\left[\Gamma(\tfrac{1}{4})\right]^2 / 1536(2\pi)^{1/2} = 0.037\,555\,511\,8, \quad (57d)$$

$$a_3 = 4697(2\pi)^{3/2} / 61440\left[\Gamma(\tfrac{1}{4})\right]^2 = 0.091\,596\,104\,8. \quad (57e)$$

Substituting

$$x = b_0 N[1 + b_1 N^{-2} + b_2 N^{-4} + \ldots] \tag{58}$$

into eq. (56) and comparing coefficients of N^j, and finally introducing the appropriate expressions for N and x we recover eq. (54) when $\varepsilon = (\tfrac{1}{2})^{1/2}$.

Bender, Olaussen and Wang [14] used the expansion in eq. (56) and determined λ by first truncating the series on the right-hand side of eq. (56) to orders $x^{-2}, x^{-4}, \ldots$ and then by iterating the values of λ until eq. (56) is satisfied. They did not write E_n explicitly as in eq. (54b).

Let us now consider the mixed quadratic and quartic oscillator with potential energy;

$$V(x) = \tfrac{1}{2}\omega^2 x^2 + \gamma x^4. \tag{59}$$

The Hamiltonian of such an oscillator is characterized by:

$$H(\omega, \gamma) = \tfrac{1}{2}(p^2 + x^2\omega^2) + \gamma x^4 = -\tfrac{1}{2}(d^2/dx^2) + \tfrac{1}{2}x^2\omega^2 + \gamma x^4. \tag{60}$$

Then, if, as has been shown by Symanzik, one introduces a new variable y by making the transformation:

$$x = y\gamma^{-1/6}, \tag{61}$$

it is found that:

$$H(\omega, \gamma) = \gamma^{1/3} H(\omega \gamma^{-1/3}, 1). \tag{62}$$

Hence, we need only find the energy levels ε_n of a new oscillator with potential:

$$v(x) = \beta x^2 + x^4, \quad \text{with} \quad \beta = \tfrac{1}{2}\omega^2 \gamma^{-2/3}. \tag{63}$$

The energy levels E_n of $H(\omega, \gamma)$ are related to the ε_n by:

$$E_n = \gamma^{1/3} \varepsilon_n. \tag{64}$$

We now apply the methods discussed above for the case of the purely quartic oscillator to the determination of the ε_n. Our basic formula is eq.

(15). The first step is to calculate:

$$\tfrac{1}{2}\oint \chi_0\, dz = \tfrac{1}{2}\oint (\lambda - \beta z^2 - z^4)^{1/2}\, dz$$

$$= \tfrac{1}{2}\oint [(\lambda - z^4) - \beta z^2]^{1/2}\, dz$$

$$= \frac{1}{2}\oint (\lambda - z^4)^{1/2}\left[1 - \frac{1}{2}\frac{\beta z^2}{(\lambda - z^4)} - \frac{1}{8}\frac{\beta^2 z^4}{(\lambda - z^4)^2} + \ldots \right] dz$$

$$= \tfrac{1}{2} I_1(\lambda) - \tfrac{1}{4}\beta S_{1,-1}(\lambda) - \tfrac{1}{16}\beta^2 S_{2,-3}(\lambda) - \tfrac{1}{32}\beta^3 S_{3,-5}(\lambda) - \ldots$$

$$= a_0 \lambda^{3/4} + b_0 \beta \lambda^{1/4} + c_0 \beta^2 \lambda^{-1/4} + d_0 \beta^3 \lambda^{-3/4}$$

$$+ e_0 \beta^4 \lambda^{-5/4} + \ldots, \qquad (65)$$

where

$$a_0 = \frac{[\Gamma(\tfrac{1}{4})]^2}{3(2\pi)^{1/2}} = 1.748\,038\,369; \qquad (66a)$$

$$b_0 = -\frac{1}{2}\frac{(2\pi)^{3/2}}{[\Gamma(\tfrac{1}{4})]^2} = -0.599\,070\,117; \qquad (66b)$$

$$c_0 = \frac{1}{32}\frac{[\Gamma(\tfrac{1}{4})]^2}{(2\pi)^{1/2}} = 0.163\,878\,597; \qquad (66c)$$

$$d_0 = -\frac{(2\pi)^{3/2}}{64[\Gamma(\tfrac{1}{4})]^2} = -0.018\,720\,941; \qquad (66d)$$

$$e_0 = -\frac{5[\Gamma(\tfrac{1}{4})]^2}{6144(2\pi)^{1/2}} = -0.004\,267\,672. \qquad (66e)$$

Next we must find:

$$-\tfrac{1}{4}\oint (\chi_1^2/\chi_0)\, dz = -\tfrac{1}{64}\oint (v')^2 \chi_0^{-5}\, dz, \qquad (67)$$

which, when $v(x) = \beta x^2 + x^4$ becomes:

$$-\frac{1}{16}\oint \frac{(4z^6 + 4\beta z^4 + \beta^2 z^2)\,dz}{(\lambda - z^4 - \beta z^2)^{5/2}}$$
$$= -\tfrac{1}{4}S_{3,-5} - \tfrac{1}{4}\beta S_{2,-5} - \tfrac{5}{8}\beta S_{4,-7} + O(\beta^2), \tag{68}$$

where $S_{n,-m}$ is determined from $S_{n,-1}$ by eqs. (44), (45), and (46). Since:

$$S_{3,-5}(\lambda) = \tfrac{1}{2}(2\pi)^{3/2}\lambda^{-3/4}/\left[\Gamma(\tfrac{1}{4})\right]^2, \tag{69a}$$

$$S_{2,-5}(\lambda) = -\tfrac{1}{12}(2\pi)^{-1/2}\lambda^{-5/4}\left[\Gamma(\tfrac{1}{4})\right]^2, \tag{69b}$$

$$S_{4,-7}(\lambda) = \tfrac{1}{24}(2\pi)^{-1/2}\lambda^{-5/4}\left[\Gamma(\tfrac{1}{4})\right]^2, \tag{69c}$$

the above sum becomes:

$$-\tfrac{1}{8}(2\pi)^{3/2}\lambda^{-3/4}\left[\Gamma(\tfrac{1}{4})\right]^{-2} - \tfrac{1}{192}(2\pi)^{-1/2}\lambda^{-5/4}\beta\left[\Gamma(\tfrac{1}{4})\right]^2 + O(\beta^2). \tag{70}$$

The terms (65) and (70), when substituted into eq. (15) yield:

$$\pi\varepsilon(n+\tfrac{1}{2}) = a_0\lambda^{3/4} + b_0\beta\lambda^{1/4} + c_0\beta^2\lambda^{-1/4} + (a_2\varepsilon^2 + d_0\beta^3)\lambda^{-3/4}$$
$$+ (b_2\varepsilon^2\beta + e_0\beta^4)\lambda^{-5/4} + \ldots \tag{71}$$

with

$$a_2 = -\tfrac{1}{8}(2\pi)^{3/2}\left[\Gamma(\tfrac{1}{4})\right]^{-2} = -0.149\,767\,529; \tag{72a}$$

$$b_2 = -\tfrac{1}{192}(2\pi)^{-1/2}\left[\Gamma(\tfrac{1}{4})\right]^2 = -0.027\,313\,100. \tag{72b}$$

If we substitute:

$$\lambda = a_0^{-4/3}N^{4/3}\left[1 + A_1N^{-2/3} + A_2N^{-4/3} + A_3N^{-6/3} + A_4N^{-8/3} + \ldots\right] \tag{73}$$

into eq. (71), and set the coefficients of each power of $N^{-2/3}$ equal to zero, the coefficients $A_1, A_2, A_3, A_4, \ldots$ can be found.

In the case $\varepsilon = \sqrt{\tfrac{1}{2}}$ and $\beta = \tfrac{1}{2}\omega^2\gamma^{-2/3}$, the eigenvalues $\lambda \equiv E_n$ of the Hamiltonian $H = \tfrac{1}{2}(-d^2/dx^2 + \omega^2 x^2) + \gamma x^4$ are given by:

$$E_n = C\gamma^{1/3}\left(n+\tfrac{1}{2}\right)^{4/3}\left\{1 + c_1(\gamma)\left(n+\tfrac{1}{2}\right)^{-2/3} + c_2(\gamma)\left(n+\tfrac{1}{2}\right)^{-4/3}\right.$$

$$\left. + c_3(\gamma)\left(n+\tfrac{1}{2}\right)^{-2} + c_4(\gamma)\left(n+\tfrac{1}{2}\right)^{-8/3} + \ldots\right\}, \qquad (74)$$

where

$$C = 3^{4/3}\pi^2\left[\Gamma(\tfrac{1}{4})\right]^{-8/3} = 1.376\,507\,40; \qquad (75a)$$

$$c_1(\gamma) = \frac{4}{9\pi}C\gamma^{-2/3} = 0.194\,735\,962\,\omega^2\gamma^{-2/3}; \qquad (75b)$$

$$c_2(\gamma) = \left\{\frac{6\pi^4}{\left[\Gamma(\tfrac{1}{4})\right]^8} - \frac{1}{32}\right\}C^{-1} = -0.008\,481\,598\,6\,\omega^4\gamma^{-4/3}; \qquad (75c)$$

$$c_3(\gamma) = \frac{1}{24}c_1(\gamma)^3 + \frac{1}{9\pi} = 0.000\,307\,699\,8\,\omega^6\gamma^{-2} + 0.035\,367\,765;$$
$$\qquad (75d)$$

$$c_4(\gamma) = \left\{\frac{1}{3\cdot 16^3} - \frac{5\pi^8}{3\left[\Gamma(\tfrac{1}{4})\right]^{16}}\right\}C^{-2}\omega^8\gamma^{-8/3}$$

$$+ \frac{1}{3^{8/3}\left[\Gamma(\tfrac{1}{4})\right]^{8/3}}\left\{\frac{4\left[\Gamma(\tfrac{1}{4})\right]^8}{3(4\pi)^4} - 1\right\}\omega^2\gamma^{-2/3} \qquad (75e)$$

$$= 0.000\,033\,587\,3\,\omega^8\gamma^{-8/3} + 0.001\,026\,946\,7\,\omega^2\gamma^{-2/3}.$$

In table 2 we compare energy levels calculated from eq. (74) to second and fourth order with exact numerical results derived in ref. [15].

The corresponding series expansions for the eigenvalues μ_n, say, of the Hamiltonians $-d^2/dx^2 + gx^4$ and $-d^2/dx^2 + \omega^2 x^2 + gx^4$ can be obtained from eqs. (54) and (74) respectively by making the substitutions $g = 2\gamma$ and $\mu_n = 2^{2/3}E_n$.

While eq. (74) is written as a series expansion in inverse powers of $(n+\tfrac{1}{2})$, it can be rearranged as a series in inverse powers of γ with

$$E_n = \gamma^{1/3}\left(\varepsilon_n + \alpha_n\gamma^{-2/3} + \beta_n\gamma^{-4/3} + \delta_n\gamma^{-6/3} + \ldots\right), \qquad (76)$$

Table 2. Comparison of the results obtained from eq. (74) with the exact results for the mixed quadratic and quartic case. For each value of n, the numbers in the first two rows give the results of the series up to the terms in $(n+\tfrac{1}{2})^{-4/3}$ and $(n+\tfrac{1}{2})^{-8/3}$ respectively and the number in the third row gives the exact result.

$n \setminus \Gamma$	0.1	1	1000	
0	0.5006123	0.7034567	5.479548	2nd order
	0.6004639	0.7850886	6.252714	4th order
	0.5591463	0.8037707	6.694221	exact
1	1.737068	2.703139	23.67073	
	1.776890	2.741465	24.04234	
	1.769503	2.737892	23.97221	
2	3.114892	5.152607	46.75456	
	3.141666	5.179697	47.01890	
	3.138624	5.179292	47.01734	
3	4.609763	7.921054	73.20981	
	4.630541	7.942632	73.42102	
	4.628883	7.942404	73.41911	
4	6.204091	10.94550	102.3384	
	6.221344	10.96371	102.5171	
	6.220301	10.96358	102.5162	
5	7.885583	14.18731	133.7212	
	7.900484	14.20322	133.8774	
	7.899767	14.20314	133.8769	
6	9.645158	17.61988	167.0728	
	9.658363	17.63410	167.2126	
	9.657840	17.63405	167.2123	
7	11.47580	21.22356	202.1844	
	11.48771	21.23647	202.3114	
	11.48732	21.23644	202.3112	
8	13.37189	24.98309	238.8948	
	13.38279	24.99497	239.0117	
	13.37897	24.99495	239.0116	

where, from eqs. (54) and (74) since ε_n is the result for a purely quartic oscillator [eq. (54)]:

$$\varepsilon_n = 1.37650740\left(n+\tfrac{1}{2}\right)^{4/3}\left\{1+0.035367\left(n+\tfrac{1}{2}\right)^{-2}\right.$$
$$\left. -0.003527\left(n+\tfrac{1}{2}\right)^{-4} - 0.001765\left(n+\tfrac{1}{2}\right)^{-6} + \ldots\right\},$$
$$\alpha_n = 0.26805549\left(n+\tfrac{1}{2}\right)^{2/3}\left\{1+0.005273534\left(n+\tfrac{1}{2}\right)^{-2} + \ldots\right\}, \quad (77)$$

Table 3. Comparison of the values of α_n from eq. (77) with those of the exact results.

n	WKB 0th order	WKB 2nd order	α_n (exact)
0	0.16886	0.17243	0.14367
1	0.35125	0.35208	0.35780
2	0.49376	0.49418	0.49397
3	0.61793	0.61819	0.61826
4	0.73063	0.73082	0.73084
5	0.83522	0.83536	0.83536
6	0.93361	0.93373	0.93373
7	1.02707	1.02716	1.02716
8	1.11644	1.11653	1.11653
9	1.20238	1.20245	1.20245
10	1.28534	1.28540	1.28540

$$\beta_n = -0.011674983\left\{1+O\left[\left(n+\tfrac{1}{2}\right)^{-2}\right]\right\}, \tag{78a}$$

$$\delta_n = 0.000423551\left(n+\tfrac{1}{2}\right)^{-2/3}\left\{1+O\left[\left(n+\tfrac{1}{2}\right)^{-2}\right]\right\}. \tag{78b}$$

The numbers ε_n, α_n, and β_n were calculated "exactly" from numerical tables [15] of E_n for $n = 0, 1, \ldots, 10$. These are compared to the analytical expressions above in table 3. Except for the ground state ($n=0$) for which the present investigation is not appropriate, our analytical expressions are in excellent agreement with the exact results. Of course eq. (76) is an asymptotic expansion for large values of γ.

5. Density of states in multidimensional systems

Another interpretation of eq. (10) is evident when written in the form:

$$n = -\frac{1}{2} + \frac{1}{2\pi\varepsilon}\oint[\lambda_n - v]^{1/2}\,\mathrm{d}z + \sum_{j=2}^{\infty}\frac{1}{2\pi\varepsilon}(\varepsilon/\mathrm{i})^j\oint\chi_j(z)\,\mathrm{d}z. \tag{79}$$

Here n may be regarded to be the number of characteristic values of n less than or equal to λ_n. Then, if the right hand side of eq. (79) is considered to be a function $f(\lambda)$, of the variable λ (with $\lambda \geqslant \min v(x)$) it is precisely the cumulative distribution function of the characteristic values λ_n. When n is large (or ε small) since as we have seen in many interesting cases the combination ε/n^η becomes the relevant expansion parameter, only one or two terms are required in eq. 79.

One would like to generalize eq. (11) to a system of many degrees of freedom, say one characterized by a differential equation:

$$\Delta u(x_1\ldots x_k) - [\lambda - v(x_1\ldots x_k)]u(x_1\ldots x_k) = 0, \tag{80a}$$

where

$$v(x_1\ldots x_k) \to \infty \text{ as } |x| \to \infty, \tag{80b}$$

$$\Delta \equiv \partial^2/\partial x_1^2 + \partial^2/\partial x_2^2 + \ldots + \partial^2/\partial x_k^2.$$

Titchmarsh has shown that $N(\lambda)$, the number of characteristic values of eq. (80), with values less than or equal to λ grows asymptotically as $\lambda \to \infty$ according to

$$N(\lambda) \sim \frac{1}{2^k \pi^{k/2} \Gamma(\tfrac{1}{2}k+1)} \int_{v<\lambda} [\lambda - v(x)]^{k/2} d^k x. \tag{81}$$

By choosing $k=1$ and setting $\varepsilon = 1$ in eq. (79) we recover the first order approximation of eq. (79), that of retaining only the first integral and neglecting all other terms on the right hand side of the equation. It would be interesting to find correction terms to eq. (81) that are analogous to the higher order terms in eq. (79). Such a program is in progress.

While more detailed investigation of the multi-dimensional case will be given elsewhere, an indication of the accuracy of eq. (81) can be seen from an analysis of a three dimensional non-linear oscillator with Hamiltonian:

$$H = -\tfrac{1}{2}\nabla^2 + r^4 \tag{82a}$$

and Schrödinger equation:

$$\left[-\frac{1}{2}\frac{d^2}{dr^2} + \frac{\tfrac{1}{2}l(l+1)}{r^2} + r^4 \right] u = Eu. \tag{82b}$$

The first 160 energy levels E_{nl} were obtained numerically by the method described in appendix 3 of ref. [17] for the corresponding two dimensional case. We have plotted some of the levels E_{nl} values in fig. 1. The corresponding comparison is made in fig. 2 for the Hamiltonian:

$$H = \sum_{j=1}^{3} \left[x_j^4 - \tfrac{1}{2}(\partial^2/\partial x_j^2) \right].$$

To apply eq. (81) one identifies $2E$ with λ, and in the case of eq. (82) one sets

$$v(x) \equiv l(l+1)r^{-2} + 2r^4, \tag{83}$$

while in the case eq. (83),

$$v(x) = 2\sum_1^3 x_j^4. \tag{84}$$

In both cases $k=3$ and one finds respectively from eq. (81) that for eq. (82),

$$N(E) \sim \frac{32\pi^{1/2} E^{9/4}}{45\left[\Gamma\left(\frac{1}{4}\right)\right]^2} = 0.0958849\ldots E^{9/4} \tag{85}$$

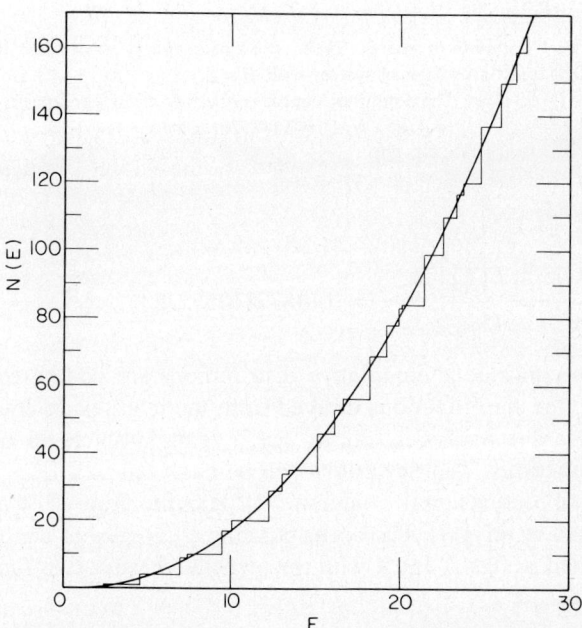

Fig. 1. The exact numbers of energy levels $N(E)$ with energy values less than E for three spherically coupled oscillators system with $H = \sum_1^3 -\frac{1}{2}\partial^2/\partial x_i^2 + (\sum_1^3 x_i^2)^2$ are shown as step function as E increases. The continuous curve is eq. (85), $N(E) = 0.0958849 E^{9/4}$.

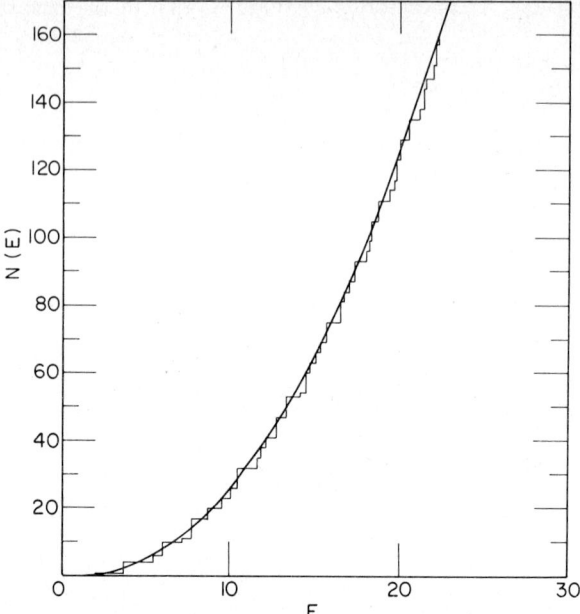

Fig. 2. The exact numbers of energy levels $N(E)$ with energy values less than E for three-independent quartic oscillators system with $H = \Sigma_1^3(-\frac{1}{2}\partial^2/\partial x_i^2 + x_i^4)$ are shown as step function as E increases. The continuous curve is obtained from Titchmarsh's formula, eq. (86), $N(E) = 0.1483781\, E^{9/4}$.

and for eq. (83),

$$N(E) \sim \frac{2^{3/2}\left[\Gamma(\tfrac{1}{4})\right]^2 E^{9/4}}{45\pi^{3/2}} = 0.148378105\ldots E^{9/4}. \tag{86}$$

These two continuous cumulative distributions are compared in figs. 1 and 2 with the step functions derived from the more exact counting of energy levels. The analytical curves are a remarkably good fit to the exact step functions. The deviations will be analyzed in a later publication. The two dimensional analogues of Hamiltonians (82) and (83) were discussed in ref. [17]. The corresponding analogues of eqs. (85) and (86) are plotted in figs. 3 and 4 with the associated exact step functions.

Appendix A. Expressions required for derivation of λ_4 and λ_6

We need the expansion of the function given in eq. (18) to order ε^6. It is

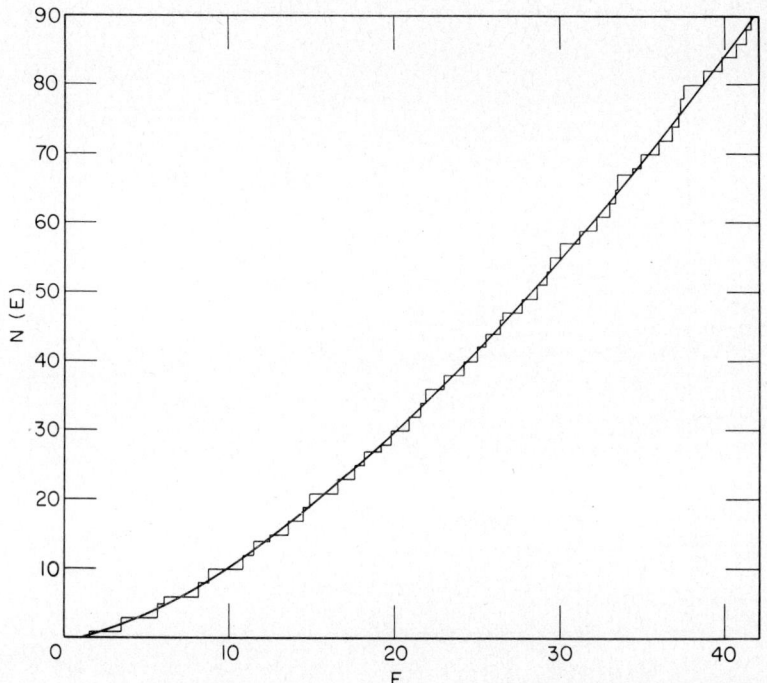

Fig. 3. The exact numbers of energy levels $N(E)$ with energy values less than E for two circularly coupled oscillators system with $H = \Sigma_1^2 - \frac{1}{2} \partial^2/\partial x_i^2 + (\Sigma_1^2 x_i^2)^2$ are shown as step function as E increases. The continuous curve is obtained from Titchmarsh's formula $N(E) = \frac{1}{3} E^{3/2}$.

found that, defining S_n by eq. (23):

$$\begin{aligned}\frac{1}{2\pi} \oint (\lambda - v)^{1/2} dz &= \frac{1}{2\pi} \oint (\lambda_0 - v + \varepsilon^2 \lambda_2 + \varepsilon^4 \lambda_4 + \ldots)^{1/2} \\ &= \frac{1}{2\pi} \oint (\lambda_0 - v)^{1/2} dz \left[1 + \frac{\varepsilon^2 \lambda_2 + \varepsilon^4 \lambda_4 + \ldots}{(\lambda_0 - v)} \right]^{1/2} \\ &= \frac{1}{2\pi} \oint (\lambda_0 - v)^{1/2} dz + \frac{1}{2} \varepsilon^2 \lambda_2 S_1 \\ &\quad + \tfrac{1}{8} \varepsilon^4 (4\lambda_2 S_1 - \lambda_2^2 S_3) \\ &\quad + \tfrac{1}{16} \varepsilon^6 (8\lambda_6 S_1 - 4\lambda_2 \lambda_4 S_3 + \lambda_2^3 S_5) + \ldots \quad (A.1)\end{aligned}$$

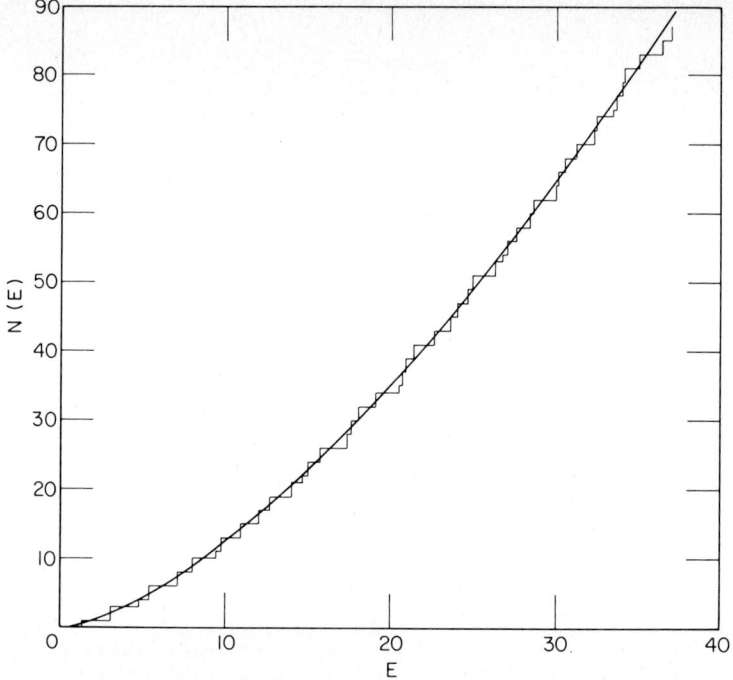

Fig. 4. The exact numbers of energy levels $N(E)$ with energy values less than E for two independent quartic oscillators system with $H = \Sigma_1^2(-\frac{1}{2}\partial^2/\partial x_i^2 + x_i^4)$ are shown as step function as E increases. The continuous curve is obtained from Titchmarsh's formula $N(E) = 0.3934469 E^{3/2}$.

The extension of eq. (19) is:

$$\oint (x_1^2/\chi_0) dz = \frac{1}{16} \oint \frac{(v')^2 dz}{(\lambda - v)^{5/2}}$$

$$= \tfrac{1}{16} J_5 - \tfrac{5}{32}\varepsilon^2 \lambda_2 J_7 - \tfrac{5}{32}\varepsilon^4 \lambda_4 J_7 + \tfrac{35}{128}\varepsilon^4 \lambda_4 J_7 + O(\varepsilon^6), \tag{A.2}$$

where

$$J_n \equiv \oint (v')^2 (\lambda_0 - v)^{-n/2} dz.$$

Furthermore:

$$-2\oint \chi_4 \, dz = \oint (\chi_2^2/\chi_0) \, dz$$

$$= \frac{1}{64} \oint \frac{dz}{(\lambda_0-v)^{1/2}} \left[\frac{v''}{(\lambda_0-v)^{3/2}} + \frac{5}{4} \frac{(v')^2}{(\lambda_0-v)^{5/2}} \right]^2$$

$$-\frac{\lambda_2 \varepsilon^2}{128} \left\{ 7\oint \frac{(v'')^2 \, dz}{(\lambda_0-v)^{9/2}} + \frac{45}{2} \oint \frac{(v')^2 v'' \, dz}{(\lambda_0-v)^{11/2}} \right.$$

$$\left. + \frac{275}{16} \oint \frac{(v')^4 \, dz}{(\lambda_0-v)^{13/2}} \right\} + \cdots \tag{A.3}$$

Appendix B. Alternative forms for λ_4

Our expression (22) for λ_4 is not the same as that of Maslov (eq. (41)). To derive Maslov's form for λ_4 we note eqs. (27) and (28). Since, upon integration by parts:

$$\oint \frac{v''(v')^2 \, dz}{(\lambda_0-v)^{9/2}} = \frac{1}{3} \oint (\lambda_0-v)^{-9/2} \, d(v')^3$$

$$= -\frac{3}{2} \oint \frac{(v')^4 \, dz}{(\lambda-v)^{11/2}}, \tag{B.1}$$

we re-express the following combination from eq. (22) for λ_4:

$$\frac{1}{64} \oint (\lambda_0-v)^{-1/2} \left[\frac{v''}{(\lambda_0-v)^{3/2}} + \frac{5}{4} \frac{(v')^2}{(\lambda_0-v)^{5/2}} \right]^2 dz$$

$$= \frac{1}{64} \left[\oint \frac{(v'')^2 \, dz}{(\lambda_0-v)^{7/2}} - \frac{35}{16} \oint \frac{(v')^4 \, dz}{(\lambda_0-v)^{11/2}} \right]$$

$$= -\frac{1}{24} \left(\frac{1}{5} \frac{d^3 I_1}{d\lambda_0^3} - \frac{1}{36} \frac{d^5 I_2}{d\lambda_0^5} \right). \tag{B.2}$$

Then, by combining eqs. (22), (23), (29), (30), (27), (28), (B.1) and (B.2) we obtain Maslov's formula:

$$\lambda_4 = \lambda_2 \frac{(d\lambda_2/dn)}{(d\lambda_0/dn)} - \frac{1}{2} \frac{\lambda_2^2 d^2\lambda_0/dn^2}{(d\lambda_0/dn)^2}$$

$$- \frac{1}{24} \frac{d}{dn} \left(\frac{1}{5} \frac{d^2 I_1}{d\lambda_0^2} - \frac{1}{36} \frac{d^4 I_2}{d\lambda_0^4} \right). \qquad (B.3)$$

The factor (1/24) in (B.2) replaces Maslov's (1/48) because our integrals are contour integrals while Maslov's are ordinary definite integrals. The two types of integrals differ by a factor of 2 here.

Note added in proof

Since this article was written, the authors have received a Ph.D. thesis by A. Voros which contains a discussion of the purely quartic oscillator which overlaps our calculation.

References

[1] G. Wentzel, Zs. f. Physik **38** (1926) 518.
[2] H. M. Kramers, Zs. f. Physik **39** (1926) 828.
[3] L. Brillouin, Comptes Rendus **183** (1926) 24.
[4] H. Jeffreys, Proc. London Math. Soc. [2], **23** (1924) 428.
[5] J. Liouville, J. Math. Pures. Appl. [1], **2** (1837) 16.
[6] G. Green, Trans. Camb. Phil. Soc. **6** (1837) 457.
[7] R. E. Langer, Bull. Am. Math. Soc. **40** (1934) 545; Trans. Amer. Math. Soc. 37 (1935) 397; Phys. Rev. **51** (1937) 669; Trans. Am. Math. Soc. **67**, (1949) 461.
[8] A. Erdélyi Asymptotic Expansions (Dover, New York, 1956).
[9] F. W. J. Olver, Asymptotics and Special Functions (Academic Press, New York, 1974).
[10] N. Fröman and P. O. Fröman, JWKB Approximation-Contributions to the Theory (North-Holland, Amsterdam, 1965).
[11] E. C. Titchmarsh, Eigenfunction Expansions Associated with Second Order Differential Equations (Oxford, 1946).
[12] J. L. Dunham, Phys. Rev. **41** (1932) 713; **41** (1932) 721.
[13] V. P. Maslov, Théorie des Perturbations et Méthod Asymptotiques Dunod, Paris, 1972 (Russian edition Moscow 1965).
[14] C. M. Bender, K. Olaussen and P. S. Wang, Phys. Rev. **D16** (1977) 1740.
[15] F. T. Hioe and E. W. Montroll, J. Math. Phys. **16** (1975) 1945.
[16] E. C. Titchmarsh, Eigenfunction Expansions, Vol. **II** (Oxford, 1958).
[17] F. T. Hioe, D. MacMillen and E. W. Montroll, Phys. Rep., **43**, (1978) 305.

CHAPTER 17

Equilibrium Density Matrix for Fluids

J. E. MAYER

*Department of Chemistry,
University of California, S.D.
La Jolla, CA 92092
U.S.A.*

© *North-Holland Publishing Company 1981*   *Perspectives in Statistical Physics
Ed. H. J. Raveché*

Contents

1. Introduction 325
2. Notation and preliminaries 325
3. Correction to J. E. Mayer and M. G. Mayer 333
4. Conclusion 341
References 341

1. Introduction

In this article we treat the coordinate representation of the density matrix for an equilibrium ensemble of monatomic molecules applicable to low temperature dense gas or liquid states. The normalization used is that the trace of the matrix, which we symbolize by Ω, be unity*, eq. (9), in which case the diagonal elements, $\omega(q,q)$, of Ω in the $3N$ dimensional cartesian coordinate q are probability densities in q.

The problem was first attacked by Wigner [1] in his now classical 1932 paper in which he gave the quantum correction to first order in h^2 as a linear addition $[1+h^2\Delta_j(q)]$ multiplying the classical probability density $W_{cl}(q)$. Later Mayer and Band [2] showed that Wigner's correction $h^2\Delta_1$ could be put as a multiplicative factor $\exp h^2\Delta_1$ times $W_{cl}(q)$.

2. Notation and preliminaries

We use $1 \leq j < k < l \ldots N$ to indicate specified numbered molecules or sometimes $1 \leq j_1 < j_2 < \ldots j_\nu$ for a subset of $\nu \leq N$ molecules. Coordinates, $q_{j\alpha}$, $\alpha = 1, 2, 3$ for x, y, z respectively of molecule j will be used, and

$$\mathcal{D}_{j\alpha} \equiv \partial/\partial q_\alpha \tag{1}$$

for the differential operator. We use the notation:

$$\sum_{j\alpha}^{3N} \mathcal{D}_{j\alpha} = \sum_{j=1}^{N} \sum_{\alpha=1}^{3} \mathcal{D}_{j\alpha} \tag{2}$$

to indicate summation over all $3N$ degrees of freedom or $\sum_{j\alpha}^{3\nu} \mathcal{D}_{j\alpha}$ for a subset of three each of ν molecules.

With

$$\beta = 1/kT, \tag{3}$$
$$\lambda^2 = \beta h^2/2\pi m, \tag{4}$$
$$\zeta = \lambda^2/2\pi = \beta h^2/4\pi^2 m, \tag{5}$$

*Other normalizations are frequently used.

the kinetic energy operator times β is:

$$\mathcal{K} = \frac{\beta}{2m}\left(\frac{h}{2\pi i}\right)^2 \sum_{j\alpha}^{3N} \mathcal{D}_{j\alpha}^2$$

$$= -\frac{\lambda^2}{4\pi}\sum_j^{3N} \mathcal{D}_{j\alpha}^2 = -\zeta \sum_{j\alpha}^{3N} \frac{1}{2}\mathcal{D}_{j\alpha}^2, \tag{6}$$

and with U_1 the potential energy times β, $\beta U(q_{1,1},\ldots,q_{N,3})$, the Hamiltonian operator times β is:

$$U_1 = \beta U, \quad \beta \mathcal{K} = \mathcal{K} + U_1. \tag{7}$$

The density matrix for an equilibrium Petite Canonical Ensemble normalized to have unit trace is, with $A = A(V, \beta, N)$ the Helmholtz free energy,

$$\Omega = (N!)^{-1} \exp\beta(A - \mathcal{K})$$
$$= e^{\beta A}(N!)^{-1} \sum_{n=0} (n!)^{-1}[-(\mathcal{K}+U_i)]^n, \tag{8}$$

in which, for a dense gas or liquid, the numerically important terms are those with $n \sim N \sim 10^{24}$. The condition

$$\text{trace } \Omega = 1, \tag{9}$$

with eq. (8) determines A. To evaluate the trace we use a unitary matrix Θ of elements $\theta_{\nu\mu}$ with its adjoint matrix $\Theta^\dagger$ of elements $\theta_{\mu\nu}^\dagger = \theta_{\nu\mu}^*$ for which, with I the unit matrix,

$$\left.\begin{array}{ll} \Theta^\dagger\Theta = I, & \Theta\Theta^\dagger = I \\ \sum_\nu \theta_{\mu\nu}^\dagger \theta_{\nu\mu'} = \delta(\mu-\mu'), & \sum_\mu \theta_{\nu\mu}\theta_{\mu\nu'} = \delta(\nu-\nu') \end{array}\right\} \tag{10}$$

with $\delta(0) = 1$ and zero otherwise.

The simplest formal unitary matrix for this problem is the complete set, $\Theta = \Psi$ of real normalized eigenfunctions of the Hamiltonian operator:

$$\mathcal{K}\Psi_K\left(\sum_{j\alpha}^{3N} q_{j\alpha}\right) = E_K\Psi_K\left(\sum_{j\alpha}^{3N} q_{j\alpha}\right), \tag{11}$$

with K a $3N$ dimensional quantum number and $\Sigma q_{j\alpha} = q$ a $3N$ dimensional coordinate set Θ the unitary matrix of elements $\theta_{q'K'}$ and its real adjoint $\Theta^\dagger$ of elements $\theta_{Kq'}$. The equilibrium density matrix Ω will be diagonal in the K, K' representation of elements $\omega(K, K) = \exp -\beta(A - E_K)$, $\omega(K, K') = 0$ if $K \neq K'$. In the coordinate qq' representation the elements will be:

$$\omega(q,q') = \sum_K \theta_{q,K} \omega(KK) \theta_{Kq'}$$
$$= \sum_K \Psi_K(q) \exp[-\beta(A - E_K)] \Psi_K(q'), \tag{12}$$

with diagonal values,

$$\omega(q,q) = \sum_K \Psi_K^2(q) \exp[\beta(A - E_K)]. \tag{12.1}$$

Unless the Hamiltonian $\mathcal{H}$ of eq. (11) is separable in operations on single molecules or in independent non-overlapping subsets of small numbers of molecules the solutions $\Psi_K(q)$ are inaccessible and eqs. (12) and (12.1) are not numerically useful.

We will follow Wigner's [1] procedure of using for Θ the product of $3N$ integrals of the normalized continuous eigenfunctions $\theta_{pj\alpha}(q_{j\alpha})$ of the single cartesian momentum operator $\mathcal{P}$. The momentum operator $\mathcal{P}_{j\alpha}$ is:

$$\mathcal{P}_{j\alpha} = (h/2\pi i)\partial/\partial q_{i\alpha} = (h/2\pi i)\mathcal{D}_{j\alpha}, \tag{13}$$

and the eigenfunctions in a macroscopic volume are continuous (purists set $V \to \infty$). The functions are:

$$\left.\begin{array}{l}\theta_p(q) = h^{-1/2} \exp[2\pi i h^{-1} pq], \\ \theta_q^\dagger(p) = \theta_p^*(q) = h^{-1/2} \exp-[2\pi i h^{-1} pq],\end{array}\right\} \tag{14}$$

so that

$$\mathcal{P}\theta_p(q) = p\theta_p(q),$$
$$\mathcal{P}\theta_q^\dagger(p) = -p\theta_q^\dagger(p) = -p\theta_p^*(q), \tag{15}$$
$$\tfrac{1}{2}\left[\theta_q^\dagger(p)\mathcal{P}^{2n-1}\theta_p(q) - \theta_p(q)\mathcal{P}^{2n-1}\theta_q^\dagger(p)\right] = 0, \tag{16}$$

but

$$\tfrac{1}{2}\left[\theta_q^\dagger(p)\mathcal{P}^{2n}\theta_p(q) + \theta_p(q)\mathcal{P}^{2n}\theta_q^\dagger(p)\right] = \theta_q^\dagger(p)\theta_p(q)p^{2n}. \tag{17}$$

For any single coordinates $q_{j\alpha}=q$ the complete set Θ of all functions $\theta_p(q)\theta_q^\dagger(p)$ obey the requirements of a unitary matrix Θ of elements $\theta_p(q)$ and its adjoint $\Theta^\dagger$ of elements $\theta_q^\dagger(p)$ namely that:

$$\Theta^\dagger\Theta = \int dp\, \theta_q^\dagger(p)\theta_p(q') = \delta(q-q'), \\ \Theta\Theta^\dagger = \int dq\, \theta_p(q)\theta_q^\dagger(p') = \delta(p-p'), \quad (18)$$

so that $\Theta^\dagger\Theta$ and $\Theta\Theta^\dagger$ are both unit matrices with Dirac delta functions on the diagonal and zero off-diagonal elements.

The diagonal elements, $\omega(q,q)$ of the Hamiltonian matrix Ω of eq. (8) are real and positive and give the probability density in the $\Sigma^{3N} q_{j\alpha} = q$-space. With:

$$\text{trace } \Omega = \int \cdots \int_V dq\, \omega(q,q) = 1, \quad (19)$$

determining the value of A. Using the symmetric operation of half the sum of that on θ plus that on $\theta^\dagger$ displayed in eqs. (16) and (17) and summing by integration over all $p_{j\alpha}$'s, with eqs. (8) and (14), we have:

$$\omega(q,q) = (N!)^{-1} \exp(\beta A) \prod_{j\alpha}^{3N} \int_{-\infty}^{+\infty} dp_{j\alpha} h^{-1} \\ \times \frac{1}{2} \left[\exp(-2\pi i p_{j\alpha} q_{j\alpha}) \sum_{n=0}^{\infty} \frac{[-(\mathcal{K}+U_1)]^n}{n!} \exp(2\pi i p_{j\alpha} q_{j\alpha}) \right. \\ \left. + \exp(2\pi i p_{j\alpha} q_{j\alpha}) \sum_{n=0}^{\infty} \frac{[-(\mathcal{K}+U_1)]^n}{n!} \exp(-2\pi i p_{j\alpha} q_{j\alpha}) \right]. \quad (20)$$

The terms $[-(\mathcal{K}+U_1)]^n/n!$ in eq. (8) for large n, since $\mathcal{K}U_1$ and $U_1\mathcal{K}$ do not commute, $\mathcal{K}U_1 \neq U_1\mathcal{K}$ contain 2^n different single terms from U_1^n to $\mathcal{K}^m$ operating on Θ or $\Theta^\dagger$. Of these for fixed m there are $\binom{n}{\nu} = n!/\nu!(n-\nu)!$ having ν powers of $\mathcal{K}$ and $\mu = n-\nu$ powers of U_1. Now from eq. (16) if for any *single* $j\alpha$ the operation $\mathcal{D}_{p_{j\alpha}}$ on Θ is to an odd power the term is zero*. In the sum of terms $[(\mathcal{K}+U)^m/n!]\Theta$ then, from

---

*The artifice of writing half the sum of operating on Θ and on $\Theta^\dagger$ is unnecessary. Integration from $-\infty$ to $+\infty$ over p removes the odd terms whichever is used.

eq. (6), only terms in which the complete operator $(\beta\mathcal{K}) = [-\zeta\Sigma_{j\alpha}^{3N}\frac{1}{2}\mathcal{D}_{j\alpha}]$ acts on Θ are non-zero, so that in the sum of powers of the operator $-\zeta\Sigma_{j\alpha}^{3N}\frac{1}{2}\mathcal{D}_{j\alpha}^{2}$, eq. (6), acting on powers of U in $-\beta(\mathcal{K}+U)^m/n!$, eq. (8), only even powers of the operator survive.

In the second edition [3] of Mayer and Mayer, a treatment based on the conventional model that

$$\beta U = \sum\sum_{1 \leq j < k \leq N} \beta u(r_{ik}), \tag{21}$$

is discussed in detail. Counting of the number of terms in eq. (8) for Ω shows that we can write Ω as a product of two commuting real matrices, one of which is independent of U,

$$\Omega = (N!)^{-1}\Omega(\mathcal{K},U)\Omega(\mathcal{K}) = (N!)^{-1}\Omega(\mathcal{K})\Omega(\mathcal{K},U), \tag{22}$$

where

$$\Omega(\mathcal{K},U) = \sum_{\nu > 0}^{\infty} (\nu!)^{-1}\left[-(\mathcal{K}+U_1(\boldsymbol{q}))\right]^{\nu}\mathbf{1}, \tag{23}$$

in which the $\mathcal{K}$ operator acts only on U, $\mathcal{K}\mathbf{1} = 0$ and the terms in eq. (8) where $\mathcal{K}^{n-\nu}$ do not operate on U or powers of U^{μ}, $\mu \geq 2$ are in $\Omega(\mathcal{K})$, which is:

$$\Omega(\mathcal{K}) = \prod_{j\alpha}\left\{\int_{-\infty}^{+\infty}\mathrm{d}\left(\frac{p_{j\alpha}}{h}\right)\theta_{qj\alpha}^{\dagger}(p_j)\sum_{\mu=0}^{\infty}\frac{(-\mathcal{K})^{\mu}}{\mu!}\theta_{pj\alpha}(q_{j\alpha})\right.$$

$$= \prod_{j\alpha}^{3N}\delta(q_{j\alpha}-q_{j\alpha}')\int_{-\infty}^{+\infty}\mathrm{d}\left(\frac{p_{j\alpha}}{h}\right)\sum_{\mu=0}^{\infty}\frac{(-\beta(2m)^{-1}p_{j\alpha}^2)^{\mu}}{\mu!}$$

$$= \prod_{j\alpha}^{3N}\delta(q_{j\alpha}-q_{j\alpha}')\int_{-\infty}^{+\infty}h^{-1}\mathrm{d}p_{j\alpha}\exp\left[-\beta(2m)^{-1}p_{j\alpha}^2\right]$$

$$= \prod_{j\alpha}^{3N}\lambda^{-1}\delta(q_{j\alpha}-q_{j\alpha}') = \lambda^{-3N}\delta(\boldsymbol{q}-\boldsymbol{q}'), \tag{24}$$

and of course the same result is found if the symmetric form of operation on half the sum of that on θ and $\theta^{\dagger}$ of eqs. (16) and (17) is used. Since $\int_V\int \mathrm{d}\boldsymbol{q}\,\mathrm{d}\boldsymbol{q}'\,\delta(\boldsymbol{q}-\boldsymbol{q}') = V^N$ the "matrix" $\Omega(\mathcal{K})$ is a pure number (V/λ^3) multiplying $\Omega(\mathcal{K},U)$ of eq. (23) so that

$$\Omega = \exp\left[\beta A + N(\ln VN^{-1}\lambda^{-3}+1)\right]\Omega(\mathcal{K},U). \tag{25}$$

With $U=0$ in the perfect gas limit $\Omega(\mathcal{K}, U=0)=1$ since only the first term, $n=0$ of eq. (23) is non-zero and

$$\Omega(U=0) = \exp\left[\beta A + N(\ln VN^{-1}\lambda^{-3} + 1)\right], \tag{25.1}$$

which is the semiclassical equation for the unsymmetrized Petite Canonical Ensemble determining βA by the condition $\Omega = 1$,

$$\beta A = -N\left[\ln VN^{-1}\lambda^{-3} + 1\right] \quad \text{(perfect gas)}. \tag{25.2}$$

The thermodynamic relation:

$$\beta\left(\frac{\partial \beta A}{\partial \beta}\right)_{V,N} = \beta\left[A + \beta\frac{\partial A}{\partial \beta}\right] = \beta\left[A - T\frac{\partial A}{\partial T}\right] = \beta[A + TS] = \beta E \tag{25.3}$$

combined with eq. (4) for λ,

$$\beta \partial \ln \lambda / \partial \beta = \tfrac{1}{2}, \tag{25.4}$$

yields

$$\beta E = \tfrac{3}{2} N \quad (U \equiv 0, \text{ perfect gas}), \tag{25.5}$$

which is the classical result, and correct for the unsymmetrized ("Boltzmann statistics") perfect gas for which E is the kinetic energy.

For the general case the $\beta U \neq 0$ and eq. (9) that trace $\Omega = 1$ we have that.

$$\begin{aligned}\beta A &= -N\left[\ln VN^{-1}\lambda^{-3} + 1\right] \ln V^{-N} \quad \text{trace } \Omega(\mathcal{K}, U) \\ &= -N\left[\ln VN^{-1}\lambda^{-3} + 1\right] - \ln\langle\Omega(\mathcal{K}, U)\rangle,\end{aligned} \tag{26}$$

where $\langle\Omega(\mathcal{K}, U)\rangle$ is the average in V of the real diagonal elements of the matrix,
and

$$\beta E = \frac{3}{2} N + \beta\left(\frac{\partial}{\partial \beta}\right)_{V,N} \ln\langle\Omega(\mathcal{K}, U)\rangle. \tag{27}$$

The term $\tfrac{3}{2} N = \beta E = \langle\sum_{j\alpha}^{3N}\beta(2m)^{-1}p_{j\alpha}^2\rangle$ is the kinetic energy in the perfect gas limit, but *not* in general. The $p_{j\alpha}$'s in the single functions $\theta_{pj\alpha}(q_{j\alpha})$ and $\theta^*_{pj\alpha}(q_{j\alpha})$ are quantum numbers of a set of orthonormal functions. They have the dimensions of momenta and are continuous, and contribute additively to a kinetic energy term but by no means constitute the total kinetic energy of the ensemble.

There is a relation for all ensembles relating the average of the square minus the average squared of an extensive quantity to a derivative of it. For our V, N, T ensemble it is, with $C_V = \beta E/\partial T)_{N,V}$,

$$\langle (\beta E)^2 \rangle - \langle \beta E \rangle^2 = k^{-1} C_V = -\beta \frac{\partial \beta E}{\partial \beta}, \tag{28}$$

which we record here for future reference.

In reference [3] it was erroneously argued that for a sum of pair functions of distances $f(r_{jk})$ such as eq. (21) for U_1 since $\partial f(r_{jk})/\partial r_{j\alpha} = -\partial f(r_{jk})/\partial r_{k\alpha}$ and, since the second derivatives are also a sum of function of pair distances, only even derivatives, $\mathcal{K}^\nu U_1$ would survive in $\Omega(\mathcal{K}, U)$. With our eqs. (1) to (6) inclusive we find that:

$$-\mathcal{K} f(r) = \zeta \left(\frac{d^2 f(r)}{dr^2} + \frac{2}{r} \frac{d f(r)}{dr} \right), \tag{29}$$

so that with

$$-u_\kappa(r) = \left[\zeta \left(\frac{d^2}{dr^2} + \frac{2}{r} \frac{d}{dr} \right) \right]^{\kappa-1} \beta u(r), \tag{30}$$

the matrix $(\mathcal{K}, U_1)$ of eq. (23) would consist solely of products for all $\frac{1}{2} N(N-1) jk$ pairs, $1 \le j < k \le N$ of sums of products of $u_\kappa(r_{jk})$'s times combinatorial factors, with $1 \le \kappa \le \infty$.

For large κ the $\kappa - 1$ power of the operator in eq. (23) looks complicated. A simpler form:

$$u_\kappa(r) = \zeta^{\kappa-1} \left[\frac{d^{2(\kappa-1)}}{dr^{2(\kappa-1)}} + \frac{2(\kappa-1)}{r} \frac{d^{2\kappa-3}}{dr^{2\kappa-3}} \right] \beta u(r), \tag{30.1}$$

is the same as eq. (30) for the classical first term of no operator with $u_1(r) = \beta u(r)$ and for the first quantum correction of $\kappa = 2$. One can then show by algebraic calculation that:

$$\left[\frac{d^2}{dr^2} + \frac{2}{r} \frac{d}{dr} \right] \left[\frac{d^{2(\kappa-1)}}{dr^{2(\kappa-1)}} + \frac{2(\kappa-1)}{r} \frac{d^{2\kappa-3}}{dr^{2\kappa-3}} \right]$$
$$= \frac{d^{2\kappa}}{dr^{2\kappa}} + \frac{2\kappa}{r} \frac{d^{2\kappa-1}}{dr^{2\kappa-1}}, \tag{30.2}$$

and hence by induction that eq. (30.1) is valid for all $1 \le \kappa < \infty$.

In reference [3] the combinatorial problem of counting the terms in $\Omega(\mathcal{K}, U)$ which are products of the $\frac{1}{2}N(N-1)$ pair terms $u_\kappa(r_{jk})$ was solved. We shall designate this matrix by $\Omega_p(\mathcal{K}, U_1)$. The result was that the diagonal q,q elements

$$[\Omega_p(\mathcal{K}, U)]_{q,q} = \prod\prod_{1 \leq j < k \leq N} \omega(r_{jk}), \tag{31}$$

were given by

$$\omega(r) = \sum_{\nu_1 > 0} \sum_{\nu_2 > 0} \cdots \prod_{\nu_x > 0} \prod_{\kappa > 1} (\nu_k!)^{-1}(u_\kappa(r)/\kappa!)^{\nu_\kappa}$$

$$= \exp \sum_{\kappa > 1} u_\kappa(r)/\kappa!. \tag{32}$$

The conventional realistic pair potential with minimum $-u_0$ at $r = r_0$ is often assumed to be the Lennard–Jones 6–2 potential:

$$u_{LJ}(r) = u_0\left[(r_0/r)^{12} - 2(r_0/r)^6\right], \tag{33}$$

for which, however, due to the 6th and 12th order poles at $r = 0$, the series $\Sigma \beta u_\kappa/\kappa!$ of eq. (32) diverges. The Morse potential:

$$u_M(r) = u_0\left[\exp 12(1 - r/r_0) - 2\exp 6(1 - r/r_0)\right], \tag{34}$$

has the same second derivative as U_{LJ} at r_0,

$$\left[r(d^2 u_M/dr^2)\right]_{r=r_0} = \left[r(d^2 u_{LJ}/dr^2)\right]_{r=r_0} = 72u_0, \tag{35}$$

and mercifully eq. (32) converges. One would assume that numerical evaluation would best be done with u_M of eq. (34).

Before proceeding to the next section discussing added terms concerning those in reference [3] we make one parenthetic comment justifying the discussion following eq. (27) that the *kinetic* energy includes more than the term $\frac{3}{2}N$ in eq. (27). This is obvious from eq's (27) and (32) with (33) or (34) for u_κ. The first term, $u_1(r)$, in eq. (33) is the classical potential u times β and summed over all molecular pairs is the total potential energy of the system as a function of the coordinates q. The contribution of all other terms u_κ, $\kappa \geq 2$ to eq. (27) are corrections to the *kinetic* energy ensemble average as a function of q analogous to the $\frac{1}{2}h\nu$ term in an oscillator potential.

3. Correction to J. E. Mayer and M. G. Mayer [3]

With U_1 the dimensionless total potential energy times β, and $\mathcal{K}$ the dimensionless kinetic energy operator of eq. (6) we seek to evaluate the trace of the matrix $\Omega(\mathcal{K}, U_1)$ of eq. (23), or actually the trace of

$$\Omega_E(\mathcal{K}, U_1) = \beta (\partial/\partial \beta)_{V,N} \Omega(\mathcal{K}, U), \tag{36}$$

which from eq. (27) leads directly to the thermodynamic energy E. Since $\mathcal{K}$ and U_1 are both proportional to β we find from eq.'s (23) and (36):

$$\Omega_E(\mathcal{K}, U_1) = \sum_{\nu=1}^{\infty} (\nu - 1!)^{-1} \left[-(\mathcal{K} + U_1) \right]^{\nu} \cdot \mathbf{1}, \tag{37}$$

of which, since $\mathcal{K} \mathbf{1} = 0$, the first, $\nu = 1$, term is $-U_1 = -\beta U$, and in the term with $n = \nu - 1$, $0 \leq n \leq \infty$,

$$\left[-(\mathcal{K} + U_1) \right]^{n+1} \mathbf{1},$$

only terms ending in U_1 will be non-zero. For fixed n there will be a total of 2^{n-1} terms with $\mathcal{K}$'s and U_1's in all permutations of their orders. Of these, with $0 \leq \kappa \leq n-1$ a total of $\binom{n-1}{\kappa}$ will have κ operator $\mathcal{K}$'s and $n - \kappa = \mu + 1$ U_1's, but always with (at least) one U_1 at the right-hand end, so that the non-zero permutations are between κ objects of one kind, ($\mathcal{K}$'s), and μ of the other, (U_1's). Of these are $\binom{\kappa+\mu}{\kappa} = \binom{\kappa+\mu}{\mu}$ and the sum over κ from 0 to $n-1$ is 2^{n-1}.

Now U_1 is a sum of $\frac{1}{2}N(N-1) = \binom{N}{2}$ terms $u_1(r_{jk})$ and with $\mathcal{D}_j u(r_{jk}) = -\mathcal{D}_k u(r_{jk})$ but $\mathcal{D}_j^2 u_1(r_{jk}) = \mathcal{D}_k^2 u_1(r_{jk}) \neq 0$ we have that with eq. (30.1) for $u_\kappa(r)$ one finds that $\mathcal{K} U_1$ is a sum of $\binom{N}{2}$ terms $u_2(r_{jk})$:

$$-\mathcal{K} U_1 = \sum\sum_{i<j<k<N} -\mathcal{K}_{jk} \frac{\zeta}{2} u_1(r_{jk}) = \sum\sum_{i<j<k<N} u_2(r_{jk}). \tag{38}$$

Since $u_2(r_{jk})$ is a function of r_{jk} alone and for any $f(r_{jk})$ we have $\mathcal{D}_j f(r_{jk}) = \mathcal{D}_k f(r_{jk})$ we have that:

$$(-\mathcal{K})^n U_1 = \sum\sum_{1<j<k<N} u_{n+1}(r_{jk}). \tag{39}$$

The square, U_1^2, contains a sum of $\binom{N}{2}$ terms $u^2(r_{jk})$, plus $\binom{N}{3}$ terms $u(r_{jk})u(r_{jl}) + u(r_{jk})u(r_{kl}) + u(r_{jl})u(r_{kl})$ and $\binom{N}{4}$ terms $u(r_{jk})u(r_{lm})$ with all four molecules different $j \neq k \neq l \neq m$, etc.

We defer discussion of the operation of $(-\mathcal{K})^n$ on terms $u^\nu(r_{jk})$. Operation of $(-\mathcal{K})^n$ on unconnected products $u(r_{jk})u(r_{lm})$ simply gives us terms:

$$\zeta^n \sum_{\mu=0}^{n} u_{n+1-\mu}(r_{jk}) u_{\mu+1}(r_{lm})$$

and similar terms for higher powers of U_1. The operation of $-\mathcal{K}$ on the terms like $u(r_{jk})u(r_{jl})$ leads, however, to terms not included in ref. [3], namely to terms involving the cosine of an angle:

$$\mathcal{D}_j u_1(r_{jk}) \mathcal{D}_j u_1(r_j)$$

$$= \zeta \cos \frac{(r_j - r_k) \cdot (r_j - r_l)}{r_{jk} r_{jl}} \frac{d}{dr_{jk}} u_1(r_{jk}) \frac{d}{dr_{jl}} u_1(r_{jl})$$

$$\equiv \zeta \cos(kjjl) u_1'(r_{jk}) u_1'(r_{jl}), \tag{40}$$

where we use

$$\cos(kjjl) = \cos \frac{(r_j - r_k) \cdot (r_j - r_l)}{r_{jk} r_{jl}} \tag{40.1}$$

and

$$u_1'(r) = du_1(r)/dr. \tag{40.2}$$

Now we will need to examine the operation of $(-\mathcal{K})^n$ on terms in U_1^m which in general involve various products of functions $f(jk)$ not necessarily identical in functional form but all functions of r_{jk}'s. Assume there are ψ numbered molecules j, $1 \leq j \leq \psi$ having $f(j,k) \neq 1$ so that there are $\frac{1}{2}\psi(\psi-1)$ non-trivial products in $F(\psi)$,

$$F(\psi) = \prod\prod_{1 \leq j < k \leq \psi} f(j,k),$$

and $\frac{1}{2}\psi(\psi-1)(\psi-2)$ involving sets of three pairs, jk, jl, kl. For a given numbered set j, k, l, define:

$$F_3(jkl) = f(jk) f(jl) f(kl). \tag{41}$$

The remaining product:

$$F(\psi - jkl) = F(\psi) - F_3(jkl), \tag{41.1}$$

will not be a function of j, k, or l and will not be acted on by

$$(-\mathcal{K}_{jkl})^n = \zeta^n \tfrac{1}{2}(\mathcal{D}_j^{2n} + \mathcal{D}_k^{2n} - \mathcal{D}_l^{2n}). \tag{41.2}$$

Now

$$\mathcal{D}_j F_3(jkl) = \mathcal{D} f(jk) f(jl) f(kl)$$
$$= f(kl)[\mathcal{D}, f(jl)] f(jk) + f(jl) \mathcal{D}_j f(jk),$$

and

$$\mathcal{D}_j^2 F_3(jkl) = f(k,l) \{ [\mathcal{D}_j^2 f(jk)] f(jl) + f(jk) [\mathcal{D}_j^2 f(jl)]$$
$$+ 2[\mathcal{D}_j f(jk)][\mathcal{D}_j f(jl)] \}. \tag{41.3}$$

With

$$\tfrac{1}{2}\zeta^n(\mathcal{D}_j^{2n} + \mathcal{D}_k^{2n}) f(jk) \mathbf{1} = f_n(jk), \tag{41.4}$$

analogous to eq. (30.1) for u_κ but with $\kappa + 1 = n$ we find with $f'(r) = df/dr$

$$-\mathcal{K}_{jkl} F_3(jkl) = f_1(jk) f(jl) f(kl) + f(jk) f_1(jl) f(kl)$$
$$+ f(jk) f(jl) f_1(kl) + \zeta [\cos(kjjl) f'(jk) f'(jl) f(kl)$$
$$+ \cos(jk \cdot kl) f'(jk) f(jl) f'(kl)$$
$$+ \cos(jllk) f(jk) f'(jl) f'(kl)]. \tag{42}$$

Now

$$\zeta \mathcal{D}_j^2 \cos(kjjl) f'(jk) f'(jl) = f_1(jk) f_1(jl) \tag{42.1}$$

plus terms that involve cosine products of $f_1(jk) f(jl)$ etc. so that

$$(-\mathcal{K}_{jkl})^2 F_3(jkl) = f_2(jk) f(jl) f(kl) + f(jk) f_2(jl) f(kl)$$
$$+ f(jk) f(jl) f_2(kl) + 3 f_1(jk) f_1(jl) f(kl)$$
$$+ 3 f_1(jk) f(jl) f_1(kl) + 3 f(jk) f_1(jl) f_1(kl)$$
$$\tag{42.2}$$

plus other cosine terms.

Introduce a symbol $\{\lambda_3\}_n$ for a set of 3 non-negative integers:

$$\{\lambda_3\}_n = \lambda(jk), \lambda(jl), \lambda(kl), (\lambda(jk) \geq 0),$$

$$\lambda(jk) + \lambda(jl) + \lambda(kl) = n \tag{43}$$

and

$$\sum_{\{\lambda_3\}_n} \equiv \text{summation over all } \{\lambda_3\}_n \text{ sets.} \tag{43.1}$$

The pair notation is clumsy and can readily be obviated.

Label the three pair sets jk, jl, kl in that order by a single index γ, $\gamma = 1, 2,$ or 3 so that with $f(jk) = f_0(1)$; $f(jl) = f_0(2)$; $f(kl) = f_0(3)$

$$F_3(jkl) = \prod_{\gamma=1}^{3} f_0(\gamma), \tag{44}$$

$$\prod_{\gamma=1}^{3} f_{\lambda(\gamma)} = f_{\lambda(jk)}(jk) f_{\lambda(jl)}(jl) f_{\lambda(kl)}(kl), \tag{44.1}$$

$$\cos(kjjl) = \cos(1,2), \quad \cos(jkkl) = \cos(1,3), \quad \cos(jllk) = \cos(2,3). \tag{44.2}$$

We use a single symbol for the rather complicated looking sum of the three cosine products in eq. (42) with $f_{\lambda(\gamma)}(\gamma)^3$, namely:

$$\{\lambda_3\}'_n = \cos(1,2) f'_{\lambda(1)}(1) f'_{\lambda(2)}(2) f_{\lambda(3)}(3)$$
$$+ \cos(1,3) f'_{\lambda(1)}(1) f_{\lambda(2)}(2) f'_{\lambda(3)}(3)$$
$$+ \cos(2,3) f_{\lambda(1)}(1) f'_{\lambda(2)}(2) f'_{\lambda(3)}(3), \tag{44.3}$$

with

$$\sum_{\gamma} \lambda(\gamma) = n, \tag{44.4}$$

and from eq. (42.1):

$$-\mathcal{K}_{jkl}\{\lambda_3\}'_n = \sum_{\gamma'=1}^{3} \prod_{\gamma \neq \gamma'}^{2} f_{\gamma(\gamma)+1}(\gamma) f_{\lambda(\gamma')}. \tag{44.5}$$

Using this notation we at least manage to compress the expression for $(-\mathcal{K}_{jkl})^n F_3(j,k,l)$ to one line:

$$(-\mathcal{K}_{jkl})^n F_3(j,k,l) = \sum_{\{\lambda_3\}_n} \prod_{\gamma=1}^{3} \kappa_n(\gamma) f_{\lambda_n(\gamma)}(\gamma) + \{\lambda_3\}'_{n-1}, \qquad (45)$$

where

$$\{\kappa_3\}_n \equiv \kappa_n(1), \kappa_n(2), \kappa_n(3), \qquad (45.1)$$

is to be determined from the recursion relation:

$$(-\mathcal{K}_{jkl})^{n+1} F_3(jkl) = (-\mathcal{K}_{jkl})\left[-(\mathcal{K}_{jkl})^n F_3(j,kl)\right]. \qquad (45.2)$$

We turn back to eq. (37) for Ω_E which, with the discussion following the expression, we write as:

$$\begin{aligned}
\Omega_E(\mathcal{K}, U_1) &= \sum_{n>0} (n!)^{-1} \left[-(\mathcal{K}+U_1)\right]^n (-U_1) \cdot \mathbf{1} \\
&= \sum_{\kappa>0}^{\infty} \sum_{\mu>0}^{\infty} \frac{(-\mathcal{K})^\kappa}{\kappa!} \frac{(-U_1)^\mu}{\mu!} (-U_1) \cdot \mathbf{1} \\
&= \sum_{\kappa=0}^{\infty} \frac{(-\mathcal{K})^\kappa}{\kappa!} [\exp - U_1](-U_1) \cdot \mathbf{1} \\
&= \sum_{\kappa<0}^{\infty} \zeta^\kappa (\kappa!)^{-1} \sum_{j=1}^{N} \mathcal{D}_j^{2\kappa} [\exp - U_1](-U_1) \cdot \mathbf{1} \\
&= \frac{1}{2} \sum_{j=1}^{N} \left[\sum_{\mu=0}^{\infty} \frac{\zeta^\mu}{\mu!} \mathcal{D}_j^{2\mu}(-U_1) \cdot \mathbf{1} \right] \left[\sum_{\lambda=0}^{\infty} \frac{\zeta^\lambda}{\lambda!} \mathcal{D}_j^{2\lambda} e^{-U_1} \cdot \mathbf{1} \right] \\
&\quad + \mathcal{D}_j^2 \left[\sum_{\mu=1}^{\infty} \frac{\zeta^\mu}{\mu!} \mathcal{D}_j^{2\mu-1}(-U_1) \right] \cdot \left[\sum_{\lambda=1}^{\infty} \frac{\zeta^\lambda}{\lambda!} \mathcal{D}_j^{2\lambda-1} e^{-U_1} \right] \cdot \mathbf{1} \\
&= \sum_{j=1}^{N} \left\{ \left[\sum_{\mu=0}^{\infty} \mathcal{D}_j^{2\mu}(-U_1) \cdot \mathbf{1} \right] \left[\sum_{\lambda=0}^{\infty} \frac{\zeta^\lambda}{\lambda!} \mathcal{D}_j^{2\lambda} e^{-U_1} \cdot \mathbf{1} \right] \right. \\
&\quad \left. + \sum_{\mu=1}^{\infty} \frac{\zeta^\mu}{\mu!} \mathcal{D}_j^{2\mu-1}(-U_1) \cdot \left[\sum_{\lambda=0}^{\infty} \frac{\zeta^\lambda}{\lambda!} \mathcal{D}_j^{2\lambda+1} e^{-U_1} \right] \cdot \mathbf{1} \right\},
\end{aligned}$$
$$(46)$$

where the "feed back" of the scalar product of the odd derivative terms just doubles the product of even derivatives and completely dissociates the two terms as independent additions to Ω_E. However, similar other scalar product terms arise in the expansion of $\mathcal{D}_j^{2\lambda}e^{-U_1}$.

In the cluster expansion of the classical imperfect gas one introduces functions*:

$$f(jk) = \exp[-\beta u(r_{jk})] - 1 = f(kj), \qquad (47)$$

so that

$$e^{-\beta u} = \sum_{n>0} \frac{1}{n!}\left[1 + \sum_j \sum_k f(jk)\right]^n$$

$$= 1 + \sum_j \sum_k \frac{f(jk)}{1!} + \sum_j \sum_k \sum_l \frac{1}{2!} f(kj)f(jl)$$

$$+ \sum_i \sum_j \sum_k \sum_l \frac{1}{2!} f(ij)f(kl) + \ldots, \qquad (47.1)$$

where we drop the awkward limiting notation under the summations by requiring that in each product all indices, i, j, k, l refer to different molecules: $\frac{1}{2}N(N-1)$, $\frac{1}{2}N(N-1)(N-2)$ and $\frac{1}{2}N(N-1)(N-2)(N-3)$ terms in the double triple and quadruple sums respectively. A one to one identification is then convenient to make between products of $f(jk)$'s and diagrams of N numbered vertices with lines connecting j and k for every $f(jk)$ in the product, and the vertices of a product of m molecules which are connected by lines are said to form a "cluster" of products of f's, $C_m[\Pi_{jk}(f(r_{jk}))]$.

In such a connected product there may be "nodal" vertices or "articulation points" namely vertices which if suppressed (i.e. erased) divide the connected cluster into two independent clusters in which case integration in coordinate space leads to a product of integrals. We refer to products with no nodal vertices as irreducible clusters.

The operation of $(-\mathcal{K})^\kappa$ outlined in eq. (46) with the discussion of eqs. (40) to (45) leads to similar relations requiring distinction between unconnected and connected clusters and between clusters with nodal vertices and irreducible clusters. The operation of $(-\mathcal{K})^\kappa$ on the products of a single diagram gives a product of separate terms, identical after

---

*The development of powers of $-U_1$ as products of sums of $-\beta u(r_{jk})$'s includes sums of powers of single pairs $[-\beta u(r_{jk})]^\nu/\nu!$ and then directly to the same development as we make here.

averaging in the ensemble, of operations by $(-\mathcal{K}_{j_1, j_2, \ldots j_m})$ on the products of f's in connected clusters, with zero for $\kappa \geq 1$ for the operation on a molecule, j, which does not appear in any $f(jk)$. The scalar product of odd derivatives on j is proportional to the cosine of the angle between $r_{jk} r_j$, etc. and if j is a nodal vertex the ensemble average of the cosine is zero, $\int_0^\pi d\theta \cos\theta = 0$, so the term survives in Ω_E diagonal elements only for irreducible clusters, the simplest of which is the ring diagram:

$$R_n = f(r_{12}) f(r_{23}) f(r_{34}) \cdots f(r_{n-1;n}) f(r_{n,1}), \tag{48}$$

which can be integrated and averaged by "folding."

We now examine, with $\beta u(r)$ the Morse pair potential of eq. (34), the operation on it of $(-\mathcal{K})^m/n!$ summed to infinity. With r_0 the distance of the minimum, $-\beta u_0$,

$$R = r r_0^{-1}, \tag{49}$$

$$g(R) = e^{6(1-R)}, \tag{49.1}$$

$$-\beta u(r) = -\beta u_0 [g^2(R) - 2g(R)], \tag{49.2}$$

$$\xi = 6\zeta r_0^{-2} = 3\lambda^2/\pi r_0^2 = 3\beta h^2/2\pi^2 m r_0^2, \tag{49.3}$$

and if $R = R_{jk}$, $-\mathcal{K}_{jk} = \xi \mathcal{D}_R^2$,

$$-\sum_{n>0}^{\infty} (n!)^{-1} (-\mathcal{K})^n \beta u(R)$$

$$= -\beta u_0 \sum_{n=0}^{\infty} (n!)^{-1} \xi^n \left[(-12)^{2n} g^2(R) - 2(-6)^{2n} g(R) \right]$$

$$= -\beta u_0 [\exp 12(1 + 2\xi - R) - 2 \exp 6(1 + \xi - r)], \tag{50}$$

which is larger in magnitude than $-\beta u(r)$ but finite and goes exponentially to zero at $R \to \infty$. Its derivative which respect to R is:

$$d/dR \left[-\sum_{n<0}^{\infty} (n!)^{-1} (-\mathcal{K})^n \beta u(R) \right] = -12\beta u_0 [\exp 12(1 + 2\xi - R) - \exp 6(1 + \xi - R)], \tag{50.1}$$

which will be zero when the two exponents are equal, namely at

$$R = R_m = 1 + 3\xi. \tag{50.2}$$

The maximum (positive) value of $-\sum_n (n!)^{-1} (-\mathcal{K})^n \beta u(R)$ will then be at

$R = R_m$ and will have the numerical value:

$$e^{-12\xi\beta u_0} \quad \text{at} \quad R = R_m.$$

From eq. (47) for $f(jk)$ we need to evaluate:

$$\sum_{n=0}^{\infty} [(-\mathcal{K})^n/n!] e^{-\beta u(r)} = \exp \sum_{n=0}^{\infty} (-\mathcal{K})^n (n!)^{-1} [-\beta u(r)], \quad (51)$$

which with eq's (49) to (50) leads us for the Morse potential to a double exponential:

$$\sum_{n=0}^{\infty} [(-\mathcal{K})^n/n!] \exp{-\beta u(R)} = \exp\{-\beta u_0 [\exp 12(1+2\xi-R) \\ -2\exp 6(1+\xi-R)]\}. \quad (51.1)$$

Since, however, eq. (27), gives the thermodynamic energy E in terms of the ensemble average of the logarithm of the diagonal elements of Ω_E the first exponential vanishes.

Define:

$$\phi(R) = -\beta u_0 [\exp 12(1+2\xi-R) - 2\exp 6(1+\xi-R)], \quad (51.2)$$

so that

$$(\zeta^\mu/\mu!)[\mathcal{D}_j^{2\mu} + \mathcal{D}_k^{2\mu}](-U_1) \cdot \mathbf{1} = \sum_j \sum_k -\phi(R_{jk}), \quad (51.3)$$

with $\frac{1}{2}N(N-1)$ terms, and

$$\sum_{n=0}^{\infty} (-\mathcal{K})^n/n!) \exp{-\beta u(R_{jk})} = \exp{-\phi(R_{jk})}. \quad (51.4)$$

With $f(jk)$ given by eq. (47):

$$\sum_{n=0}^{\infty} (-\mathcal{K})^n f(jk) = [\exp{-\phi(R_{jk})}] - 1,$$

and in any connected product of $f(jk)$'s in the expansion of $\exp(-\beta U)$ the operation of $\sum_{n>0}(-\mathcal{K})^n f(jk)$ simply replaces the classical imperfect gas or liquid equations with $\beta u(r_{jk})$'s by $\phi(R_{jk})$ but for products of ϕ's with no nodal vertices the scalar product terms at the end of eq. (46) must be added. Since, however the order of differentiation commutes, summing the even derivatives first and we always have only terms which are single first derivative of $\phi(R_{jk})$'s given in detail in eq. (42).

4. Conclusion

With the replacement of $\beta u(R)$ by $\phi(R)$ for the dimensionless pair potential the numerical evaluation of the thermodynamic functions E, A...etc. still has all the difficulties of the classical equations for the very dense gas or liquid, but not apparently much greater ones. The inclusion of the scalar product terms do not seem to be formidable. The requirement of symmetrization is difficult but can be done by a modification of Uhlenbeck's [4] which was adapted to a dense medium by Baldursson [5].

The Morse potential is more realistic than the Lennard–Jones at R near zero for which the potential must go to that of an atom of nuclear charge $2z$ which is finite, but is unrealistic for large R values going exponentially to zero rather than as R^{-6}. A single analytic potential rather than a fit for different R values seems to be required to sum to infinity the operation of $(-\mathcal{H})^n/n!$. Quite possibly after $\phi(R_{jk})$ is obtained a more realistic "fitted" two part form could be used rather than the exponential to infinite R.

Finally one comment is rather obvious. Averaging in coordinate space can be used with any classical Gibbs ensemble with the $\phi(R)$ replacing $\beta u(R)$ and paying attention to the scalar product terms.

References

[1] E. Wigner, Phys. Rev. **40** (1932) 749.
[2] J. E. Mayer and W. Band, J. Chem. Phys. **15** (1947) 141.
[3] J. E. Mayer and M. G. Mayer, "Statistical Mechanics," 2nd Ed., (John Wiley, New York, 1977) p. 447–457.
[4] B. Kahn and G. E. Uhlenbeck, Physica **5** (1938) 399.
[5] S. Baldursson and J. E. Mayer with H. Aroeste, J. Chem. Phys. **31** (1959) 814.

CHAPTER 18

The Mechanisms of Stochasticity in Classical Dynamical Systems

A. S. WIGHTMAN

Joseph Henry Laboratories of Physics
Princeton University
Princeton, New Jersey 08544
U.S.A.

Contents

1. Introduction — 345
2. Stability and instability — 348
3. Homoclinic points and stochastic behavior — 358
4. Conclusion: open problems — 360
References — 361

1. Introduction

In the century which has elapsed since the creation of the statistical theory of matter, classical Hamiltonian dynamical systems have been among the main test cases for the effectiveness of the concepts of statistical mechanics. Boltzmann proposed to reconcile the macroscopic behavior of such systems with their microscopic dynamics by the *Ergodic Hypothesis* which said that, in the course of time, the point of phase space representing the state of the Hamiltonian system wanders everywhere on its hypersurface of constant energy. In its modern form, the ergodic hypothesis is replaced by the notion of *metric transitivity*: every subset of a hypersurface of constant energy that is carried into itself by the time development of the system is either of measure zero or is the complement of a subset of measure zero. The measure referred to is that given by the so-called *micro-canonical ensemble* i.e. if S is a subset of the hypersurface of energy, E, then there is assigned a non-negative real number $\mu_{\text{micro}}(S)$, the measure of S:

$$\mu_{\text{micro}}(S) = \int_S dq\, dp\, \delta(E - H(q, p)),$$

where $q = \{q_1 \ldots q_n\}$ and $p = \{p_1 \ldots p_n\}$ are the canonical variables and H is the Hamiltonian. One can think of the temporal evolution of the system as given by a family of measure-preserving mappings of the hypersurfaces carrying a point $\{q, p\}$ into its time translate after time t: $T^t\{q, p\} = \{q(t), p(t)\}$, $-\infty < t < \infty$. T^t then satisfies:

$T^0 = 1$, the identity mapping,

$T^t T^s = T^{s+t}$,

and the measure-preserving character of T^t is expressed by:

$$\mu_{\text{micro}}(T^t S) = \mu_{\text{micro}}(S).$$

Such a measure-preserving family of mappings is called a *flow*. Any flow induces a one parameter group of transformations of functions, f, defined on the hypersurface, Ω_E, of energy E:

$$(U^t f)(q,p) = f(T^t\{q,p\}).$$

One can give an alternative characterization of metric transitivity of a flow in terms of this action on functions: A flow is called *ergodic* if the only functions f satisfying

$$U^t f = f \quad \text{for all } t, \quad -\infty < t < \infty,$$

except possibly on a set of μ_{micro}-measure zero are constant almost everywhere. A flow is metrically transitive if and only if it is ergodic.

When the functions, f, are chosen in the Hilbert space, $L^2(\Omega_E, d\mu_{\text{micro}})$, of square integrable functions with respect to the measure μ_{micro}, then $\{U^t | -\infty < t < \infty\}$ turns out to be a continuous unitary one-parameter group:

$$U^s U^t = U^{s+t}, \quad U^0 = 1, \quad (U^t)^* = U^{-t}.$$

When, in addition $\mu_{\text{micro}}(\Omega_E) < \infty$, von Neumann's ergodic theorem says that

$$\lim_{T \to \infty} \frac{1}{2T} \int_{-T}^{T} f(T_t x) \, d\mu_{\text{micro}}(x)$$

exists in $L^2(\Omega_E, d\mu_{\text{micro}})$ and equals

$$\int_{\Omega_E} f(x) \, d\mu_{\text{micro}}(x),$$

i.e. time averages equal phase averages. This was where the mathematical theory stood in the 1930's. Professional statistical mechanicians were not much impressed. They asked: how does one verify that a concretely given dynamical system is ergodic? The answer is that it isn't so easy to do and at that time rather few ergodic flows were known.

We now know many ergodic flows and also many which are *not* ergodic. It is natural to ask what are the distinguishing features of a Hamiltonian that gives rise to an ergodic flow; that is the problem referred to in the title of this article. The first point to recognize is that ergodicity is only the beginning of the story.

In the early development of statistical mechanics usually no distinction was made between what has just been defined as an ergodic flow, and flows with even more irregular behavior. The later development of ergodic theory led to precise definitions of a whole chain of such properties:

Weak Mixing;
Mixing;
Mixing of nth order;
K system;
Bernouilli system;

each more restrictive and random than the preceding. I won't repeat all these definitions here especially since I have already given a popular account of most of them before [1]. Suffice it to say that each tries to capture some essential feature of the stochastic microscopic behavior of a dynamical system which is supposed to give rise to smooth macroscopic thermodynamic behavior. What will be attempted here is an expository account of part of what is now known about the converse: how can randomness fail? how can statistical mechanics be justified in the presence of this failure?

The starting point is Poincaré's analysis of the behavior of orbits in the neighborhood of a periodic orbit. In passing, let me remark that it is really a pedagogical scandal that after more than three quarters of a century the simple and enlightening ideas of Poincaré do not appear in elementary mechanics books written for physics students. Will we have to wait for the third millenium to see an analytical mechanics book for undergraduates in which there is described what really happens when two harmonic oscillators are coupled together with a non-linear coupling [2]? All the people who really need to have their students know this, teach it – the celestial mechanicians, the accelerator designers, the plasma physicists, the quantum chemists – sometimes without mentioning that it comes from Poincaré. Why this Victorian prudishness on the part of the physics textbook writers? One might almost think, in the spirit of the day, that there is a conspiracy against Poincaré. In any case, we have to be thankful to Abraham and Marsden [3] to Arnold and Avez [4], to Arnold [5], to Moser [6] and to Thirring [7].

Poincaré's analysis of the linear approximation to a flow in the neighborhood of a periodic orbit led to a number of conjectures about the possible non-ergodic behavior of the exact flow which were only established in the 1950's and 1960's, especially in the work of Kolmogorov, Arnold and Moser. The full complexities involved are still only partly disentangled but the work of the 1970's has shown at least that the typical smooth Hamiltonian flow is neither ergodic nor completely integrable. It is with results of this kind that the end of sect. 2 of this article will deal primarily. Section 3, on the other hand, deals with some results on the part of a Hamiltonian flow which is random.

Finally, in sect. 4 I have tried to state what the preceding implies for classical statistical mechanics.

It is sad that this volume has become a memorial to Mel Green rather than a celebration of his work and influence. We were graduate students together in his Princeton days and friends from then on. For sentimental reasons, I dedicate this contribution to him and the spirit of his work: trying to understand what is really going on.

2. Stability and instability

This section is divided into three parts. The first discusses some stability phenomena, which can occur in Hamiltonian systems. The second treats some instability phenomena. Finally, the third part sketches some of what is known about the actual occurrence of such stabilities and instabilities. It turns out that for most smooth Hamiltonian systems both kinds of phenomena occur. In particular, we have the result of Markus and Meyer which says that generic behavior for Hamiltonian systems is neither ergodic nor completely integrable.

In his great treatise, *Les Nouvelles Méthodes de la Méchanique Céleste* (1892–9) Poincaré studied the behavior of orbits in the neighborhood of a given periodic orbit [8]. He considered a *transverse section* of such a periodic orbit at a point P of the orbit i.e. a piece of smooth hypersurface in the phase space passing through P but not tangent to any orbit (see fig. 1).

What is now called the *Poincaré mapping* associated with the flow in phase space is the mapping, ϕ, of the transverse section into itself which associates to a point P' the point $\phi(P')$ where P' first crosses the section again in the same direction (the point of first return). Clearly, P is a fixed point of ϕ. The next step in Poincaré's analysis is to consider the

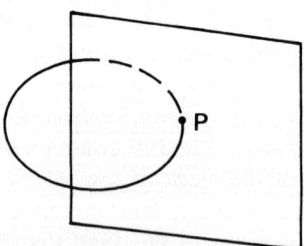

Fig. 1. Transverse sector of a flow at a periodic orbit.

linear approximation to ϕ obtained by expanding it in Taylor series on the transverse section at P, and throwing away all but the linear terms. It is given by the matrix T:

$$T = (\nabla \phi)_P.$$

An indication of the qualitative behavior of the orbits near the given periodic orbit through P is obtained by studying the iterates, T^n, $n = \pm 1$, $n = \pm 2$, ... of T. If T has an eigenvalue λ with eigenvector y, then

$$T^n y = \lambda^n y,$$

so points separated from P by a vector in the direction y will run away if $|\lambda| > 1$ and approach P if $|\lambda| < 1$. If $|\lambda| = 1$ and $\lambda \neq 1$ they will circle around P.

The possible eigenvalues of T are restricted by the fact that it is a symplectic transformation. That the time evolution $x \to T^t x$ defines a symplectic mapping of phase space is simply the expression of the fact that the Hamiltonian form of the equations of motion is preserved under time translations. It is customarily expressed analytically by the invariance of the differential form:

$$\sum_{j=1}^{n} dq_j \wedge dp_j,$$

under time evolution. It is not quite obvious that the symplectic property for ϕ follows from this because distinct points of the transverse section, in general, return to the section at different times. Nevertheless, it is true [9]. In fact, the complex eigenvalues occur in quadruplets $\lambda, \bar{\lambda}, \lambda^{-1}, \bar{\lambda}^{-1}$ of equal multiplicity and only when λ is real or on the unit circle may these quadruplets reduce to doublets λ, λ^{-1}. It turns out that $\lambda = 1$ always has odd multiplicity ≥ 1 [10]. When all the eigenvalues are of absolute value 1, the fixed point is called elliptic and the orbits lie (in linear approximation) on *invariant tori* whose intersection with the transverse section is indicated schematically in fig. 2. For a system of n degrees of freedom, the invariant tori have dimension n and lie in the $2n - 1$ dimensional hypersurfaces of constant energy. Thus, in linear approximation the orbits near an elliptic fixed point show a certain stability.

What happens to these invariant tori when one replaces the linear approximation by the exact flow? This was a problem posed by Poincaré and studied by a long line of distinguished mathematicians.

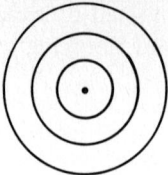

Fig. 2. Qualitative behavior of the orbits in a neighborhood of an elliptic fixed point. The successive hits $T^k x$, $k = 1, 2, 3, \ldots$ of a phase point x circle the fixed point. For a system of two degrees of freedom the invariant torus is a two-dimensional submanifold of the three-dimensional energy surface.

Very important results were obtained by Kolmogorov, Arnold and Moser in the 1950's and 1960's. They showed that close to the periodic orbit enough invariant tori survive (possibly distorted) to comprise a set of positive measure. The resulting picture has become familiar by repetition since Arnold and Avez first had the courage to draw it for their book [11]. It is indicated schematically in fig. 3.

The importance of the KAM results from our present somewhat restricted point of view is that they guarantee that the presence of an elliptic periodic orbit destroys the ergodicity of a flow, the set of

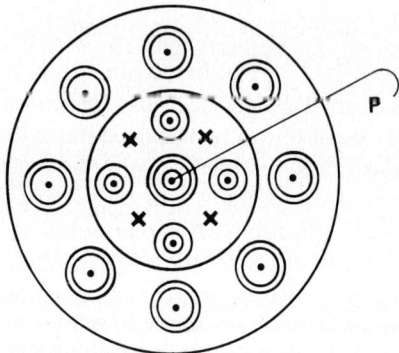

Fig. 3. Schematic version of the effect of a perturbation on the orbits around an elliptic fixed point, P (compare with fig. 2). The point in the middle is a fixed point, P, of the Poincaré mapping, ϕ. The circles surrounding it are the invariant tori of ϕ. They get denser and denser as one approaches P, filling a fraction of the area that approaches one. P is surrounded by an infinite number of other fixed points of ϕ and its powers, some elliptic, some hyperbolic. As one goes away from P the surviving invariant tori thin out and stochastic seas lie between them. I will return to the problem of the structure of these seas later.

invariant tori is not of measure zero nor is its complement of measure zero. Thus it is of the greatest importance to know when elliptic orbits occur in a Hamiltonian system. This question will be put aside for the moment. First I want to recall some of the basic facts about the instabilities of flows which lead to ergodicity.

One of the important developments of ergodic theory in the 1930's was the proof that certain special flows are indeed ergodic. The prime example is that of a mass point sliding on a surface of constant negative curvature. (This is referred to in another terminology as *geodesic flow* on the surface. A recognition of the basic instability of geodesics involved in the example goes back to Hadamard [12].) In their study of this example, Hedlund and Hopf succeeded in isolating a mechanism which produces ergodicity [13, 14]. Using this deeper understanding they showed that ergodicity also holds for surfaces whose curvature is non-constant provided the curvature is negative and bounded away from zero. Hedlund and Hopf's ideas were taken up by Krylov, who tried to argue that the relaxation processes in a hard sphere gas are governed by this same mechanism [15]. Eventually, proofs of some of Krylov's assertions were supplied by Sinai for the hard sphere gas [16] and by Anosov for geodesic flows on a large class of smooth manifolds [17].

The key to the behavior of Anosov systems is that the non-vanishing tangent vectors to the transverse section of a flow at a fixed point P should be classified into stable and unstable, the former spanning a subspace E_S, the latter a subspace E_U, such that the whole tangent space, $T_P M$, may be written:

$$T_P M = E_S + E_U \quad \text{with} \quad E_S \cap E_U = \{0\}.$$

The properties stable and unstable are defined by the behavior of the linear approximation to the nth iterate of the Poincaré mapping:

$$|(\nabla(\phi^n))|_P v| \geq c\lambda^n |v| \quad \text{for some} \quad c, \lambda > 0$$

and all $n = 1, 2, 3, \ldots$ if $v \in E_U$,

$$|(\nabla(\phi)^n)|_P v| \leq c\lambda^{-n} |v| \quad \text{for some} \quad c, \lambda > 0 \tag{1}$$

and all $n = 1, 2, 3 \ldots$ if $v \in E_S$. Clearly nearby orbits in unstable directions tend to run away, nearby orbits in stable directions tend to approach the given orbit. The | | in these equations is a norm which is introduced somewhat arbitrarily into the tangent space. (For compact phase spaces the results on Anosov systems are independent of the choice of this norm.)

Now I turn to the question of what actually happens in typical Hamiltonian systems. The first question that has to be answered is what the adjective "typical" should be defined to mean. There are quite a few answers. One, familiar to physicists and Alice is "all, with the exception of those that I hereby declare to be uninteresting, if they happen to exist". That definition has the advantage that it focuses attention on the set of exceptions which should be "small" in some sense. One notion of "small" is measure zero with respect to some measure, but so far no one has found any natural measure to put on the family of Hamiltonian dynamical systems. On the other hand, for Hamiltonian systems whose phase space is a given manifold there is a natural family of topologies, the so-called Whitney topologies. Two Hamiltonians on the phase space, M, are close in the C^r Whitney topology, if all the derivatives, $\partial H/\partial q$ and $\partial H/\partial p$ appearing in Hamilton's equations, together with all their derivatives up to order r are uniformly close all over M. These topologies can be defined in terms of norms which won't be given here [18].

Now one straightforward definition of a small set in a topological space is: a set whose closure contains no interior points, a so-called *nowhere dense* set. (Recall that the closure of a set is the set together with all its limit points and, the interior of a set, S, consists of all points, P, of S such that there is some neighborhood of P also in S.) If the exceptional points in our space of Hamiltonian dynamical systems on a phase space M are a nowhere dense set, then the non-exceptional points are a dense open set i.e. a dense set for which each point is an interior point. Properties which hold on an open set have a certain stability: if one makes small changes in the Hamiltonian, sufficiently small in the sense of the given topology, one does not lose the property. On the other hand, when the non-exceptional points are open and dense, every point has a non-exceptional point in its neighborhood.

Although it would be lovely if the notion of exceptional set that we have just defined could be used in the theory of dynamical systems, it has turned out that a much weaker notion is the useful one: the exceptional set may be a countable union of nowhere dense sets, a so-called *meager* set. Therefore, the non-exceptional set contains a countable intersection of dense open sets; it is a so-called *residual* set. Residual sets are also called sets of the second category because in the early part of this century Baire divided sets into two categories and proved a remarkable theorem based on the distinction. (As earnest students of the methods of modern mathematical physics know [19], Baire's theorem says that a complete metric space contains a residual set.) A property which holds on a residual set is said to be *generic*.

A proof that a property of a Hamiltonian system is generic does not usually establish that any particular Hamiltonian has the property, and therefore such proofs are very unsatisfying if what you really want to know is whether the property holds for a particular Hamiltonian. Nevertheless, genericity is really a natural notion from the physical point of view, for two reasons at least: (a) the Hamiltonian systems used to model Nature usually have some arbitrariness in their formulation and it is comforting to know that an observed property holds not only for one particular Hamiltonian but for typical Hamiltonians; (b) when a property is generic it means that it is essentially i.e. with a meager set of possible exceptions, a consequence of the Hamiltonian structure of the dynamical system alone. When enough generic properties of Hamiltonian dynamical systems are known one will have a general survey of what to expect from such systems. Of course, it will not reduce the importance of establishing specific properties for individual systems of special physical interest.

Having tried to make clear why I think the determination of generic properties of Hamiltonian systems is interesting, if not the whole story, I want to go into a few more technical details. The first has to do with the smoothness of the dynamical systems under study. There is certainly a big difference between the typical behavior of dynamical systems which are merely continuous and those which are smooth. It can be grossly summarized in the slogan: *rough systems are more likely to be chaotic*. A more precise illustration of the principle involved is a famous result of Oxtoby and Ulam [20] on measure preserving homeomorphisms of a connected polyhedron of dimension ≥ 2; it asserts that the subset of such transformations that are ergodic is residual, so ergodicity is a generic property for such continuous transformations. On the other hand, as we will see later, for smooth dynamical systems this result is false.

One way to deal systematically with the distinctions introduced by smoothness assumptions is to distinguish between systems with C^r-generic properties for different r: a property is C^r-*generic* if it holds for a residual set of dynamical systems in the C^r Whitney topology. There are several properties of Hamiltonian dynamical systems known to be C^r-generic for $r \geq 2$ but not known to be C^r-generic for $r=1$. (See for example, [3] pp. 587–592 for a review.) Furthermore, the basic Kolmogorov–Arnold–Moser theory of invariant tori has been established for three or more degrees of freedom only for C^r, $r \geq 5$ Hamiltonians, and for two degrees of freedom for $r \geq 4$ [21]. On the other hand, there are properties which are known to be C^1-generic but not known to be C^r-generic for $r \geq 2$. For example the density of periodic orbits is C^1-generic but the C^r-genericity for $r \geq 2$ is an open problem [22]. These are outstanding problems in analytical mechanics, whose solution will be necessary before we can have a clear view of what is the typical behavior in Hamiltonian systems.

In passing, it should be noted that one can introduce the C^r Whitney topology in the set of C^r vector fields on a manifold. Each such vector field defines a differential equation whose solutions describe a dynamical system (but not a Hamiltonian system, in general). Thus, one can also talk about C^r-generic properties of dynamical systems including dissipation. The last few years have seen very interesting developments of the theory with applications to turbulence and other chaotic phenomena. It turns out that generically orbits in such systems approach some limit sets: limit points, limit periodic orbits, or limit higher dimensional sets. (That limiting behavior is impossible for a Hamiltonian system, because the flow has to preserve volume in the phase space, according to Liouville's theorem.) The behavior of a dynamical system is strongly dependent on the dimension of its manifold. For dimension two, a smooth dynamical system has only limit points and limit periodic orbits; there is no "turbulence" in a two-dimensional system. For dimension three one has, in addition to limit points, limit periodic orbits and limit two-dimensional tori, limit sets of a highly irregular structure which have been called "strange attractors". Their occurrence is generic for dimension three or more [22]. Orbits which approach such highly irregular sets will appear "turbulent". When strange attractors are present, they also play an important role in the computation of time averages. For a large class of dynamical systems (the so-called Axiom A systems of Smale) an analogue of the Birkhoff ergodic theorem has been proved:

$$\frac{1}{T}\int_0^T f(x_t)\,dt$$

converges to a limit as $T \to 0$ for almost all x. (Almost all is defined relative to any smooth measure on the phase space locally equivalent to Lebesgue measure. The family of sets of measure zero is independent of the choice of the measure at least if the phase space is compact.) The idea is that, after a possible set of exceptional points have been dropped, the phase space splits up into a sum of sets (a finite sum if the phase

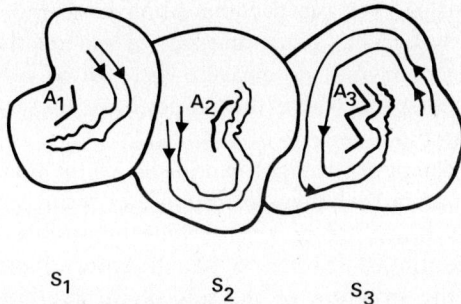

Fig. 4. Schematic diagram of the asymptotics of a general Axiom A flow. The phase space is, up to a set of measure zero the sum of the basins $S_1, S_2, S_3, \ldots S_n$. Within the basin S_i the flow approaches the attractor A_i.

space is compact), S_i, called *basins of attractors*. All points within a basin, S_i, approach the attractor, A_i, of the basin as $t \to \infty$, a behavior indicated schematically in fig. 4. The time average of a function then splits into a sum of contributions from the different basins. The contribution from a single basin is phase space average with respect to a measure, ν_i, concentrated on the attractor, A_i:

$$\int d\nu_i(x) f(x),$$

ν_i is expected to be invariant under the flow. This statement:

$$\lim_{T \to \infty} \frac{1}{T} \int_0^T f(x_t) dt = \sum_i \int d\nu_i(x) f(x) \qquad (2)$$

for almost all x, is for Axiom A dynamical systems, a theorem of Bowen and Ruelle [23]. It is the starting point of a new chapter in the rigorous classical statistical mechanics of dissipative systems. It should be compared with the *Birkhoff Ergodic Theorem* for measure preserving flows which says the limit:

$$\lim_{T \to \infty} \frac{1}{T} \int_0^T f(x_t) dt$$

exists for almost all x and, for all f integrable with respect to μ, is equal to:

$$\int d\rho(a) \int d\mu_a(x) f(x). \qquad (3)$$

Here $d\mu = \int d\rho(a) d\mu_a(x)$ is the decomposition of μ into ergodic measures $d\mu_a$. If the system has extra integrals of motion the variable, a, will include their values and the measure $d\mu_a$ will be concentrated on the subset of phase space where the points have those values of the integral of motion.

It is natural to compare the right-hand sides of the formulae (2) and (3). They are similar in that they correspond to decompositions of the flow into subflows which are in a sense indecomposable. Whether this analogy can be continued further to the structure of the attractor in comparison with the structure of the sets supporting ergodic flows is unclear because very little is known about the structure of strange attractors. Thom has proposed that the KAM tori around an elliptic periodic orbit be regarded as the Hamiltonian analogue of an attractor; he suggested the name *vague attractor* [24]. There are several reasons which support this terminology

(a) For two degrees of freedom the invariant tori, having dimension two, separate the (three dimensional) hypersurfaces of constant energy into two disjoint regions; orbits inside an invariant torus cannot get outside.

(b) For n degrees of freedom, $n \geqslant 3$ the invariant tori have dimension n but the hypersurfaces of constant energy have dimension $2n-1$ so the invariant tori have no inside and outside. Arnold showed how an orbit not on an invariant torus can go from a point arbitrarily close to one invariant torus to a point arbitrarily close to any other, a process customarily called *Arnold diffusion* [25]. However, there is still some stability even in the presence of Arnold diffusion: the diffusion time is long. More precisely, suppose the Hamiltonian is of the form:

$$H = H_0(I_1, \ldots I_n) + \varepsilon H_1(I_1 \ldots I_n, \phi_1 \ldots \phi_n),$$

where the $I_1 \ldots I_n$ and $\phi_1 \ldots \phi_n$ are action and angle variables for H_0. For the perturbed Hamiltonian H, the action variables become functions of t, and one can ask for how long the inequality:

$$|I_j(t) - I_j(0)| \leqslant \varepsilon^a$$

will hold. The answer is, if H is analytic and its first two derivatives satisfy certain inequalities, for $0 < t < T$ where,

$$T = \frac{1}{\varepsilon^b} \exp \frac{1}{\varepsilon}.$$

Here, a and b are certain numbers computable from H_0. If H is only C^r

instead of analytic the exponential is replaced by a power law where the power increases with r [26].

Now I return to the main issue for Hamiltonian systems. As we have seen the occurrence of elliptic periodic orbits and their associated invariant tori destroys ergodicity. Is the occurrence of elliptic periodic orbits a generic property for Hamiltonian systems? Markus and Meyer said yes for C^∞ Hamiltonian systems with a compact phase space and therefore concluded that for such Hamiltonian systems ergodicity is not generic [27]. The idea of their proof is simple. They look at the absolute minimum of the Hamiltonian, H, on the compact phase space. For a generic Hamiltonian the Taylor series of H must begin with quadratic terms that define non-degenerate symplectic form. Thus, the situation is precisely that which is the starting point of the KAM theory. (The usual application of KAM theory is to the Poincaré mapping in a neighborhood of a periodic orbit; here it is being applied at an equilibrium point instead.) Markus and Meyer's argument uses the existence of one elliptic equilibrium point. The next natural question is how common periodic orbits of various kinds are in generic Hamiltonian systems.

As far as hyperbolic periodic orbits are concerned one has a remarkable result of Pugh and Robinson: for a C^1-generic Hamiltonian flow hyperbolic periodic orbits are dense [28]. It is not known at the moment whether this result is true with C^1 replaced by C^r. To state what is known for elliptic periodic orbits we have to introduce the notion of k-elliptic periodic orbits. A fixed point, P, of a Poincaré mapping, ϕ, on a hypersurface of constant energy (and the associated periodic orbit) is called k-elliptic if the tangent mapping $T = (\nabla \phi)_P$ has k non-real eigenvalues of absolute value 1, and the rest of absolute value different from 1. Notice that what we formerly called elliptic periodic orbits for a Hamiltonian system of n degrees are $(n-1)$ elliptic in this terminology. Then for a C^1-generic Hamiltonian each energy surface either contains only hyperbolic periodic orbits and the flow is Anosov or is in the closure of the set of all 1-elliptic orbits [29]. This statement is very strong for $n=2$. If it were not for the gap between the C^1 smoothness of this statement and the C^5 currently needed for the KAM theory we could say that the C^1-generic Hamiltonian for two degrees of freedom either yields a transverse section like that of fig. 3 with a dense set of elliptic fixed points or no elliptic fixed points at all. Actually, this result of Newhouse was preceded by a detailed study of $n=2$ by Birkhoff [30], who already established that an elliptic fixed point is a limit point of both elliptic and hyperbolic periodic points.

Clearly, the preceding results represent great progress in unraveling the structure of Hamiltonian flows, even though there are many difficult mathematical questions still unanswered. If, to try and assess the prospects for classical statistical mechanics we simply assume that no new complications will arise, it would seem that for a given Hamiltonian we can expect the flow on an energy surface either to be ergodic and indeed Anosov, or that there will be a dense set of elliptic periodic orbits and KAM tori breaking out like measles all over the place. To understand what happens in the latter case I turn to the study of another phenomenon discovered by Poincaré: homoclinic points.

3. Homoclinic points and stochastic behavior

A transverse section of a flow in the neighborhood of a hyperbolic fixed point P is shown in fig. 5. The indicated stable and unstable manifolds pass through the points satisfying the inequalities (1) describing the successive hits of an orbit on the transverse section. Clearly, the orbits on the unstable manifold run away from P, while those on the stable manifold approach P. It can happen that the stable and unstable manifolds of a hyperbolic point, P, cross each other at some point Q. If so, Q is called a *homoclinic* point. Continued beyond Q the stable manifold thrashes wildly crossing the unstable manifold an infinite number of times; a similar statement holds with stable and unstable interchanged. A genuine picture of this phenomenon based on numerical calculations of a concrete mapping is to be found in [31]. It is known that, generically, every hyperbolic point has a homoclinic point [32].

Poincaré already recognized that in the neighborhood of a homoclinic point the action of a Hamiltonian flow must be very complicated.

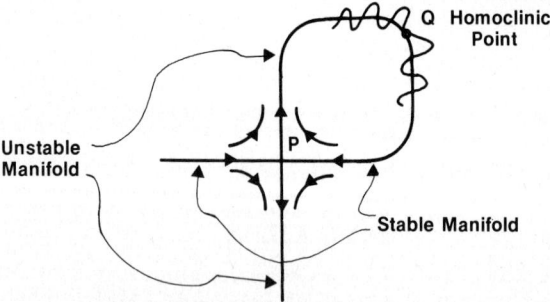

Fig. 5. Schematic indication of a transverse section of a flow in the neighborhood of a hyperbolic fixed point. Successive hits of an orbit on the unstable manifold travel away from P, while those on the stable manifold approach P.

Exactly how complicated is still not understood but there is a very nice result of Smale, generalized by Conley and Moser which gives one precise account of the randomness involved [33], and our next goal is to describe it.

To explain the Smale–Conley–Moser result, I have to introduce the notion of a *Bernouilli shift* which for my purposes can be defined as follows. Consider the set, S, of all doubly infinite sequences α,

$$\alpha = \ldots \alpha_{-2}\alpha_{-1}\alpha_0\alpha_1\alpha_2\ldots$$

where the α_k belong to some fixed denumerable set A. (S is sometimes written $A^{\mathbb{Z}}$, where $\mathbb{Z}$ stands for the integers.) The shift T is a transformation mapping S into itself defined by:

$$(T\alpha)_k = \alpha_{k-1}.$$

In probability theory, one assumes that a probability measure, μ, is given on the set A. Then one constructs a probability measure on S as the infinite product measure of μ with itself, one for each integer. It turns out that the space S and the shift T then define a model for a roulette wheel spun an infinite number of times. For our present purposes this probabilistic construction plays no role. What matters is T as a topological transformation. S can be defined as a topological space by giving a base for the neighborhoods of a point α, S,

$$U_j(\alpha) = \{\beta \mid \alpha_k = \beta_k, \quad \text{for all } k, \text{ such that } |k| < j\} \quad j = 0, 1, 2, 3, \ldots$$

With this definition of neighborhoods for points of S given, T becomes a continuous one-to-one mapping of S onto itself with a continuous inverse

$$(T^{-1}\alpha)_k = \alpha_{k+1}.$$

Now comes the connection with homoclinic points. Suppose that one is given a homoclinic point, Q, on a transverse section of a hyperbolic periodic orbit, and the unstable and stable manifolds intersect transversely at Q. The theorem then says that there is an embedding of the space S in a neighborhood of Q so that a power of the Poincaré mapping becomes the Bernoulli shift. Crudely speaking, the Poincaré mapping the chosen subset of points acts like a roulette wheel. (As stated, the theorem is in the form given by Smale where one has to restrict the alphabet A to be a finite set. In the Conley–Moser version the power of the Poincaré mapping is allowed to vary and the set A is allowed to be denumerably infinite.)

How does one choose the embedding of the space S in a neighborhood of Q? That is one of the deeper tricks of the theory associated with the phrases "symbolic dynamics" and "Smale's horseshoe". I will not go into them here but will recommend Moser's exposition in [6].

The preceding theorem throws some light on the stochastic nature of the flow but it gives little insight into the situation in the neighborhood of an elliptic periodic orbit. The most general and elementary question we can ask there is: how does the stochastic part of the flow, the part not on invariant tori decompose into ergodic components? A systematic theory of such decompositions has been developed by Pesin [34]. He uses systematically the notion of *characteristic Lyapunov exponent*. It is a real-valued function of a point, x, on the phase space and a vector v belonging to the tangent space at x; it is defined by:

$$\chi^+(x,v) = \varlimsup_{T\to\infty} \frac{1}{T}\ln|\nabla U^T|_x v|.$$

Thus the condition of hyperbolicity (1) says precisely that $\chi^+(x,v)<0$ for $v\in E_S$ and $\chi^+(x,v)>0$ for $v\in E_U$. It turns out that the Lyapunov exponent can take at most $2n$ values at a given point x. These $2n$ values in turn give rise to a decomposition of the flow into a denumerable set of ergodic subflows. The construction is difficult and long. It runs parallel in part to the construction which was used in Anosov's treatment of Anosov flows [17]. (In the latter case, the flow is ergodic and there is only one ergodic component.) Pesin's analysis is carried out for any dynamical system on a smooth manifold with a smooth measure invariant under the flow; this includes smooth Hamiltonian systems as a special case.

The use of the Lyapunov characteristic exponents to characterize the ergodic decomposition of flows deserves a detailed study from a physical point of view. Numerical experiments have already given strong indications of ergodic components distinguished by different Lyapunov exponents, but it would be very instructive to understand how a complete decomposition looks in some concrete cases [35]. It is really a striking fact that all one needs to know are these exponents in order to make the complete decomposition of the flow.

4. Conclusions: Open problems

The preceding sections have, I hope, made clear that unless one considers Anosov systems, one can expect non-ergodic behavior for typical

Hamiltonian systems because of the appearance of elliptic periodic orbits. Thus, the applicability of the traditional statistical mechanics to such non-ergodic systems can be justified only on the grounds that the elliptic periodic orbits and their accompanying invariant tori "don't matter". Why not? There are several possible reasons which have been part of the folklore of statistical mechanics for a long time:

(a) the measure of the regions on the constant energy hypersurfaces which is occupied by the invariant tori becomes small compared to the measure of the total hypersurfaces in the thermodynamic limit.

(b) the macroscopic observables with which the standard statistical mechanics concerns itself are insensitive to the presence (or absence) of invariant tori, especially in the thermodynamic limit.

It seems plausible that both (a) and (b) will play a role in a satisfactory version of the foundations of classical statistical mechanics. Each poses interesting problems. For example, in connection with (a), is there some computable auxiliary function whose behavior reflects the relative importance of the elliptic and hyperbolic parts of a flow? In connection with (b), suppose that one knows about a type of dynamical system only its correlation functions in the thermodynamic limit, is that an essential restriction on the observables which would make them insensitive to the occurrence of invariant tori?

Of course, if you can prove that the system you wish to study is an Anosov system you will be in splendid shape, as far as classical statistical mechanics is concerned. Not only that but your system will remain Anosov under small perturbations.

Clearly, there are plenty of basic questions still to be answered in classical statistical mechanics.

References

[1] A. S. Wightman, Statistical Mechanics and Ergodic Theory—An Expository Lecture, in: Statistical Mechanics at the Turn of the Decade, ed. E. G. D. Cohen (Marcel Dekker, Inc., New York, 1971) pp. 1–32.
[2] In preparation for this article I looked at 18 physics text books on mechanics in Fine Hall Library at Princeton without finding one which discussed the stability of periodic orbits in Hamiltonian systems in the linear approximation and beyond, not to speak of the notions of stable and unstable manifolds and homoclinic and heteroclinic points.
[3] R. Abrahams and J. E. Marsden, Foundations of Mechanics, Second Edition (Benjamin/Cummings Publishing Co. Reading, Mass., U.S.A., 1978).
[4] V. Arnold and A. Avez, Ergodic Problems of Classical Mechanics (Benjamin, New York, 1968).

[5] V. Arnold, Mathematical Methods of Classical Mechanics (Benjamin Springer-Verlag, New York, 1978).
[6] J. Moser, Stable and Random Motions in Dynamical Systems With Special Emphasis on Celestial Mechanics, Annals of Math. Studies No. **77** (Princeton University Press, 1973).
[7] W. Thirring, A Course in Mathematical Physics 1, Classical Dynamical Systems (Springer-Verlag, New York, 1978).
[8] See especially vol. I, ch. VII and vol. III, ch. XXXIII.
[9] See [3], pp. 521–5.
[10] See [3], p. 573 and pp. 169–71.
[11] See [4] p. 91.
[12] J. Hadamard, Les surfaces à courbures opposées et leurs lignes géodésiques, J. Math. Pure and Appl. **4** (1898) 27–73.
[13] G. Hedlund, The Dynamics of Geodesic Flows, Bull. Amer. Math. Soc. **45** (1939) 241–260.
[14] E. Hopf, Statistik der geodätischen Linien in Mannigfaltigkesten negativer Krümmung, Ber. Verh. Sächs. Akad. Wiss. Leipzig **91** (1939) 261–304.
[15] N. S. Krylov, Works on the Foundations of Statistical Physics (Princeton Press, 1979).
[16] Ya. Sinai, Ergodicity of Boltzmann's Gas Model, in: Statistical Mechanics Foundations and Applications, ed. T. A. Bak (Benjamin, New York, 1967) pp. 559–573.
[17] D. V. Anosov, Geodesic Flows on Closed Riemannian Manifolds of Negative Curvature, Proc. Steklov Inst. **90** (1967) 1–235.
[18] R. Abraham and J. Robbin, Transversal Mappings and Flows ch. II (Benjamin/Cummings Reading, Mass., 1967).
[19] M. Reed and B. Simon, Methods of Modern Mathematical Physics I (Academic Press, New York, 1972) p.80.
[20] J. C. Oxtoby and S. M. Ulam, Measure Preserving Homeomorphisms and Metrical Transitivity, Annals of Math. **42** (1941) 874–920*.
[21] H. Rüssmann, Kleine Nenner I Über invariante Kurven differenzierbarer Abbildungen eines Kreisrings, Nachr. Akad. Wiss. Gött. Math. Phys. Kl II (1970) 67–105.
[22] D. Ruelle and F. Takens, On the Nature of Turbulence, Commun. in Math. Phys. **20** (1971) 167–192; **23** (1971) 343.
[23] R. Bowen and D. Ruelle, The Ergodic Theory of Axiom A Flows, Invent. Math. **29** (1975) 181–202.
[24] R. Thorn, Stabilité structurelle et morphogenèse, (Benjamin Reading, Mass., 1972).
[25] V. Arnold, Instability of Dynamical Systems with Several Degrees of Freedom, Sov. Math. Dokl **156** (1964) 581–585. See also [4] pp. 109–114.
[26] N. N. Nekhoroshev, An Exponential Estimate of the Time of Stability of Nearly Integrable Hamiltonian Systems, Russ. Math. Surveys **32**:6 (1977) 1–65.
[27] L. Markus and K. R. Meyer, Generic Hamiltonian Systems are neither Integrable nor Ergodic, Mem. Amer. Math. Soc. No. **144** (1974).
[28] C. Pugh and R. C. Robinson, The C^1 Closing Lemma Including Hamiltonians. See Clark Robinson, Introduction to the Closing Lemma, in: The Structure of Attractors in Dynamical Systems, Lecture Notes in Math. No. **688** (Springer–Verlag, Berlin, 1978) 225–230.

*A homeomorphism is a one-to-one continuous mapping of the polyhedron onto itself with a continuous inverse. Such mappings need not be associated with a flow and a differential equation.

[29] S. Newhouse, Quasi-elliptic periodic points in conservative dynamical systems, Amer. Jour. Math. **99** (1977) 1061–1087.
[30] G. D. Birkhoff, Nouvelles recherches sur les systèmes dynamiques, Mem. Pont. Acad. Scient. Novi Lyncei Ser. III Vol. **1** (1934).
[31] R. W. Easton, Perturbed Twist Maps, Homoclinic Points and Ergodic Zones, in: Instabilities in Dynamical Systems Applications to Celestial Mechanics, ed. V. G. Szebehely (D. Reidel, Dordrecht, 1979) pp. 41–47.
[32] F. Takens, Homoclinic Points in Conservative Systems, Invent. Math. **18** (1972) 267–292.
[33] See [6] chs. III and VI.
[34] Ya B. Pesin, Characteristic Lyapunov Exponents and Smooth Ergodic Theory, Russ. Math. Surveys **32**:4 (1977) 55–114.
[35] G. Contopoulos, Instabilities in Systems of Three Degrees of Freedom, in: Instabilities in Dynamical Systems Proc. NATO Advanced Study Institute, Cortina d'Ampezzo, 1978, ed. V. G. Szebehely (D. Reidel Pub., Dordrecht, 1979) pp. 25–9; N. Saito and A. Ichimura, Ergodic Components in the Stochastic Region in a Hamiltonian System, in: Stochastic Behavior in Classical and Quantum Hamiltonian Systems Como 1977, eds. G. Casati and J. Ford (Springer-Verlag, Berlin, 1977) pp. 137–144;
The reader interested by the above references will perhaps find the following useful: D. Ruelle, Dynamical Systems with Turbulent Behavior, in: Mathematical Problems in Theoretical Physics, Rome, 1977, eds. G. Dell'Antonio, S. Doplicher and G. Jona-Lasinio (Springer-Verlag, Berlin, 1978) pp. 341–360.

Subject Index

Anderson model, 145
anomalous diffusion, 43
Anosov systems, 351
Antonow's rule, 278, 280
approximate constants of the motion, 126
Arnold diffusion, 356
articulation points, 338
attractor, 355
autocorrelation function, 5, 6

Baire's theorem, 353
basins of attractors, 355
BBGKY hierarchy, 63, 64, 73
Bernoulli shift, 359
Bethe–Guggenheim approximation, 287, 288, 289
binary
  collisions, 29, 30, 66, 70
  liquid mixtures, 253
Birkhoff ergodic theorem, 354, 355
Bohr–Wilson–Sommerfeld condition, 297, 304
Boltzman
  equation, 5, 6, 9, 34, 103
  statistics, 330
Bragg–Williams theory, 184, 189
Brownian motion, 87
Burnett
  coefficients, 8, 13, 18, 53
  equations, 38

Cahn transition, 280, 283, 284
Callan–Symanzik equation, 140
Chapman–Kolmogoroff equation, 78, 82, 93, 104
cluster expansion, 338
coarse-grained free energy, 206, 213, 218, 219
coarse-graining, 203, 204, 205, 213

coefficient of
  self-diffusion, 10, 34, 37, 158
  shear viscosity, 34, 37
  thermal conductivity, 34, 37, 113, 117, 119
coherence length, 191, 278, 288
collision sequence, 32
composition profiles, 277
connected clusters, 338
cooperative phenomena, 142
correction to scaling, 249, 250, 255, 260
correlation length, 204
Couette flow, 53
coupling constant, 141, 152, 205, 227, 230, 237
critical
  concentrations, 157
  end point, 276, 278, 279
  exponents, 140, 176, 186, 188, 190, 237, 244, 247, 250, 255, 297
  phenomena, 139, 176
  point, 181, 241, 276
  point universality, 176, 237, 245, 255
  solution point, 276
crossover, 263, 264
Curie temperature, 181, 186
Curie–Weiss law, 181

decimation, 151
density
  matrix, 325
  of states of multidimensional systems, 297, 315
  profile, 285, 286

ε-expansion, 140, 204, 213
Ehrenfest wind-tree model, 8, 9, 42
eight vertex model, 196
Enskog's theory, 34

ergodic hypothesis, 345
Euler equations, 38

Fick's law, 20, 159, 171
field
  theory, 140
  variables, 250
fixed point, 194, 227, 231, 235, 236, 237
flow, 345
fluctuation formulae, 65
fluctuation–dissipation theorem, 5, 98
Fokker–Planck equation, 79, 84, 93, 94, 98, 128
four-component mixtures, 277
Fourier's law, 113
functional integral, 147

generic set, 353
geodesic flow, 351
gradient expansions, 45
gravitational stratification, 244
Green–Kubo formulae, 61, 80, 84, 86

H-theorem, 103
hard
  disks, 5, 36
  spheres, 34, 36, 66, 67, 69
Heisenberg model, 187
He^3–He^4 mixture, 148, 149
homoclinic point, 358
Hubbard model, 145, 146
hypercell, 77, 78

imperfect wetting, 280, 283
interfacial tension, 290
interferometric experiments, 242
invariant
  critical subspace, 235
  tori, 349, 350, 351
irreducible clusters, 338
Ising model, 187

KAM theory, 350, 357
kinetic theory, 5, 62, 130
Kramers–Moyal expansion, 81

λ-point transitions, 185
Landau theory, 195
Langevin equation, 87, 128, 181
lattice-gas, 181
Lennard–Jones 6–12 potential, 332
Lienard equation, 163
limit cycle oscillations, 157, 162, 171
linear-response theory, 92
linear-model, 254, 255
Liouville's equation, 26, 27, 36, 61
Liouville's theorem, 354
long-time
  behavior, 10, 14
  tails, 45, 48, 49, 51, 52
long-range order, 142
Lorentz gas, 5, 6, 7
Lorentz–Lorenz law, 182
Lyapunov exponent, 360

macroscopic equations, 91
Markovian, 5
Maslov's method, 304
Mayer–Band theory of quantum statistics, 325
meager set, 352
mean field theory, 143, 149
mean-free-path damping, 39, 49
measure preserving flows, 345, 355
memory effects, 126
metal–insulator transitions, 142, 144
metastable states, 168, 169, 170, 171, 203, 204
metric transitivity, 345, 346
microscopic variables, 91
mode–mode coupling, 77, 86, 130
Morse potential, 332, 339, 341

N-particle distribution function, 27, 28, 38
n-vector ferromagnetic model, 197
Navier–Stokes equations, 29, 30, 38, 47, 125
nearest-neighbor harmonic forces, 113
nodal vertices, 338, 341
non-classical behavior, 190, 263
non-critical interface, 276

non-equilibrium fluctuations, 69
non-linear
  Langevin equation, 95
  projection operator, 86
  transport theory, 125, 127, 130

1-component ϕ^4 model, 247
"Onsager revolution", 178
order parameter, 141, 143, 186, 203, 204
Ornstein–Zernike theory, 182, 183, 187, 227, 228

Padé approximant, 188
parametric representation, 250, 253, 254, 263
penetrable-sphere model, 290
perfect wetting, 283
phase separation, 276
Poincaré mapping, 348, 351, 357, 359
polymer gels, 143
projection operators, 92, 131

Q-component Potts model, 197, 232, 237
quartic anharmonic oscillators, 297

Rayleigh equation, 162, 163
real space renormalization, 195, 227, 228
refractive index, 244
relaxation oscillations, 159, 168
renormalization group, 140, 141, 146, 149, 152, 176, 179, 192, 204, 206, 242, 250, 255
residual set, 352
Ricatti equation, 298
ring graphs, 7

scaling, 139, 140, 190, 242, 244, 245
scaling fields, 250
Schrödinger equation, 297, 298

second-order transitions, 146, 185
series expansion techniques, 247
spherical-model, 189, 191
spin-glasses, 142
star-triangle transformation, 231, 232, 233, 235, 236
Stosszahlansatz, 26, 27, 29, 92
strange attractors, 354
super-Burnett self-diffusion coefficient, 46
sympletic transformation, 349

three-body collision, 31
three-dimensional Ising model, 247
three-phase equilibrium, 280, 283
time correlation functions, 45, 52, 132
tricritical point, 149, 281
two-dimensional ferroelectric models, 196
two-component
  lattice gas, 285
  mixtures, 276, 279

unconnected clusters, 338
universality class, 247
unstable states, 203, 204

vague attractor, 356
Van der Pol equation, 163
Van der Waals equation, 182
velocity autocorrelation functions, 7, 10, 13, 18, 47, 51
Von Neumann's ergodic theorem, 346
vortex pattern, 48

Wegner expansion, 242, 251, 253, 256, 259, 262, 264
wetting transition, 283
Whitney topology, 352, 353, 354
Wigner theory of quantum statistics, 325, 327
WKB method, 297

ND H. FOGLER LIBRARY
ATE DUE